AN INTRODUCTION
TO CIRCUITS
AND ELECTRONICS

AN INTRODUCTION TO CIRCUITS AND ELECTRONICS

J.R. COGDELL

Department of Electrical and Computer Engineering
The University of Texas at Austin

PRENTICE-HALL, INC., *Englewood Cliffs, NJ 07632*

Library of Congress Cataloging-in-Publication Data

COGDELL, J. R.
    An introduction to circuits and electronics.

    Includes index.
    1. Electric circuits. 2. Electronics. I. Title.
TK454.C63 1986   621.319′2   85–12037
ISBN 0-13-479346-3

Editorial/production supervision and
        interior design: *Eileen M. O'Sullivan*
Cover design: *20/20 Services, Inc.*
Manufacturing buyer: *Rhett Conklin*

Printed in the United States of America

10 9 8 7 6 5 4 3 2 1

ISBN 0-13-479346-3 01

PRENTICE-HALL INTERNATIONAL (UK) LIMITED, *London*
PRENTICE-HALL OF AUSTRALIA PTY. LIMITED, *Sydney*
PRENTICE-HALL CANADA INC., *Toronto*
PRENTICE-HALL HISPANOAMERICANA, S.A., *Mexico*
PRENTICE-HALL OF INDIA PRIVATE LIMITED, *New Delhi*
PRENTICE-HALL OF JAPAN, INC., *Tokyo*
PRENTICE-HALL OF SOUTHEAST ASIA PTE. LTD., *Singapore*
EDITORA PRENTICE-HALL DO BRASIL, LTDA., *Rio de Janeiro*
WHITEHALL BOOKS LIMITED, *Wellington, New Zealand*

CONTENTS

2   THE ANALYSIS OF DC CIRCUITS   21

6 INTRODUCTION TO ELECTRONICS   209

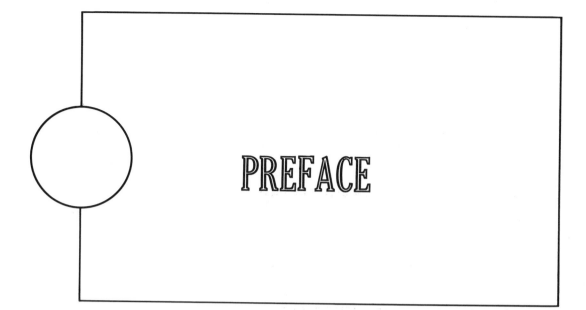

PREFACE

I agree with the philosophers: that to understand the Greeks, you must understand the ideas that dominated Greek thought; that to understand Martin Luther, you have to understand the religious and political ideas of his time; that to understand the Beatles, you must understand the thinking of the 1960s; and so forth. I am sure that the philosophers would agree with me: that to understand electrical engineering, you must understand the fundamental ideas that shape the thinking and methods of electrical engineers. In this book we examine many ideas that permeate electrical engineering at this time. These are not transitory ideas, here today and gone tomorrow; for ideas develop slowly and their influence penetrates deeply into the mentality of believers. For this reason, you can be confident that what you learn from this book will remain current. The transistor may follow the vacuum tube to obsolescence, and the computer go the way of the mechanical calculator; but the ideas that led to the transistor and the computer will persist and continue to bear fruit.

The ideas of electrical engineering, to my thinking, come from two sources: the mind of man and the mind of God. God thought up matter and gave to it useful and interesting electrical properties. Men have studied the electrical properties of matter, understood them in measure, and brought forth useful and interesting devices. To our understanding of the physical creation we should add the fruits of mathematical reasoning. Mathematics offers us an orderly means for reasoning about an orderly creation; hence mathematical methods and concepts play an important role in this book.

One warning: do not judge the importance of an idea or technique by the number of pages given to it in this book. I have sought to make every section understandable, and that means giving greater place to those subjects that are harder to understand.

ACKNOWLEDGMENTS

I appreciate the assistance of my colleagues who have used this book in draft form (Lee E. Baker, Loyd Dreher, and Bill Hamilton), several generations of students who took the trouble to tell me where they stumbled over the errors, and most of all, my wife, who has borne with almost perfect patience my occupation and preoccupation with this project.

TO THE INSTRUCTOR

Thank you for using my book. I have designed this book to ease your task: so that you do not have to spend most of your class time explaining the basic material (provided that the students have read the book, of course). This will leave you class time to work examples and show how the ideas and material in the book relate to realistic situations. For me, offering examples and applications is more rewarding than explaining the basics.

This is a one-semester introduction to circuits and electronics. I assumed at the outset that most students will balk at having to read more that about 10 pages per class period; hence my goal was about 400 pages of text, not counting the homework problems. To keep the size down, I have focused on the basic ideas, offering examples and elaborations primarily to clarify these ideas. The applications are too numerous, and they evolve too rapidly, to permit exploration of applications in a book of this length. I accepted this limitation to free you to use class time to present applications from your own experience.

You will find a few conventions in this book which are not standard:

1. I have avoided introducing voltage *rises* and *drops*, but have simply referred to the voltage across a resistor, a source, and so on. My definition of voltage is the definition of a voltage drop.

2. I always use peak values for phasor magnitudes. This is done to reinforce the concept that the time-domain voltage is related to the phasor by rotation in the complex plane and projection on the real axis. The phasor concept is very difficult for the beginning student, and I judge it wise to retain the simplest possible relationship between the phasors and the time domain quantities. The convention of this book requires putting some $\sqrt{2}$'s here and there and also some factors of $\frac{1}{2}$ in various power formulas, but in my view that is a small price to pay for the pedagogical benefits mentioned above.

3. I have included many examples in this book, but they are incorporated into the text. The recent custom of highlighting examples through a change in format tempts a lazy student to study only the examples just enough to work the homework problems. I fail to see any advantage in this shortcut to you or to the student. In this book, students have to read the text to identify the examples.

This book can be used with prerequisites of freshman physics and calculus. A simultaneous course in differential equations is desirable, if only for the mathematical maturity it develops. Additional background in mechanics is also desirable because I introduce many mechanical analogies.

The circuits section of this book is developed with sufficient care so that students should be well prepared to progress to a rigorous course in electromechanical energy conversion. The electronics section runs less deep, surveying as it does the main ideas in a subject that grows more vast and complicated every year.

TO THE NONELECTRICAL ENGINEERING READER

Let's assume the worst—that during your formative years you stuck your finger into a light socket—and to the terrible pain which resulted your mother added strong words of rebuke and warning. From that day on, you have avoided electricity, and you were careful to choose a major other than electrical engineering when you began college. So why do you have to take this course? The reason is easy to find; your faculty requires it.

From your point of view, at some point in your college career you chose a major subject—mechanical engineering, aerospace engineering, civil engineering, whatever. But from the university point of view, you chose to study under a certain faculty, and this faculty prescribes your degree requirements. They, not the electrical engineering department or your instructor in this course, require you to study electrical engineering.

Your faculty requires you to take this course because its members are convinced that you would otherwise be hampered in your career. They know that you will be working side by side with electrical engineers in design teams and other groups. They know that you need to know the vocabulary of the electrical engineer and, beyond that, you need to know how the electrical engineer thinks, how he or she solves problems. Your faculty also anticipate that you will be required to perform simple design tasks and perhaps field repairs of electrical or electronic equipment. This course will help you immeasurably in these tasks.

To these reasons for your taking this course, your author adds his personal hope that you will be enriched intellectually by this course and this book. Electrical problems offer challenges to the problem solver. The ideas and methods that have evolved from solving such problems are both practical and fascinating. If you interact with the ideas presented in this book in an honest and intelligent way, you will find yourself better able to solve problems in all your studies.

J.R. Cogdell

1

ELECTRIC CHARGE AND ELECTRICAL ENERGY

1.1 INTRODUCTION TO ELECTRICAL ENGINEERING

1.1.1 The Place of Electrical Engineering in Modern Technology

What Is Electrical Engineering? Electrical engineering is, in one sense, the opposite of lightning. Lightning unleashes electrical energy unpredictably, seldom to the benefit of human society. Electrical engineering harnesses electrical energy for human good—for transporting energy and information, for lifting the burdens of toil and tedium. When electrical energy is important for its own sake, to turn motors or illuminate department stores, we think of the electrical power industry. When energy is important for its symbolic (information) content, we think of the electronics industry. Both use electrical energy for useful purposes.

This book explains the fundamental ideas and techniques of electric circuits and electronics. Our goal is to provide you with a strong foundation to solve basic and practical problems and to furnish you with a repertoire of words and concepts for clear thinking and clear communication. With this course for preparation, you may advance to further study of circuits, electronics, or electrical energy conversion.

Electrical engineering is considered one of the more difficult subjects on most college campuses. Electrical phenomena are invisible for the most part: you cannot see the current in a wire, the magnetic field causing a motor to turn, the radio waves passing through your body at this moment. You must, therefore, develop the ability to visualize the invisible in order to master electrical concepts.

Although the mathematical demands of this book are modest, electrical engineering, like physics, uses advanced mathematical methods to solve its problems. Another

1

reason for the deserved reputation of electrical engineering as a tough curriculum originates in this mathematical sophistication.

Despite these difficulties, electrical engineering is increasingly prevalent in
modern technology. Just as electronic watches and calculators have replaced mechanical
watches and calculators, electrical parts will in the future replace many mechanical parts
in appliances, automobiles, manufacturing processes, instrumentation, and other areas.
This technological revolution, still in its infancy, originates in social, historical, and physical causes.

Why Electrical Engineering Increasingly Dominates Technology. Electrical
science has developed in recent times. Perhaps it was the invisibility of electrical phenomena which retarded scientific investigation; whatever the reason, knowledge of electrical phenomena was vague and fragmentary until about one hundred years ago when
Maxwell unified electrical theory with his famous equations. Understanding of the electrical properties of matter, particularly matter in the solid state, is of course a twentieth-
century triumph.

Significant technology usually precedes scientific understanding. Excellent swords
were manufactured before scientific metallurgy developed; violins came before acous-
tics; thoroughbreds preceded genetics. Except for the magnetic compass, however, little
significant technology predated electrical science. Again we might blame the invisibility
and resulting abstraction of electrical things for retarding the intuitive invention of useful
gadgets. It is historical fact that electrical science and technology have marched side by
side. At the present time, continued scientific advances are producing steady and spectacular progress in electrical tools and toys.

Consider next the physical factors underlying this electrical dominance. Matter,
the stuff making up the entire physical creation, is fundamentally electrical, being comprised of charged nuclei surrounded by shells of charged electrons. Most of the macroscopic properties of matter arise from electrical properties. When one bumps a toe
against a chair in the dark, does matter touch matter? No: when matter touches matter,
we get nuclear reactions, not sore toes. What we experience as toe-striking-chair is basically the electrons in our toe being repelled, without physical contact, by the electrons in
the chair. Similarly, stress–strain relationships in materials are largely macroscopic mani-
festations of the stretching of the electrical bonds holding the matter in crystalline order.
As mentioned above, recent gains in the understanding of matter in the solid state have
opened the door to new electrical technologies: transistors, lasers, integrated circuits,
liquid crystal displays.

Consider further that electric charge is bipolar, that is, that electrical engineers
work with positive and negative charges. Think of the wonderful machines one could
invent with balls that rolled uphill in addition to ordinary balls that roll downhill. That is
in effect what exists in electricity—positive charges move one way, negative charges the
other. Transistors depend on this type of action, and modern electronics exploits fully the
bipolarity of charged matter.

Consider as a third physical factor the number 1.76×10^{11}. This enormous number is the charge-to-mass ratio for the electron in the MKS system, as if the electric part dominates the mechanical part of the electron by this enormous ratio. Because of this ratio, we can control electrons electrically and move them about without supplying much mechanical energy; hence electric charges respond rapidly to electrical forces. The enormous range of frequencies which electrical engineers employ in communication networks, and the amazing speed with which electronic computers operate, testify to the significance of the charge-to-mass ratio of the electron.

The other large number favoring electrical technology over alternatives is 2.998×10^8. This you might recognize as the speed of light in the MKS system of units. This is as fast as things can go in this world, and this is exactly the speed at which electrical energy travels. When you turn on the light switch at the wall, your action is transmitted to the light bulb with the speed of light. More to the point, when the electrons in a radio antenna in Paris, France, are shaken by a radio transmitter, the electrons in the repeater satellite midway over the Atlantic respond after a delay equal to the distance (22,000 miles) divided by the velocity of light, a delay of about 0.118 second. No wonder, is it, that our communication systems depend on electrical means?

Finally, there are social factors pushing electrical technology toward dominance. There are more people all the time, and an increasing amount of information is compiled, processed, communicated, and stored per person in all modern societies. Of course, the fruits of electrical engineering, computers and communication systems, process and transport all this information.

Summary.   There exist historical, physical, and social factors which enliven electrical engineering at this time. The scientific understanding of the electrical nature of matter, the charge-to-mass ratio of the electron and the velocity of light, the bipolarity of electric charge, and the population and information explosions—all these interact to press electrical engineering to the fore. And only the tip of the iceberg shows. This is the beginning of the age of electricity.

1.1.2 Why Start with Circuit Theory?

The first half of this book is about electric circuit theory. There are three reasons why we start with circuit theory. It is neither as abstract nor as mathematically sophisticated as other branches of electrical engineering, so it is an easy place to start. More important, practically every electrical device is a circuit. A radio is a circuit, as is the power distribution system that runs your air conditioner. A computer is also a vastly complicated circuit. Hence an understanding of the language and methods of circuits is basic to everything else.

Another reason why circuits provide a logical starting point is that circuit theory has generated the language of electrical engineering. Even devices which are more sophisticated than an electrical circuit (an aircraft radar, for example), are described by electrical engineers in the language of circuits. Hence we study circuits as an introduction to the language and ideas underlying much of electrical engineering.

1.2 THE PHYSICAL BASIS OF CIRCUIT THEORY

1.2.1 Energy and Charge

Charge Is a Fundamental Physical Quantity.   When something electrical happens, a traffic light changes or an electric motor begins turning, it means that electric charges are in motion. Charge is a property of matter, like mass; indeed, charge joins mass, length, and time as one of the fundamental units from which all scientific units are derived. The unit of electric *charge* is the coulomb (abbreviated C), named in honor of Charles Augustin de Coulomb (1736–1806). There are two types of charge, *positive* and *negative*. The names derive from the fact that the two types of charge produce opposite effects. Thus equations describing the effects of charges encompass both types of charges if we associate a positive number with one type and a negative number with the other. Traditionally, the electron has been assigned a negative sign and the proton a positive. The charge of the electron is the smallest possible charge; in the MKS system of units this is

$$e = -1.602 \times 10^{-19} \text{ C} \qquad (1.1)$$

Since the mass of the electron is 9.11×10^{-31} kg, the charge-to-mass ratio of the electron is 1.76×10^{11} in the MKS system. The charge on the proton is positive in sign and equal in magnitude to that of the electron, but the proton mass is 1846 times greater. Since the charge-to-mass ratio of the fundamental bits of matter is so great, electrical effects usually dominate mechanical inertia effects. Hence we usually talk about charge as if it were not tied to mass—as if it were massless.

Forces between Charges.   We know about electric charges because charges exert forces on other charges. There are two types of forces between charges. The equation

$$\vec{F}_2 = \underbrace{\frac{Q_1 Q_2 \vec{a}_{12}}{4\pi\epsilon_0 R^2}}_{\vec{F}_e} + \underbrace{\frac{\mu_0 Q_2 \vec{u}_2 \times (Q_1 \vec{u}_1 \times \vec{a}_{12})}{4\pi R^2}}_{\vec{F}_m} \qquad (1.2)$$

gives the force on charge Q_2 caused by charge Q_1 when they are moving with velocity $\vec{u}_2$ and $\vec{u}_1$, as shown by Fig. 1.1. The constants ϵ_0 and μ_0 are required to make the units correct. The vector $\vec{a}_{12}$ has unit magnitude and is directed from Q_1 to Q_2. The first type of force, $\vec{F}_e$, describes repulsion or attraction, depending on whether the charges are alike or opposite. This type of force is called an *electrostatic* force because it depends only on the position and not the motion of the charges. This type of force is responsible for lightning, because charges are separated in clouds by droplet separation; and electrostatic forces are used in photocopy machines to form images on glass drums with charged bits of dry ink.

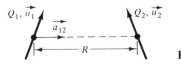

Figure 1.1 Two charges in motion.

The second type of force, $\vec{F}_m$, is motional, that is, it depends on the location and motion of the charges; these are called *magnetic* forces. Magnetic forces turn motors, deflect electron beams in TV tubes, and effect energy conversion in generators. Electrical engineers thus have both electrostatic and magnetic effects to use in manipulating electrical energy.

The Importance of Energy.   Energy is the medium of exchange in a physical system, like money in an economic system. Energy is exchanged whenever one physical thing affects another. In mechanics it takes force and movement to do work (exchange energy), and in electricity it takes electrical force and movement of charges to do work (exchange energy). The electrical force is represented by the voltage and the movement of charge by the current in an electrical circuit.

1.2.2 What Is Circuit Theory?

A Circuit Problem.   To see how voltage and current describe energy exchanges, and to illustrate what circuit theory involves, imagine that we remove the battery and one headlight from a car and connect them together with wire, as shown in Fig. 1.2. From our experience with batteries, wire, and headlights, we expect that the light will glow and also get hot, which suggests that the battery is supplying energy to the headlight. Most people assume that the energy flows through the wire, but it is more accurate to say that the wire guides the energy from the battery to the headlight. In Fig. 1.3 we show a circuit representing the physical situation depicted in Fig. 1.2: a 12-volt (V) battery symbol and a 4.8-ohm (Ω) resistor (modeling the headlight), with lines representing the wire. From Ohm's law, we calculate the current to be 2.5 amperes (A). The power is voltage multiplying current, or 30 watts (W). This power indicates the rate at which energy is flowing from the battery into the headlight, where it is converted to heat and light energy.

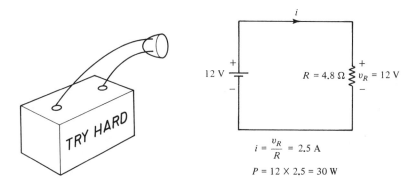

Figure 1.2   Physical circuit.   **Figure 1.3**   Circuit theory model.

The Nature of Electrical Circuit Theory.   Soon we will carefully define current, voltage, and Ohm's law. Here we intend to show only what circuit theory encompasses. The solution of an engineering problem normally proceeds through four stages: first, the problem is identified; second, the problem is modeled; third, the model is analyzed; and fourth, the results are applied to the original physical problem. In the case of the battery and headlight, we skipped the first and last steps, but we did model a physical situation (battery and headlight) with common circuit symbols and we did analyze the circuit model (2.5 A and 30 W). Circuit theory consists only in the third step: taking a given circuit model and solving for certain results through the application of known circuit laws such as Ohm's law. This third step is what you will learn to do in the first half of this book. In the second half you use circuit theory in learning the principles of electronics. We begin with the definitions of current and voltage.

1.3 CURRENT AND KIRCHHOFF'S CURRENT LAW

1.3.1 Definition of Current

Current Is Charge in Motion.   In the experiment with the battery and headlight described above, we recognize that the headlight glows because of charges moving through the wires. If we had connected our circuit with a nonconductor of electricity, say, a piece of string, the headlight would not glow. What makes an electrical conductor is the presence of conduction electrons in the material; a string has no such conduction electrons, whereas a copper wire does. A nonconductor has plenty of electrons but its electrons are bound to the nuclei and cannot move.

The movement of charge in a wire comprises an electric current. Consider a wire with a cross section of A square meters, and an electron concentration n_e of conduction electrons per cubic meter, moving with a $\bar{\text{v}}$elocity u from left to right, as pictured in Fig. 1.4. Recalling that e is the charge on the electron, we define the current to be

$$i = n_e e A u \qquad \text{C/s} \quad \text{or} \quad \text{ampere}$$

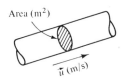

Area (m^2)

$\bar{u}$ (m/s)

Figure 1.4   Wire with current.

Note that the units of current are coulombs per second, but to honor André Ampère (1775–1836) we give this unit a special name, the ampere (A). The electron concentration for copper is about 1.13×10^{29} electrons/m^3 and hence the charge density is $n_e \times e$ or -1.81×10^{10} C/m^3. Traveling through No. 12 wire (0.081 in. in diameter) at a snail's pace of 0.1 mm/s, the charges constitute a current of $i = -5.8$ A (left to right). We could also express this result by saying that the current is $i = +5.8$ A (right to left). In this book we use a lowercase italic letter such as i to represent a quantity that may vary in time and uppercase italic letter such as I to indicate a quantity that must be constant.

Reference Directions.   That we can express the current two ways reveals the importance of reference directions. When we write "left to right" we are indicating the direction to which we are referring current flow. To specify a current, we require a reference direction plus a numerical value, which may be positive or negative. We will call current a *signed* quantity, since it can be numerically positive or negative. In this sense charge is also a signed quantity. In this book the current reference directions are indicated by arrowheads drawn on the lines representing the conductors as in Fig. 1.5, or with arrows beside the wires. The relationship between the reference direction and the sign of the current is shown in Fig. 1.5, which expresses the same physical current in two ways and with both symbols for current reference directions. The problem solver is responsible for assigning current reference directions as a beginning step in analyzing a circuit. These reference directions may be assigned without regard for the direction of physical charge flow; they are assigned for bookkeeping purposes. This freedom in assigning reference directions will be clarified below when we state Kirchhoff's current law.

Same   **Figure 1.5**   The same physical current.

To summarize, current describes the movement of charges in conductors. To specify the current in a conductor, we need both a reference direction and a numerical value, which can be positive or negative.

A Mechanical Analogy.   For most of us, an understanding of mechanics comes easier than an understanding of electrical phenomena. We have all experienced forces and motion, springs and inertia. Also, mechanics is taught early in most curricula, so we have had more instruction and more time to ponder. Because our intuition for mechanics is relatively well developed, in this book we present mechanical analogs for many electrical quantities and phenomena.

Velocity is the simplest analog for electric current, and displacement would therefore be analogous to charge. By these analogies, we mean only that the equations relating these quantities are similar.

For example, if we know the velocity of an object, $u(t)$, the change in displacement, Δx, during the period $t_1 < t < t_2$ would be

$$\Delta x = \int_{t_1}^{t_2} u(t)\ dt$$

Similarly, if we know the current in a wire, $i(t)$, the charge Δq, passing a cross section of the wire would be

$$\Delta q = \int_{t_1}^{t_2} i(t)\ dt \tag{1.3}$$

Velocity and displacement in general are vector quantities. Charge motion is also in general a vector quantity; but in circuits, the wires channel the current in established directions, and a plus or minus sign is all that remains of the "vectorness" of the current. In other words, charge flow in circuit theory corresponds to linear motion in mechanics.

An Example. A steady current of $+10^{-6}$ A flows toward the right in a copper wire of 0.001 in. diameter. Find the speed and direction of the electron flow. How many electrons pass a cross section of the wire in 1 μs?

Because the current to the right is positive, the electrons must be traveling to the left. Their velocity would be

$$u = \frac{I}{An_e|e|} = \frac{10^{-6}}{\pi(0.0005 \times 0.0254)^2 \times 1.81 \times 10^{10}} = 1.09 \times 10^{-8} \text{ m/s}$$

The total charge passing a cross section in 10^{-6} s would be

$$\Delta q = \int_0^{10^{-6}} i \, dt = 10^{-6} \text{ A} \times 10^{-6} \text{ s} = 10^{-12} \text{ C}$$

The number of electrons would be 10^{-12} C$/1.61 \times 10^{-19} = 6.21 \times 10^6$.

1.3.2 Kirchhoff's Current Law

Conservation of Charge and Charge Neutrality. All the evidence suggests that the universe was created charge-neutral; that is, there exists somewhere a positive charge for every negative charge. As was implied in our discussion on electrostatic forces [Eq. (1.2)], positive and negative charges can be separated by natural causes (leading to lightning) or man-made causes (TV tubes), but most matter does not contain surplus charge; that is, most matter is charge-neutral. This fact, combined with the fact that charge is conserved, leads to a constraint on the currents at a junction of wires.

KCL at a Junction of Wires. The junction of two or more wires is called a *node*. The constraint imposed by conservation of charge and charge neutrality is known as *Kirchhoff's current law* (KCL) and can be stated as follows: Because charge is conserved, the sum of the currents leaving a node is zero at all times. We could have written "currents entering a node" just as well. Another equivalent statement would be that the sum of the currents entering a node is equal to the sum of the currents leaving that node. Figure 1.6 illustrates KCL at node *a*. Note that a + sign is used for currents referenced departing from the node and a − sign is used for currents referenced toward the node. In general form, KCL is

$$\sum_{\text{node}} \pm i_n = 0 \tag{1.4}$$

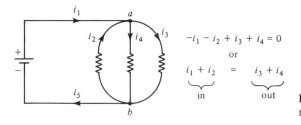

$-i_1 - i_2 + i_3 + i_4 = 0$

or

$\underbrace{i_1 + i_2}_{\text{in}} = \underbrace{i_3 + i_4}_{\text{out}}$

Figure 1.6 Kirchhoff's current law at node *a*.

The signs come from the reference directions; the i's are counted $+$ if referenced departing from the node and $-$ if referenced entering the node.

KCL Applied to Groups of Nodes.   Kirchhoff's current law is one of the fundamental laws of electric circuits because it expresses the conservation of charge. Not only does it apply to a simple node, but it applies also to groups of nodes. For example, if in Fig. 1.6 we add KCL for node a to KCL for node b, the i_2 term must appear with opposite signs in the two equations; thus the two i_2 terms will cancel. For the same reason, the i_3 and i_4 terms will also cancel. Clearly, this cancellation will always occur for the internal currents in a group of nodes. Thus in Fig. 1.6 we can easily show that i_1 and i_5 are equal by adding KCL for nodes a and b.

1.4 VOLTAGE AND KIRCHHOFF'S VOLTAGE LAW

1.4.1 Definition of Voltage

Voltage and Energy Exchanges.   Voltage expresses the potential of an electrical system for doing work. *Voltage* is defined to be the work done per unit charge by the electrical system in moving a charge from one point to another in a circuit. In Fig. 1.7 we repeat the circuit modeling the battery–headlight experiment described earlier. In the figure we have identified points in the circuit with the letters a, b, c, and d. The motion of charges around the circuit effects the transfer of energy from the battery to the headlight, which is represented by the resistor. The work done by the electrical system in moving a charge from a to b is measured by the voltage: the higher the voltage, the more the work that will be done. Specifically, the voltage from a to b is defined to be

$$v_{ab} = \frac{\text{work done by } q \text{ in moving } a \longrightarrow b}{q} \tag{1.5}$$

The units of voltage are energy/charge, joules per coulomb in the MKS system, but to honor Count Alessandro Volta (1745–1827) we use the special name volt (V) for this unit. In our hypothetical experiment, the voltage between a and b will be 12 V because of the car battery.

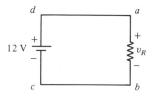

Figure 1.7 Battery circuit marked for definition of voltage.

Voltage as Potential.   Although our example implies that charges must move for the voltage to exist, such motion is not required. If the bulb were burned out and no current existed, the voltage would still be present between a and b. The voltage expresses the *potential* for doing work; that is, it measures how much work would be done if a charge were moved from a to b. It is similar to a gravitational potential, which expresses the work that would be done in a gravitational field if a mass were moved from one point to another. In other words, the voltage is defined to be the work on a hypothetical unit charge in moving between two points in a circuit. As for a gravitational potential, the path of the charge does not matter. The charge can be moved from a to b through the bulb or it can be moved outside the bulb, or it can be moved from a to the moon and then from the moon to b. In all cases the work, and hence the voltage, would be the same.

Voltage Is a Signed Quantity.   Both the work and the charge are signed quantities; hence voltage must also be a signed quantity. In the case where the battery turns on the headlight, energy is delivered by the electrical circuit to the thermodynamic system (heat and radiation); and in this case the work done by the electrical system would be considered positive. On the other hand, chemical processes in the battery are delivering energy to the electrical circuit; in this case we would consider the work done by the electrical system as negative in our definition of the voltage involving the battery. Voltage is thus the signed work done by the electrical system divided by the signed charge involved in doing that work.

There is yet a third way in which the sign of the voltage can be affected, because the direction in which the charge is moved also must be considered. If we were to move a hypothetical positive charge from a to b in Fig. 1.7, positive work would be done and the voltage would be positive. If we were to move the hypothetical positive charge from b to a, negative work would be done and the voltage would be negative. To speak meaningfully of voltage as a signed quantity, we must have a clearly defined reference direction to indicate the assumed direction of travel.

An Example.   For our battery in Fig. 1.7, $V_{dc} = +12$ V. Hence if we move a charge of $+1$ C from d to c with our hand, the electrical system (the battery) would do $+12$ joules (J) of work (on us). This means that the charge would pull on us. But if we moved a charge of -1 C from d to c, the work done by the electrical system would be -12 J, revealing that we would have to push on the charge to move it from d to c.

Voltage Reference Directions.   Because voltage is a signed quantity, we need a way to assign reference directions in writing voltage equations. We shall use two conventions for defining the reference directions of voltages. The more explicit convention uses subscripts, as in Eq. (1.5), to define the beginning point (a), the ending point (b), and thus the direction traveled (from a to b). By this convention V_{gh} would be the work done by the electrical system per charge in moving a charge from g to h. Often in a circuit, however, we desire to express the voltage across a single element, as in the present case with the battery or the headlight. In this case we can mark both ends of the circuit element with polarity symbols: a $+$ at one end and a $-$ at the other end. With this simpler notation, the convention is that you move from the $+$ to the $-$; that is, the $+$ would be the first subscript and the $-$ the second subscript. Figure 1.8 shows the circuit marked in this manner, and Fig. 1.9 shows an equally legitimate notation. Theoretically, we could also reverse the polarity mark on the battery, but we shall always consider batteries as having a numerically positive voltage.

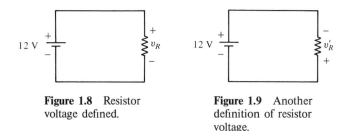

Figure 1.8 Resistor voltage defined.

Figure 1.9 Another definition of resistor voltage.

A Mechanical Analog.   Force is a mechanical analog for voltage, in two senses. As external forces impart energy to mechanical systems and thus make things happen, so voltage sources (batteries, for example) impart energy to electrical circuits. Beyond this subjective analogy, however, we will show later in this chapter that equations involving voltage are similar in form to those involving force in mechanical systems.

1.4.2 Kirchhoff's Voltage Law

Conservation of Energy.   Kirchhoff's voltage law expresses conservation of energy in electrical circuits. The charges traveling around a circuit effect the transfer of energy from one circuit element to another but do not receive energy themselves on the average. This means that if you were to move a hypothetical test charge around a complete loop in a circuit, the total energy exchanged would add to zero. During part of the loop you would have to push on the charge to move it, but during other parts of the loop it would pull on you.

Forms of KVL.   Because the energy sum is zero, it follows from the definition of voltage that the voltage sum around a closed loop is zero also. This is *Kirchhoff's voltage law* (KVL):

$$\sum_{\text{loop}} \text{voltages} = 0 \qquad (1.6)$$

We can apply KVL to the circuit in Fig. 1.7 with the result

$$v_{cd} + v_{da} + v_{ab} + v_{bc} = 0 \qquad (1.7)$$

The direct connections between d and a and between b and c imply ideal connections, and hence no voltage, between these points. Therefore, $v_{da} = 0$ and $v_{bc} = 0$, and Eq. (1.7) reduces to

$$+v_{cd} + v_{ab} = 0 \qquad (1.8)$$

When we apply KVL to voltages whose reference directions are indicated with the $+$ and $-$ polarity convention, we find that we cannot guarantee, as in Eq. (1.7), that all the signs in the resulting equation will be positive. Thus we must rewrite KVL in the form

$$\sum_{\text{loop}} \pm v\text{'s} = 0 \qquad (1.9)$$

In writing KVL equations with $+$ and $-$ polarity symbols, we write the voltage with a positive sign if the + is encountered before the − and with a negative sign if the − is encountered first. Applying this rule to Fig. 1.8 we express KVL as

$$-12 + v_R = 0$$

An Example.   As an application of KVL, let us determine the unknown voltages, v and v_{dc}, in Fig. 1.10. Here we have deliberately mixed the reference direction conventions. The boxes with the ?'s might be resistors or batteries or some other circuit element which we have not yet introduced. These boxes do not concern us, for we can write KVL from the given information without knowing what is in the boxes. We note that all the voltages around the leftmost loop are specified. Let us confirm that the given voltages satisfy KVL. We write a KVL loop equation going clockwise around the loop starting with the bottom of the battery:

$$-(10) + (+6) - (-4) = 0$$

In this equation, the signs outside the parentheses come from the reference direction and the signs inside come from the numerical value of the voltages. Thus the minus sign before the first term is there because we encounter first the − polarity symbol on the battery, and the next sign is positive because we then encounter the + polarity marking on the topmost box. We note that KVL is satisfied. What would it mean if KVL were not satisfied around this loop? It would mean that your author, or the printer, had made an error; for a circuit in which KVL is not satisfied is not a circuit at all—it is nonsense, like $1 + 1 = 3$.

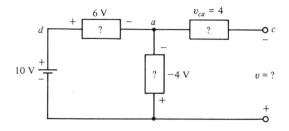

Figure 1.10   Example with mixed polarity conventions.

Let us proceed to determine the unknown v by writing KVL around the rightmost loop. What loop? There is indeed a loop even though no path exists for current, for we could still move our hypothetical test charge around this closed path. Thus KVL must be satisfied even when there is no closed path for charge to flow. Let us start at point a and go counterclockwise:

$$-(-4) + (v) + v_{ca} = 0$$
$$v = -v_{ca} - 4 = -4 - 4 = -8 \text{ V} \tag{1.10}$$

In Eq. (1.10) we have written the v_{ca} term with a + sign because we are moving from the first subscript (c) to the second (a). Finally, we can determine v_{cd} by writing KVL from c

to d to a to c:

$$v_{cd} + v_{da} + v_{ac} = 0$$

$$v_{cd} = -v_{da} - v_{ac} = -(+6) + (4) = -2 \text{ V}$$

(1.11)

Note that in the second form of Eq. (1.11) we wrote v_{da} as $+6$ because d and the $+$ symbol mark the same point in the circuit. That is, the convention for the $+$ and $-$ marking is that we move from the $+$ to the $-$, which in this case is the same as moving from d to a; hence v_{da} is $+6$ V. Note that $v_{ac} = -v_{ca}$, as shown below.

Some Rules for Subscripts.   As a second application of KVL we shall demonstrate two properties of the subscript notation for voltages. For this purpose we refer to Fig. 1.11. One application of KVL is to go from a to b and then return to a. The resulting equation is

$$v_{ab} + v_{ba} = 0 \text{ or } v_{ab} = -v_{ba}$$

Thus we see that reversing the subscripts changes the sign of a voltage. This reflects that the physical work done in moving a charge between two points is the negative of the work done in moving the same charge the opposite direction.

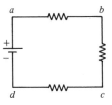

Figure 1.11   Circuit marked for subscript notation.

Let us now determine v_{ac} by moving from a to c to b and back to a:

$$v_{ac} + v_{cb} + v_{ba} = 0$$

$$v_{ac} = -v_{ba} - v_{cb}$$

$$= v_{ab} + v_{bc}$$

We eliminated the minus signs in the next-to-last form by reversing the subscripts. Note the pattern in the subscripts: the second subscript (b) of the first voltage (v_{ab}) is identical to the first subscript of the second term (v_{bc}). The final form of v_{ac} suggests that the middle point drops out, that is, does not matter. You can verify this for yourself by writing v_{ac} in terms of v_{ad} and v_{dc}. You will get the equation

$$v_{ac} = v_{ad} + v_{dc}$$

Again the middle point drops out. This makes sense: it is like saying that the distance from New York to Los Angeles is the distance from New York to St. Louis plus the distance from St. Louis to Los Angeles. Clearly, "St. Louis" is a variable; we could have just as easily said Amarillo or Tucson. These two properties of the subscript notation will prove useful in later chapters.

1.5 ENERGY FLOW IN ELECTRICAL CIRCUITS

1.5.1 Voltage, Current, and Power

Power from Voltage and Current. At the beginning of this chapter we described electrical engineering as the useful control of electrical energy. Usually, energy is being exchanged between parts of a circuit and the rate of energy flow, the power, must be calculated. We will now show that knowledge of voltage and current everywhere in an electrical circuit allows computation of energy flow (or power) throughout that circuit. Indeed, merely restating the definitions of voltage and current reveals their relationship to power flow. Voltage is the energy exchanged per charge and current is the rate of charge flow. Thus the rate of energy exchanged, the power, is by definition the product of voltage and current:

$$\frac{\text{work}}{\text{charge}} \times \frac{\text{charge}}{\text{time}} = \frac{\text{work}}{\text{time}} = \text{power}$$

Thus if we know the voltage across a circuit component and the current through that element, we can determine the power into (or out of) that component by multiplying voltage and current.

Load Sets and Source Sets. Power is a signed quantity and we must again consider reference directions. The sign in the power formula depends on the combination of the voltage and current reference directions. We show both possibilities in Fig. 1.12. In a *load set* the current reference direction is directed into the + polarity marking (or the first subscript) of the voltage reference direction. This is the convention normally used with a resistor or some other element which receives power from the circuit. In a *source set* the current reference direction is directed out of the + polarity marking (or the first subscript) of the voltage. This combination is normally used for a source of energy to the circuit such as a battery. However, in either case in Fig. 1.12 we can speak of the power out of the component and the power into the component as meaningful quantities.

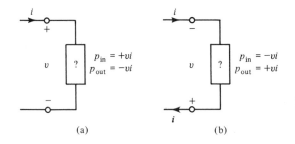

Figure 1.12 Sign conventions for power calculations: (a) load set; (b) source set.

We must stress that the formulas given in Fig. 1.12 relate solely to reference directions. Any of the numerical values of the voltages, currents, or powers can be positive or negative. The direction of energy flow at a component will be established only after the sign depending on the reference directions is combined with the numerical signs of the v's and i's.

An Example.   The use of load and source sets will be clarified in the example in Fig. 1.13. With the battery we have a source set since the current reference arrow is out of the + polarity marking. Hence the power out of the battery is $+vi$ or $+20$ W. In the preceding sentence, the + of the "$+vi$" is from the source set of reference directions and the + of the "$+20$" comes from the combination of the reference directions and the numerical values of the voltage and current. The interpretation of the positive power out of the source is that the battery is indeed delivering energy to the circuit. If the power out of the battery had been negative, we would learn that the battery is being recharged by some other source in the circuit. Box 1 has a load set and proves to be receiving energy from the circuit. Box 2 is also receiving energy from the circuit because a positive sign is produced by the combination of the minus sign from the reference directions and the minus sign of the voltage. Thus in calculating power we must continue to distinguish between signs arising from the reference directions and signs arising from the numerical values of the voltages and currents. Only the meaning of the power variable ("into" or "out of") and the numerical sign of the result allow the final conclusion as to which way energy is flowing at a given instant.

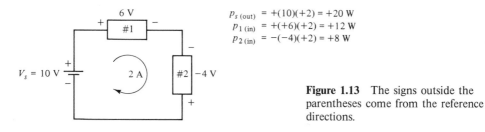

$$p_{s\,(out)} = +(10)(+2) = +20 \text{ W}$$
$$p_{1\,(in)} = +(+6)(+2) = +12 \text{ W}$$
$$p_{2\,(in)} = -(-4)(+2) = +8 \text{ W}$$

Figure 1.13 The signs outside the parentheses come from the reference directions.

Mechanical Analogies.   We have presented mechanical force, velocity, and displacement as counterparts of voltage, current, and charge, respectively. Here we shall review some of the relationships relating these mechanical variables to energy exchanged in a mechanical system. Then we will show the corresponding equations for the electrical variables.

The amount of energy exchanged, dW, by a force F operating through a distance, dx, would be

$$dW = F \, dx$$

When continuous motion is involved, the rate of energy exchanged, the power (p), would be

$$p = \frac{dW}{dt} = F \frac{dx}{dt} = Fu$$

where u is the velocity. Finally, the energy exchanged in a period of time $t_1 < t < t_2$ can be computed by integration:

$$W = \int_{t_1}^{t_2} p \; dt = \int_{t_1}^{t_2} Fu \; dt$$

These relationships have electrical counterparts. When a charge dq is moved from a to b, the energy given by the electrical circuit would be

$$dW = v_{ab} \; dq$$

When continuous charge flow, a current, is involved, the rate of energy exchanged, the power (p), would be

$$p = \frac{dW}{dt} = v_{ab} \frac{dq}{dt} = v_{ab} i \tag{1.12}$$

where i is the current referenced from a to b. Finally, the energy exchanged in a period of time $t_1 < t < t_2$ can be computed by integration:

$$W = \int_{t_1}^{t_2} p \; dt = \int_{t_1}^{t_2} v_{ab} i \; dt \tag{1.13}$$

An Example. A battery has a voltage of 1.1 V when in a discharged state but after being recharged at 0.1 A for 2 hours (T) has a full-charge voltage of 1.3 V. We shall compute the total charge given to the battery and the energy required to recharge the battery.

The charge received can be computed by integration:

$$i = \frac{dq}{dt} \Rightarrow q = \int_0^T i \; dt = iT = 0.1 \text{ A} \times 2 \text{ h} \times 3600 \text{ s/h} = 720 \text{ C}$$

We shall assume that the voltage increases linearly with time:

$$v(t) = 1.1 + 0.2\left(\frac{t}{T}\right)$$

The energy required would thus be

$$W = \int_0^T \left[1.1 + 0.2\left(\frac{t}{T}\right)\right](0.1) \; dt = 0.1\left(1.1t + \frac{0.1}{T} t^2\right)\Big|_0^T$$

$$= 864 \text{ J}$$

1.5.2 Summary

Voltage and current allow us to monitor energy flow in an electrical circuit. Voltage and current also prove to be easily measured in electrical circuits. Voltage and current also obey simple laws, KVL and KCL. For these reasons voltage and current are universally used by electrical engineers in describing the state of an electrical network.

In the next chapter we will define resistors and state Ohm's law. This addition will allow us to develop many techniques for analyzing electrical circuits. Although the knowledge gained in that chapter will allow us to analyze many practical problems, our main interest lies in the techniques of solution; for these are general enough to permit assault of many circuit problems of great complexity.

PROBLEMS

The exercises are of two types: practice problems and application problems. The practice problems are straightforward, often similar to the examples in the text and are placed in the order of the chapter discussion. Often the answers are given. The application problems follow and are intended to integrate the concepts in a realistic situation.

Practice Problems

SECTIONS 1.1–3.2

P1.1. In a copper wire 0.006 in. in diameter, electrons are moving 1 ft every minute. What is the magnitude of the current in the wire?

P1.2. The allowed safe current for a No. 10 copper wire, 0.01019 in. in diameter, is 30 A. At 30 A direct current, how fast are the electrons traveling in mm/s?

Ans: $u = 3.16$ mm/s

P1.3. For the circuit in Fig. P1.3:
  (a) Write KCL for node a.
  (b) Write KCL at node b and add to KCL at node a to show that i_1 and i_4 are equal.

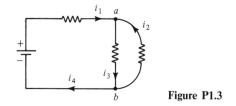

Figure P1.3

P1.4. Use KCL to determine i_1, i_2, and i_3 in the circuit shown in Fig. P1.4.

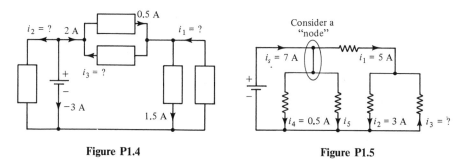

Figure P1.4   Figure P1.5

P1.5. For the circuit in Fig. P1.5, compute the unknown currents using KCL.

SECTION 4.1

P1.6. How much work is done by a 12-V battery if one electron is pulled off the negative terminal with a pair of tiny, nonconducting tweezers and placed on the positive terminal?

P1.7. In a cathode ray tube (CRT), a beam of electrons is accelerated through a certain voltage (V_c), then allowed to drift through an evacuated space until it strikes the tube surface, making a small spot of light. If $|V_c| = 10,000$ V, what velocity do the electrons acquire from the voltage source? *Hint:* Equate the final kinetic energy of an individual electron with the electrical energy delivered by the source to one electron.

Ans: $u = 5.93 \times 10^7$ m/s

SECTION 4.2

P1.8. Write KVL to determine v in the circuit in Fig. P1.8.

Ans: $v = -15$ V

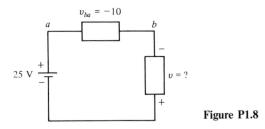

Figure P1.8

P1.9. For a circuit, $v_{ab} = 7$ V, $v_{cb} = -2$ V, $v_{dc} = 6$ V, and $v_{de} = -2$ V. Using the rules for adding subscripts, determine v_{ba}, v_{bd}, and v_{ae}. *Hint:* Draw a "circuit" showing the lettered points to aid in visualizing the patterns.

Ans: $v_{ba} = -7$ V, $v_{bd} = -4$ V, and $v_{ae} = +1$ V

P1.10. For the circuit shown in Fig. P1.10, find v_1, v_{ad}, v_{bc}, and v_{ac}.

Ans: $v_1 = +2$ V, $v_{ad} = -8$ V, $v_{bc} = +13$ V, and $v_{ac} = +6$ V

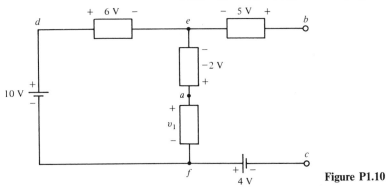

Figure P1.10

P1.11. For the circuit shown in Fig. P1.11, use KCL and KVL to determine i_1, i_2, v_{ad}, and v_x.

Ans: $i_1 = +7$ A, $i_2 = -2$ A, $v_{ad} = +5$ V, and $v_x = +12$ V

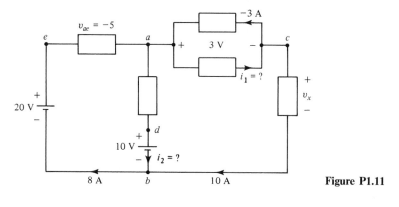

Figure P1.11

SECTIONS 5.1–5.2

P1.12. The MKS unit for energy is the joule, which is 1 watt-second, but the unit in common use by electrical utilities is the kilowatthour (kWh). How many joules are there in a kWh? At 6 cents/kWh, how many joules can one buy for a penny?

Ans: 6×10^5 J/cent

P1.13. The circuit shown in Fig. P1.13 represents a battery charger. Find the power into the battery being charged and the time it would take to impart 1 kilocoulomb (10^3 C) to the battery.

Ans: 200 s

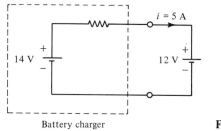

Battery charger

Figure P1.13

P1.14. For the circuit shown in Fig. P1.14:
(a) Use KCL and KVL to find V_s, v, and i.
(b) Calculate the power *into* every box and the power *out of* every source. Show that energy is conserved, that is, that $p_{out} = p_{in}$.

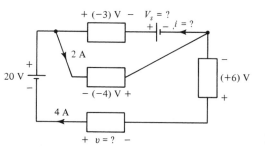

Figure P1.14

P1.15. For the circuit shown in Fig. P1.15, compute the unknown voltages and currents. (You supply the reference directions.) Compute the power *into* the two resistors and the power *out of* both sources. Show conservation of power.

Ans: $p_\text{out of sources} = 14$ W, $p_\text{into resistors} = 14$ W

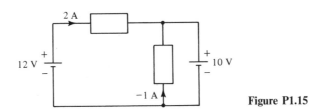

Figure P1.15

Application Problems

A1.1. In a semiconductor such as silicon, current is carried both by conduction electrons and by mobile "holes" which behave like positive charges having a charge equal in magnitude to that of the electron. Find the magnitude and direction of the current in a silicon diode of cross-sectional area 0.0001 in.^{2} with electrons (n_e of 10^{23} electrons/m^3) traveling with 0.3 mm/s to the right and holes (n_h of 1.5 $\times$ 10^{22} holes/m^3) traveling to the left at 0.1 mm/s.

A1.2. A lightning bolt might carry a current of 1000 A and last for about 1 ms. How much charge is exchanged between ground and cloud in such a lightning bolt? How many raindrops had to fall, each having a deficiency of 10 electronic charges, to create the initial charge separation neutralized by the lightning?

A1.3. A battery can be rated in "A-h," that is, ampere-hours. For example, a 60 A-h battery ought to put out 60 A for 1 h or 1 A for 60 h, or some other combination multiplying to 60. How many coulombs of charge may one hope to get out of a fully charged 80 A-h battery? How many electrons?

A1.4. A fully charged auto battery might have a voltage of 12.6 V. As the battery discharges, the voltage drops. The voltage will drop to about 10.0 V as total discharge is approached, at which point it drops to essentially zero if you try to draw current from it. Consider the case where we are discharging a 60-A-h battery over a 4-h period at a 15-A rate.

(a) Describe the voltage as a function of time during discharge.

(b) Compute the total energy output of the battery for this discharge rate.

(c) Show that the energy output is independent of the discharge rate.

(d) Change the variable of integration to be charge (instead of time) and show that the energy output is the same even if the rate of discharge varies, provided that the voltage depends only on the state of charge and not on the exact history of the output current.

THE ANALYSIS OF DC CIRCUITS

2.1 DC CIRCUIT THEORY AND OHM'S LAW

2.1.1 The Physics of Resistors

Charge, Fields, and Current.   In Chapter 1 we described electrical engineering as the utilization of electrical energy for useful purposes. We presented definitions of voltage and current and showed that we can monitor energy flow in a circuit if we know the voltage and current throughout the circuit. We also stressed that conservation of charge and conservation of energy impose constraints on the currents and voltages in a circuit, leading to Kirchhoff's current law and Kirchhoff's voltage law.

In this chapter we show the principal techniques for analyzing circuits. Although we present these for direct current (dc) circuits, these techniques are widely applicable to other types of circuits, such as alternating current (ac) circuits and pulse circuits. For this reason, we shall continue to represent voltage and current by lowercase symbols, to remind us that these could be time varying if the sources were time varying. Our sources, however, will be represented by uppercase symbols, because these are constant in time.

First we must define resistance, voltage sources, and current sources. Let us once more consider our battery–headlight experiment. We know that chemical action causes charge separation within the battery. This charge separation appears at the battery terminals, and electrostatic forces are experienced by all the charges in the vicinity of the battery, significantly by charges in the wire and the headlight. Because of these forces, electrons move in the wire and headlight, and a current exists in the circuit.

It will not surprise you that, the greater the voltage, the greater will be the resulting current. For a large class of conductors, the current increases in direct proportion to the voltage. Physical experimentation leads to the following equation, known as *Ohm's law*:

$$i = \frac{v}{R} \quad \text{or} \quad v = Ri \tag{2.1}$$

In Eq. (2.1) R is the *resistance* and has the units volts per ampere, but to honor Georg Ohm (1787–1854) we use the unit ohm, abbreviated by the uppercase Greek letter omega, Ω. The resistance of a piece of wire is directly proportional to its length and inversely proportional to its cross-sectional area. The physical material from which the wire is drawn is also a factor: copper is a good conductor; iron not so good. The resistance of a wire also depends on its temperature, but the resistance may often be considered constant. In electronic circuits we use resistors made out of a carbon-impregnated binder, high-resistance wire, or metals deposited on nonconducting surfaces. The physical behavior of such resistors leads directly to the circuit theory definition of an ideal resistance.

    The Circuit Theory of Resistance.   The circuit symbol for a resistance is given in Fig. 2.1. Note that we have arranged the reference directions of the voltage and current to be a load set. We could also write Ohm's law for a source set, which would introduce a minus sign into Ohm's law, but this is usually avoided.

    Ohm's law is presented in graphical form in Fig. 2.2. We have shown voltage to be the independent variable (cause) and current to be the dependent variable (effect). The slope of the line is the reciprocal of resistance. Customarily, Ohm's law is also stated in terms of reciprocal resistance, which is called *conductance*:

$$i = Gv \quad \text{where } G = \frac{1}{R}$$

The unit of reciprocal resistance, or conductance, is amperes per volt, but we use siemens (the official unit) or mho (the unofficial but more common term). The symbol for siemens is S, and the common symbol for the mho, "ohm" spelled backwards, is the upside-down omega (℧).

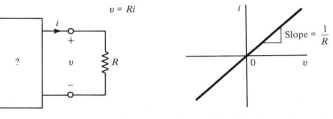

Figure 2.1   Circuit symbol for a resistance.

Figure 2.2   Ohm's law in graphical form.

    Because we have a load set, the power into the resistor is $+vi$. We can express the power into a resistor in several ways, as follows:

$$p = +vi = (Ri)i = i^2R = \frac{v^2}{R} \quad \text{watts} \tag{2.2}$$

Note that the power into a resistor is always positive. Resistance therefore always removes electrical energy from a circuit. As in the case with our battery–headlight experiment, the energy lost to the electrical circuit usually appears as heat (also some light in that case).

2.1.2 Voltage and Current Sources

Voltage Sources.   Before we can analyze circuits, we must also define voltage and current sources. In Fig. 2.3 we show the circuit symbol, mathematical, and graphical definitions for an ideal dc voltage source. The ideal voltage source maintains constant voltage, independent of its output current. In general, a voltage source may have positive or negative voltage, and it may be constant or time varying. In this chapter we will use the battery symbol for a constant (dc) voltage source and always consider the battery voltage a positive number. Note that we have used a source set for the voltage source, although this is not absolutely necessary. Normally, a voltage source would produce a positive current out of the + terminal and thus act as a source of energy for the circuit; but it is possible that some other, more powerful source might force the current out of the + terminal to be negative, thus delivering energy to the source. When this happens for a battery, we say that the battery is being charged.

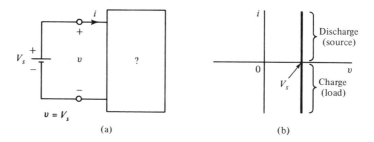

Figure 2.3   (a) Symbol for a voltage source; (b) graphical definition of a voltage source.

The dc voltage source of circuit theory models an ideal battery. We recognize that a real battery does not maintain constant voltage under heavy load—the car lights dim when the battery also has to turn the starter motor. Real batteries store a finite amount of energy and need to be recharged or replaced frequently. By contrast, the ideal voltage source can deliver energy without limit, it is inexhaustible. A physical battery can usually be represented as an ideal voltage source in a circuit problem.

Current Sources.   Figure 2.4 shows the circuit theory symbol, mathematical definition, and graphical characteristics of a current source. The current source puts out constant current, independent of its output voltage. Like the ideal voltage source, an ideal current source will deliver any required amount of energy without flinching. Unlike the voltage source, there is no physical device at your local hardware store whose electrical properties resemble a current source. In electronics, however, we can build devices which act like current sources and hence we introduce the concept here at the outset.

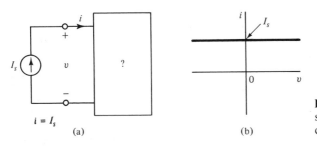

Figure 2.4 (a) Symbol for a current source; (b) graphical definition of a current source.

2.1.3 The Analysis of DC Circuits

The Battery–Headlight Circuit Solved Carefully.   We now have defined sufficient concepts and conventions to present and begin analyzing dc circuits. We will start with a careful solution of the battery–headlight problem, putting in all the steps. The circuit model is repeated in Fig. 2.5. We have used the + and − polarity notation and assigned reference directions as most people would do it, that is, such that the current reference direction is the direction in which current would be numerically positive. Shortly we will show that the choice of reference directions does not matter. Note that we have a load set for the resistor and a source set for the voltage source.

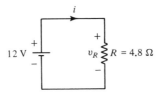

Figure 2.5 Battery–headlight circuit.

Logically, our next step is to count unknowns because we require as many equations as unknowns. We count two unknowns, i and v_R, and we therefore seek two equations. One equation comes from Ohm's law, the second from KVL:

$$v_R = 4.8i \qquad \text{and} \qquad -12 + v_R = 0$$

The solution is trivial:

$$v_R = +12 \text{ V} \qquad \text{and} \qquad i = +2.5 \text{ A}$$

The power into the resistor (headlight) is $+vi$ or $+30$ W, and the power out of the voltage source (battery) is $+30$ W.

Same Circuit, Changed Reference Directions.   We will now show that the assumed reference directions do not matter by working through the problem with reversed reference directions (Fig. 2.6). The relevant equations are still Ohm's law

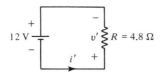

Figure 2.6 Battery–headlight circuit with different reference directions.

and KVL:

$$v'_R = 4.8i' \quad \text{and} \quad -12 - v'_R = 0$$

and the solution is again trivial:

$$v'_R = -12 \text{ V} \quad \text{and} \quad i' = -2.5 \text{ A}$$

Because the resistor still has a load set, the power into the resistor is still $+vi$, $+(-12)(-2.5)$ or $+30$ W, but this time the voltage source also has a load set. Hence the power out of the voltage source is $-vi$ or $-(12)(-2.5)$ or $+30$ W. We must accept, therefore, that the assignment of reference directions is an arbitrary matter, a preliminary we must complete in order to begin writing equations.

A Second Example.   Let us now try a harder problem. In Fig. 2.7 we show a circuit with three resistors and a current source. The resistors are called *branches* in the circuit because they connect the nodes together. Branch currents and branch voltages are already defined, including reference directions. We were careful to use load sets for the resistors and a source set for the source, but otherwise the reference directions were assigned arbitrarily. First let us count unknowns: There are four unknown v's and three unknown i's, so we require seven equations. Three of these equations come from Ohm's law:

$$v_1 = 2i_1 \quad \text{and} \quad v_2 = 4i_2 \quad \text{and} \quad v_3 = 6i_3$$

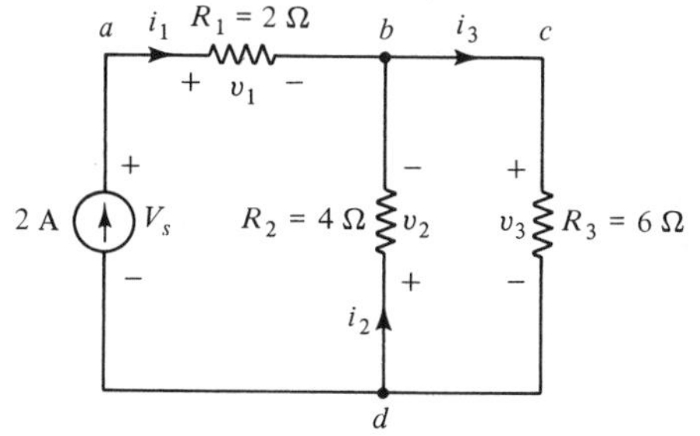

Figure 2.7   Seven equations are required to solve for the seven unknowns.

Two equations come from KVL around the two loops (*dabd* and *dbcd*):

$$-V_s + v_1 - v_2 = 0 \quad \text{and} \quad v_2 + v_3 = 0$$

We can write another KVL equation from the outer loop, but this equation can be derived from the other two and thus contains no new information. Similarly, we can write two KCL equations, but they are not independent equations (they are identical except for a minus sign); hence we get one equation from KCL:

$$-i_1 - i_2 + i_3 = 0$$

The final equation comes from the definition of a current source:

$$i_1 = 2 \text{ A}$$

The solution is mathematically trivial, but we challenge you to go through it. Unless you are very careful, you will probably make at least one mistake, not because this is a hard problem but because you become bored or impatient. The solution is

$$v_1 = 4 \text{ V} \qquad v_2 = -4.8 \text{ V} \qquad v_3 = 4.8 \text{ V}$$

$$i_2 = -1.2 \text{ A} \qquad i_3 = 0.8 \text{ A}$$

If you did attempt the solution of the seven equations in seven unknowns, whether or not you got the correct answer, you probably said to yourself, somewhere along the way, "Surely there's a better way." That is a valid hope; indeed, this chapter develops many methods which are better than the one you just used. For of all the existing methods for analyzing that circuit, you just used a nonmethod. What was wrong with this "method?" Too many equations, for one thing. But we needed all those equations because we had all those unknowns, so we conclude that defining many unknowns led to trouble. And once we had the equations, there was no system for solving them—no logical procedure to follow.

Both of these defects are addressed by the methods we present in the remainder of this chapter. We shall develop ways to reduce the number of unknowns and we shall discover systematic procedures for solving the equations that result.

2.2 SERIES AND PARALLEL RESISTORS; VOLTAGE AND CURRENT DIVIDERS

2.2.1 Series Resistors and Voltage Dividers

Resistors in Series.   Two resistors are said to be connected *in series* when the same current flows through them. Figure 2.8 shows a series connection of three resistors and a battery. It is important in this method to anticipate which way the current is positive, so we have defined the current reference direction such that the current will be positive, out of the + terminal of the voltage source; and we put the + and − polarity symbols on the resistors according to load-set conventions. Because the i's will be numerically positive, the v's across the resistors will also be positive. We can write KVL around the loop, going clockwise, as in

$$-V_s + v_1 + v_2 + v_3 = 0 \Rightarrow V_s = v_1 + v_2 + v_3 \qquad (2.3)$$

Because the v's are positive and their sum is V_s, it is reasonable to say that V_s divides between v_1, v_2, and v_3. To show how this division depends on the values of the resistors, we

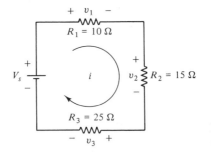

Figure 2.8   Three resistors connected in series.

introduce Ohm's law:

$$V_s = R_1 i + R_2 i + R_3 i$$
$$= (R_1 + R_2 + R_3)i \tag{2.4}$$
$$= R_{eq} i$$

That the resistors carry the same current, and thus are in series, is clearly important to the reduction of Eq. (2.4) from the first to the second form. The difference between the second and the third forms of Eq. (2.4) is that we replaced the sum of the resistors by one equivalent resistor, R_{eq}; for the current will be the same with a single 50-Ω resistor as with three series resistors totaling 50 Ω. Figure 2.9 shows a circuit which is equivalent to that in Fig. 2.8, except that the three resistors have been combined.

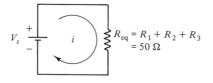

Figure 2.9   The equivalent resistor replaces three resistors in series.

Voltage Dividers.   We set out to learn how the voltage, V_s, divides between the three resistors. We have determined the current

$$i = \frac{V_s}{R_{eq}} = \frac{V_s}{50}$$

We can now obtain the voltage across the individual resistors from Ohm's law applied to the original circuit:

$$v_1 = R_1 i = R_1 \left(\frac{V_s}{R_{eq}} \right) = \frac{R_1}{R_{eq}} \times V_s = \frac{10}{50} V_s \tag{2.5}$$

We can write formulas for v_2 and v_3 similarly. For example, if V_s were 60 V, v_1 would be 12 V, v_2 would be 18 V, and v_3 would be 30 V. Notice that the 60 V divides between the series resistors. Our results are easily generalized to include an arbitrary number of series resistors; indeed, the voltage across the ith resistor would be

$$v_i = \frac{R_i}{R_{eq}} V_s \qquad \text{where } R_{eq} = R_1 + R_2 + \cdots + R_i + \cdots \text{ (all series resistors)}$$

As an intermediate result we learned that series resistors can be replaced by a single resistor whose value is the sum of the values of the series resistors. This is an important result—we will use it later to simplify complicated circuits.

2.2.2 Parallel Resistors and Current Dividers

Parallel Resistors.   Resistors are said to be connected *in parallel* when they have the same voltage across them. Figure 2.10 shows a parallel combination of three resistors and a current source. We could have defined voltages for each resistor, but because the horizontal lines represent ideal connections, wires of zero resistance, the tops of the resis-

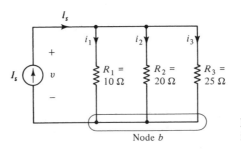

Figure 2.10   Three resistors connected in parallel.

tors are connected together and bottoms of the resistors are connected together, as we pictured them in Fig. 1.6. We were careful, however, to define the currents through the resistors with reference directions such as the current would be positive. We write KCL for node b:

$$-I_s + i_1 + i_2 + i_3 = 0 \Rightarrow I_s = i_1 + i_2 + i_3 \tag{2.6}$$

Because the current reference symbols are directed into the + end of v, we have load sets; hence we introduce Ohm's law into Eq. (2.6) in the usual form:

$$\begin{aligned}
I_s &= \frac{v}{R_1} + \frac{v}{R_2} + \frac{v}{R_3} \\
&= v\left(\frac{1}{R_1} + \frac{1}{R_2} + \frac{1}{R_3}\right) \\
&= v\left(\frac{1}{R_{eq}}\right)
\end{aligned} \tag{2.7}$$

That the resistors have the same voltage and thus are connected in parallel is important in reducing Eq. (2.7) to the second form. The third form of Eq. (2.7) introduces a resistance, R_{eq}, which is equivalent to the three parallel resistors:

$$\frac{1}{R_{eq}} = \frac{1}{R_1} + \frac{1}{R_2} + \frac{1}{R_3} \tag{2.8}$$

Current Dividers.   Equation (2.8) is important, and we will discuss it later; but now we wish to use it in determining how the current I_s divides between the three resistors. We can find the voltage from the last form of Eq. (2.7) and solve for the current in, say, R_1 from Ohm's law:

$$v = R_{eq}I_s \Rightarrow i_1 = \frac{v}{R_1} = \frac{R_{eq}I_s}{R_1} = \left(\frac{R_{eq}}{R_1}\right)I_s$$

Substituting the numbers, we learn that R_{eq} is about 5.3 Ω and that the current through R_1 is thus $0.53I_s$, or 53% of the total current through the three parallel resistors. You can confirm for yourself that 26% passes through R_2 and 21% through R_3. These results are easy to generalize: For any number of resistors which are connected in parallel, the current through the ith resistor, R_i, is

$$i_i = \frac{R_{\text{eq}}}{R_i} I_T$$

where I_T is the total current entering the parallel combination and

$$\frac{1}{R_{\text{eq}}} = \frac{1}{R_1} + \frac{1}{R_2} + \cdots + \frac{1}{R_i} + \cdots \text{ (all parallel resistors)} \qquad (2.9)$$

Equation (2.9) shows how to combine resistors in parallel. The addition of reciprocals is awkward to write in equation form but easy to accomplish on a calculator. Indeed, adding reciprocals with a calculator is just as simple as adding numbers except that you must hit the $1/x$ key on the calculator after entering the resistor values. After summing the reciprocals, you then hit the $1/x$ key to display the equivalent resistance of the parallel combination. If we desire a neat equation, we must use the conductances

$$G_{\text{eq}} = G_1 + G_2 + \cdots + G_i + \cdots \text{ (all parallel resistors)} \qquad (2.10)$$

For practice, add up the parallel combination of the three resistors in Fig. 2.10 using Eq. (2.8) to see if you get the right answer, 5.26 Ω.

Parallel resistors are common in electrical circuits, so it is important to have a simple way to indicate a parallel combination in an equation. We will use the notation $R_1 \| R_2$ to indicate that R_1 and R_2 are connected in parallel: thus

$$R_1 \| R_2 = \frac{1}{1/R_1 + 1/R_2}$$

Generalizing this notation to include more than two resistors, we calculate $12 \| 16 \| 5$ to be 2.89 Ω (check it). We note that resistors connected in parallel combine to produce a smaller equivalent resistance than any of the original resistors, whereas series resistors combine to make a larger equivalent resistance. All the lights in your house, for example, are connected in parallel since they all require the same voltage, and their combined effect is to draw more current (smaller resistance) than any one of them would individually draw.

How Voltage and Current Division Depends on Resistor Magnitude.   If you compare our solutions for series and parallel resistors, you will discover that the largest series resistor gets the most voltage in a series combination but the smallest resistor gets the most current in a parallel combination. Thus the expression "takes the path of least resistance" applies to a parallel combination, provided that you are talking about the larger part of the current, not all of the current.

Reworking Earlier Examples with Voltage Dividers.   In the preceding sections we have learned about series and parallel combinations and how voltage and current divide in them, respectively. Let us now use these concepts to solve the problem we worked in Section 2.1.3 (Fig. 2.11). We will combine resistors in order to simplify the circuit. Notice that R_2 and R_3 are connected in parallel. We can replace them with an equivalent resistance of $4 \| 6$ or 2.4 Ω. This reduces the network to that shown in Fig. 2.12a. The 2-Ω

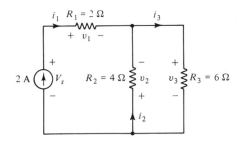

Figure 2.11 Same circuit as Fig. 2.7, to be solved by equivalent resistance method.

resistor is connected in series with the 2.4-Ω resistor and they can be combined into an equivalent resistance of 4.4 Ω (Fig. 2.12b). Clearly, V_s is 8.8 V.

We shall now restore the original circuit. Figure 12.2c is the same as Fig. 2.12a, but because we are planning to divide the 8.8 V between the two resistors we have defined voltages with the proper polarity markings for that purpose. The original v_1 already has the desired polarity marking, and we have introduced a v_4 for the voltage across the 2.4-Ω resistor. "Wait" (you might object), "that is a current source, not a voltage source you are dividing." True, but the voltage created by the current source divides in the series circuit. It does not matter whether we have a 2-A current source producing 8.8-V or a 8.8-V voltage source producing 2 A—the circuit will respond in the same way. (The type of source would matter if we changed one of the resistors in the circuit; for in one case the source current would remain constant and in the other case the source voltage would remain constant.) So we can divide the 8.8 V whether it is produced by a voltage or a current source. The results are

$$v_1 = 8.8 \times \frac{2}{4.4} = 4.00 \text{ V} \quad \text{and} \quad v_4 = 8.8 \times \frac{2.4}{4.4} = 4.80 \text{ V}$$

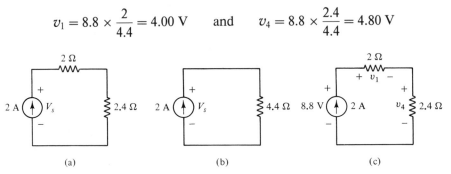

(a)   (b)   (c)

Figure 2.12 Combining resistors simplifies the network.

Because v_4 is the voltage across the parallel combination of the 4-Ω and 6-Ω resistors in the original circuit, v_4 is the same as v_3 in the original problem, and i_3 follows from Ohm's law to be 0.80 A. The voltage across the 4-Ω resistor, v_2, is also 4.8 V, but the polarity marking in Fig. 2.11 is opposite to that of v_4. Thus v_2 is -4.8 V and i_2 is -1.2 A. These complete the solution.

Reworking with Current Dividers.   We could have used current dividers instead of voltage dividers. The 2 A passes through R_1, causing 4 V across it, and then divides between the R_2 and R_3. The current through R_3 can be determined directly by current

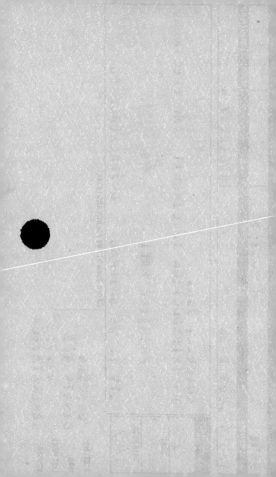

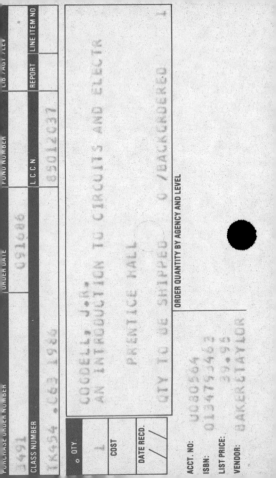

P-17

Print or type PUBLISHER's name above (and its address on back of this slip when appropriate)

Authors(s) or Editor(s) [Print or type]

Coqdull

Title in full [Print or type]

Introductor
to Circuits + Electronics

	List Price
	39.95.

Instructor's Name

Date _____ Edition _____

SPECIAL COMMENTS: _0-13-4793467-3_

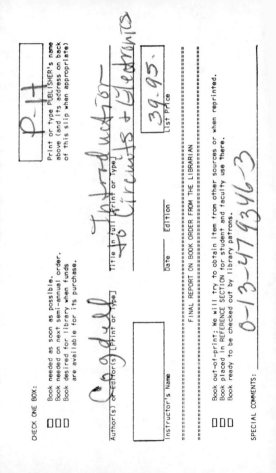

division, as also can be the current through R_2. Notice, however, that the defined reference direction for i_2 is opposite to that of the resulting current and we must account for this with a minus sign. From the currents we can calculate the voltages and complete the problem. On more complicated circuits, both voltage and current dividers might be used to learn how voltage and current distribute throughout the circuit.

A Realistic Example.   Let us work another example to show why the lights dim on a car while the starter motor is engaged. A representative circuit is shown in Fig. 2.13. We have represented the battery with a 12-V source in series with a small resistance to

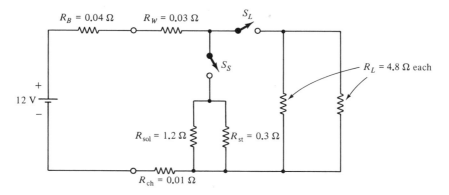

Figure 2.13 Circuit model for an automotive electrical system.

account for the internal resistance of the battery (R_B). An automotive battery cannot deliver an unlimited amount of energy and hence cannot be realistically represented here as an ideal voltage source. The resistance of the wiring is R_w and that of the ground return through the chassis is R_{ch}. The starter switch is represented by S_S and the resistances of the starter solenoid and starter motor, R_{sol} and R_{st}, are shown. Finally, S_L is the headlight switch and two R_L's represent the headlights. We will calculate the voltage across the lights with no starter and then recalculate with the starter engaged. Figure 2.14 shows the circuit, without starter engaged, after we have combined the parallel resistances of the headlights. Clearly, we have a voltage divider, and the voltage across the combined headlights will be

$$V_L = 12 \times \frac{2.4}{2.4 + 0.04 + 0.01 + 0.03}$$

$$= 11.6 \text{ V}$$

This is sufficient voltage for the lights to glow brightly.

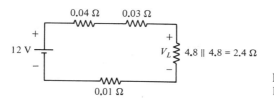

Figure 2.14 Circuit model with headlights only.

Now the starter is engaged and the circuit changes to that shown in Fig. 2.15. Both headlights are now connected in parallel with the starter circuit, and the equivalent resistance drops to 0.22 Ω, as shown. The voltage across the headlights, v'_L, will be the voltage across this parallel combination.

$$V'_L = 12 \times \frac{0.22}{0.22 + 0.04 + 0.01 + 0.03} = 8.8 \text{ V}$$

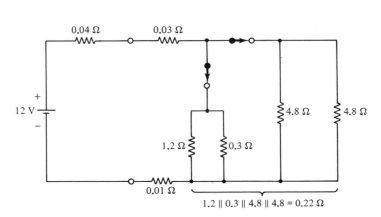

Figure 2.15 Circuit model with headlights and starter.

This reduced voltage will cause the lights to dim while the starter is engaged, as we have all observed. This effect is called *loading* because the source is weakened by the additional load of the starter. The same thing happens to the lights in a house when a heavy load such as an air conditioner is switched on the line.

Summary. Let us summarize the method we have presented in this section. We use series and parallel combinations of resistors to simplify a circuit, beginning far away from and coming toward the one source in the circuit. Ultimately, the entire circuit can be reduced to a single resistor across the source, although often the process does not have to be carried that far, as in the preceding example. The circuit is then restored, voltages and currents being divided progressively to yield the branch currents and voltages.

The Advantages of This Method. This method has many advantages. Its genius is that you are guided in the solution by the circuit connections, the shape of the circuit, so to speak. This important information is neglected in our "nonmethod"—you are faced with all those equations and no such guide. This method also avoids defining unnecessary variables. You merely introduce those you need to perform the voltage and current division. As a final advantage we might suggest that this method, after you use it a while, will develop your intuition about what is happening in circuits. The method directs your attention toward first one part of the circuit, then another; and it never forces you to consider the entire circuit at once. Before long you will find yourself possessing some feeling for how the voltage and current are going to divide. Because this method develops intuition, it will aid you in the design of simple circuits.

The Weaknesses of This Method.   Alongside these advantages we must place certain limitations. As we have presented it, the method works for circuits having only one source, although we will soon show a way around this limitation. More significantly, this method does not handle the sign for you; you are expected to bring to the problem sufficient insight to know which way the current is positive. As there is only one source, this is frequently no problem; but you do have to remember to supply the sign and not look for the mathematics to produce it. Also, this method often requires that you analyze the entire circuit, solve for virtually every branch voltage and current, to determine any single unknown. Finally, there are circuits where this method fails.

Figure 2.16 shows such a circuit. Simple though this circuit is, no two resistors are either in series or parallel. There is no place to start combining resistors. The next method we present represents a significant extension of the present method, but the revised method cannot handle the circuit in Fig. 2.16 either. However, the rest of the methods in this chapter easily solve the circuit in Fig. 2.16.

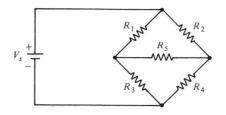

Figure 2.16 Circuit containing no parallel or series connections.

2.3 SUPERPOSITION

2.3.1 Superposition Illustrated

An Example.   The principle of superposition allows us to extend the method taught in Section 2.2. Before giving a formal definition, we shall illustrate the principle of superposition. The circuit in Fig. 2.17 will serve; we are to solve for v_2. Because the circuit has two sources, we cannot at this stage use voltage and current dividers. Thus we will use the nonmethod of Kirchhoff's voltage and current laws, plus Ohm's law.

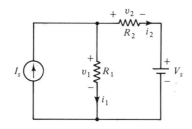

Figure 2.17 Solve for v_2 using KVL and KCL.

These are

$$\text{Ohm's law:} \quad v_1 = i_1 R_1 \quad \text{and} \quad v_2 = i_2 R_2$$

$$\text{KVL:} \quad -v_1 + v_2 + V_s = 0 \tag{2.11}$$

$$\text{KCL:} \quad -I_s + i_1 + i_2 = 0 \tag{2.12}$$

First we eliminate i_1 and i_2 with Ohm's law, so Eq. (2.12) becomes

$$\frac{v_1}{R_1} + \frac{v_2}{R_2} = I_s \tag{2.13}$$

When we eliminate v_1 between Eqs. (2.11) and (2.13), we obtain the following result:

$$v_2 = \frac{I_s - V_s/R_1}{1/R_1 + 1/R_2} = I_s(R_1 \| R_2) - V_s \times \frac{R_2}{R_1 + R_2} \tag{2.14}$$

Some Observations Based on the Result. The first form of Eq. (2.14) emerges from the algebra; the second form is more easily interpreted. Examination of the result suggests the following:

1. There are two components of v_2, one for each source. It would be reasonable to state that one part of v_2 is caused by the current source and the other part is caused by the voltage source.
2. The part of the voltage due to the current source appears to be what would be caused by the current source acting alone, provided that the voltage source is replaced by a short circuit. This is true because the two resistors are in parallel if the voltage source is a short circuit. By *short circuit* we mean an ideal connection having no voltage across it.
3. The part of the voltage due to the voltage source appears to be what would be caused by the voltage source acting alone, provided that the current source is replaced by an open circuit. By *open circuit* we mean that no path exists for current flow. Note that if there is no current flow through the current source, the two resistors are connected in series and the voltage-divider form in Eq. (2.14) is easily identified.
4. The total voltage is the sum of the separate effects of the two sources, provided that the polarity of the effects of the sources is considered. In this case the current source produces a voltage with a polarity the same as the reference direction of v_2 and it appears in the summation with a positive sign. The voltage source produces a voltage with a polarity opposite to that of v_2, and this component appears in the summation with a negative sign.

These observations suggest the general principle of superposition, which we state below.

2.3.2 The Principle of Superposition

The Principle Stated.   The response of a circuit due to multiple sources can be calculated by summing the effects of each source considered separately, all others being turned off. By "turned off" we mean that current sources are replaced by open circuits and voltage sources are replaced by short circuits.

Turned-Off Sources.   The concept of turning off a source carries importance and requires elaboration. We shall start with the graphical presentation of the i–v characteristics of a resistor in Fig. 2.18. Because the slope of the resistor characteristic is $1/R$, a zero-resistance line would have an infinite slope. The interpretation is that any amount of current can pass without creating a voltage. On the other hand, the line for an infinite resistance would have zero slope, implying zero current no matter how large the voltage. Notice that in Fig. 2.18 we have called zero resistance a short circuit and infinite resistance an open circuit. The circuit symbol for a short circuit is merely a line, and the circuit symbol for an open circuit is a break in the lines, indicating that there is no path for

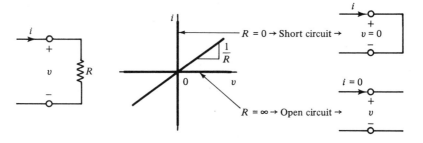

Figure 2.18   The characteristic of a short circuit is vertical and the characteristic of an open circuit is horizontal.

current. From our earlier definition of a voltage source, we know that a voltage source establishes a certain voltage at its terminals, independent of the current. In Fig. 2.19 this is represented by a vertical line at the voltage of that source (say, 12 V). It follows that a voltage source of zero volts would be represented by a vertical line of infinite slope

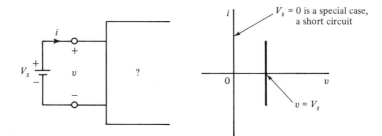

Figure 2.19   The voltage source has a vertical characteristic. Zero voltage $(V_s = 0)$ is equivalent to a short circuit.

through the origin, which is the same as the characteristic of a zero resistance. Thus a turned-off voltage source, zero volts, is a short circuit and is equivalent to zero resistance.

Figure 2.20 shows the definition of a current source. This is a horizontal line intersecting the current axis at the current equal to that of the source, because the source constrains the current, independent of the voltage. A current source of zero value produces a horizontal line through the origin, the same characteristic as an infinite resistance. Thus a turned-off current source is identical to an open circuit.

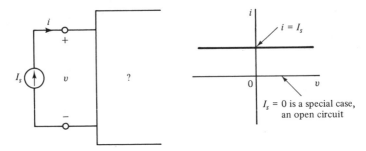

Figure 2.20   The current source has a horizontal characteristic. Zero current $(I_s = 0)$ is equivalent to an open circuit.

Superposition Restated.   Thus we can say that the response of a circuit due to multiple sources can be calculated by summing the effects of each source considered separately, all others being turned off. A turned-off voltage source is equivalent to a short circuit or zero resistance, and a turned-off current source is equivalent to an open circuit or infinite resistance.

Another Example of Superposition.   Using superposition we will solve for the voltage across the 2-Ω resistor in the circuit shown in Fig. 2.21. There are three sources, so we will have to solve three circuits which have a single source. In Fig. 2.22 we calculate the voltage due to the 4-A current source. We have turned off the 5-A current source, replacing it with an open circuit; and we have turned off the 6-V source, replacing it with

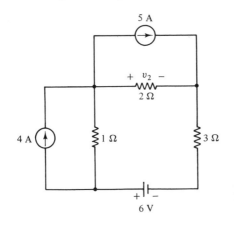

Figure 2.21   Each source will contribute to v_2.

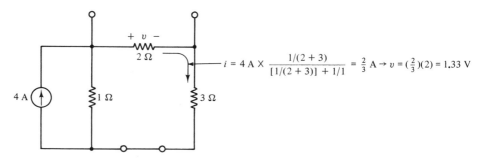

Figure 2.22 Contribution of the 4-A source with the other sources off.

a short circuit. The resulting circuit shows the 2-Ω and 3-Ω resistors connected in series with each other and together in parallel with the 1-Ω resistor. We use a current divider and then calculate the voltage across the 2-Ω resistor from Ohm's law. The polarity resulting from the 4-A source is the same as that of the total voltage across the 2-Ω resistor which we seek, and hence this contribution will appear with a + sign in the final summation.

In Fig. 2.23 we show the circuit with the 4-A and 6-V sources turned off. From the perspective of the 5-A source, the 1-Ω and the 3-Ω resistors are in series and together they are in parallel with the 2-Ω resistor. The voltage resulting from the 5-A source is thus calculated from the total equivalent resistance of the circuit as seen by the 5-A source. Here the polarity due to the source is opposite to the reference direction of the unknown total voltage, so we must use a minus sign in the final summation.

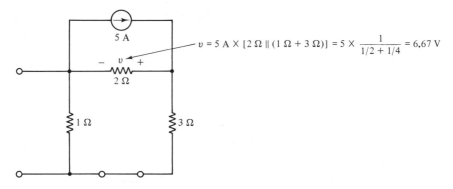

Figure 2.23 Contribution of the 5-A source with the other sources off.

Finally, we calculate in Fig. 2.24 the contribution of the voltage source. With both current sources turned off, the three resistors are connected in series and the resulting voltage due to the 6-V source is readily calculated by voltage division. Note that the sign will be positive in the final summation because the polarity of the voltage due to the volt-

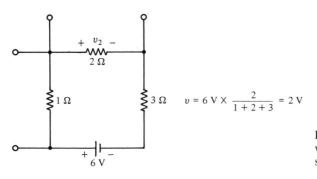

$$v = 6 \text{ V} \times \frac{2}{1 + 2 + 3} = 2 \text{ V}$$

Figure 2.24 Contribution of the voltage source with both current sources off.

age source is the same as that of the total voltage we seek. The principle of superposition states that we can now obtain the total voltage across the 2-Ω resistor by adding up the contributions of the three sources:

$$v_2 = +1.33 - 6.67 + 2 = -3.33 \text{ V} \tag{2.15}$$

2.3.3 The Limitations of Superposition

Superposition works because Kirchhoff's laws and Ohm's law are linear equations. In all KVL and KCL equations, voltage and current appear in the first power; there are no squares or square roots or functions as e^{kv}. Superposition does not work for direct power calculations, however, because power calculations involve multiplication of voltage by current. Thus you will not get the correct total power by adding the power due to each source considered separately. Superposition can be used for indirect power calculations, however, because the total current through a resistor can be computed using superposition and then this total current can be used to calculate correctly the power in that resistor. You can confirm for yourself the failure of superposition to compute power correctly by calculating the power in the 2-Ω resistor due to each source acting alone and testing if these powers add up to the true power calculated from the total voltage given in Eq. (2.15). Neither will superposition yield correct answers for circuits containing nonlinear electronic devices such as diodes or transistors.

Superposition is valid for computing the effects of individual sources on the voltage or current throughout a circuit. As such, it is a useful principle and allows us to extend the method of voltage and current dividers to circuits having multiple sources. Because this method develops intuition and thus aids in design, superposition is often used by electrical engineers.

On the other hand, this method has weaknesses. For one you often have to solve for every voltage and current, perhaps several times, to calculate a single unknown current or voltage. With the simple example we have given, there is little strain in solving the entire circuit, but in a large circuit this method becomes an ordeal. It would be much better to solve for the unknown without having to calculate every voltage and current along the way. Another weakness is the multiple solutions—much better to solve for the effects of all sources at once. Yet another weakness is that this method is still somewhat unsystematic—many decisions have to be made along the way. The methods to be presented in the remainder of this chapter eliminate these weaknesses.

2.4 NODE VOLTAGE ANALYSIS

2.4.1 The Basic Idea

Node Voltages.   The weaknesses of the preceding method are eliminated by the method of node voltage analysis, or nodal analysis. This method is both efficient and systematic. The basic idea is to write KCL at all nodes, but we write these current law equations in such a way that branch current variables are never formally defined. We avoid defining branch current variables by expressing the currents in terms of the "node voltages" in a special way.

Figure 2.25 shows a simple circuit with two voltage sources and a resistor. The point at the bottom we have denoted r for *reference node* and the points at the ends of the resistor we have called a and b. Here for clarity we use the double-subscript notation for the current: i_{ab} means the branch current flowing from node a to node b. Using KVL and Ohm's law, we may express this current thus:

$$-v_{ar} + v_{ab} + v_{br} = 0 \Rightarrow v_{ab} = v_{ar} - v_{br}$$

$$i_{ab} = \frac{v_{ab}}{R_{ab}} = \frac{v_{ar} - v_{br}}{R_{ab}} \tag{2.16}$$

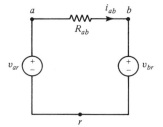

Figure 2.25   The reference node is marked r.

We will change the appearance of Eq. (2.16) by simplifying our notation. First we will drop the r in the voltage variables. The reference node that we have labeled r is like an elevation datum in surveying: We measure all elevations (voltages) relative to it. Thus when we speak of the elevation of St. Louis as being 413 ft or of the Salton Sea as being -287 ft, we are talking about the elevation relative to mean sea level, to which we assign an elevation of zero. Similarly, when we talk about the voltage at node a (v_a), we are to understand that we are talking about v_{ar}, the voltage between node a and the reference node, which is thus assigned a voltage of zero. Later we will discuss how to identify the reference node; here we are interested in the pattern of subscripts in Eq. (2.16). With the change in notation, the form becomes

$$i_{ab} = \frac{v_a - v_b}{R_{ab}} \tag{2.17}$$

The Pattern of Subscripts. Equation (2.17) can be stated in words as follows: The branch current from a to b is the voltage at a, minus the voltage at b, divided by the resistance between a and b. (This applies only when there is a direct path between a and b.) The pattern established in Eq. (2.17) is so simple and intuitive that with it we can express branch currents without defining current variables. That is, we can keep the "current from a to b" part in our heads and write on the paper the voltage at a, minus the voltage at b, divided by the resistance between a and b. This will become clear as we use this technique to solve some example problems.

2.4.2 The Node Voltage Technique

We shall find the voltage across the 4-Ω resistor and the current through the 3-Ω resistor in Fig. 2.26. We will do this through the following method.

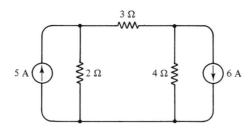

Figure 2.26 Find the voltage across the 4-Ω resistor and the current through the 3-Ω resistor.

Define a Reference Node. The circuit has three nodes. We may choose any of them as the reference node. Normally, however, the node with the most wires coming into it is chosen as the reference node. In this case the node at the bottom will be chosen. We mark the reference node with an r, as shown in Fig. 2.27.

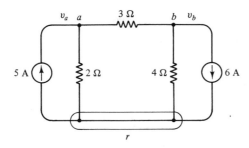

Figure 2.27 Circuit with labeled nodes. The unknowns are v_a and v_b.

Step Two. Count the independent nodes. Here we have three nodes, the reference node plus two independent nodes. This will be the number of equations to be written. A full discussion of this step follows on page 44.

Label the Independent Nodes.   We have already labeled the reference node; now we label the other nodes, as shown in Fig. 2.27. The node voltages, v_a and v_b, are our secondary unknowns from which we will calculate the primary unknowns, the voltage and current specified when the problem was defined. In nodal analysis, we first calculate the node voltages; then we calculate from these the specified unknown in the problem, a current or voltage or power somewhere.

Write KCL in a Special Form.   The special form is

$$\sum \text{currents leaving in } R\text{'s} = \sum \text{currents entering from sources}$$

Note that we avoid defining current variables; rather we express currents with the node voltages:

$$\frac{v_a - (0)}{2} + \frac{v_a - v_b}{3} = 5 \tag{2.18}$$

In Eq. (2.18) the left-hand side represents the currents leaving node a in the two resistors connected directly to node a and the right-hand side represents the current entering from the 5-A source. The two terms on the left side come from the two resistors. Note that the current flowing through the 2-Ω resistor to the reference node is merely the node voltage, v_a, divided by the resistance between node a and the reference node. We wrote the (0) for the voltage of the reference node as a reminder. Similarly, we can write KCL for node b in terms of the node voltages:

$$\frac{v_b - v_a}{3} + \frac{v_b - (0)}{4} = -6 \tag{2.19}$$

where the first term on the left-hand side represents the current flowing from node b to node a. Note that this is the negative of the second term in Eq. (2.18). This change in sign is appropriate because we are now expressing the current referenced in the opposite direction from before. Note also that the current source term on the right side has a negative sign because we are summing the currents *entering* the node from the sources.

Equations (2.18) and (2.19) are rewritten below in standard form.

$$(\tfrac{1}{2} + \tfrac{1}{3})v_a - (\tfrac{1}{3})v_b = 5$$
$$-(\tfrac{1}{3})v_a + (\tfrac{1}{3} + \tfrac{1}{4})v_b = -6$$

We may solve these equations by any method, such as Cramer's rule for determinants, with the result

$$v_a = 2.44 \text{ V} \qquad \text{and} \qquad v_b = -8.89 \text{ V}$$

Compute the Primary Unknowns.   We can now calculate the original unknowns from the resulting node voltages. The voltage across the 4-Ω resistor was sought, but we cannot calculate it without a polarity marking. If we wish the + polarity symbol at node a, we have the marking shown in Fig. 2.28. We can determine v_3 by writing KVL around the loop *r-a-b*:

$$v_{ra} + v_3 + v_{br} = 0$$
$$v_3 = -v_{ra} - v_{br} = v_{ar} - v_{br} = v_a - v_b \tag{2.20}$$

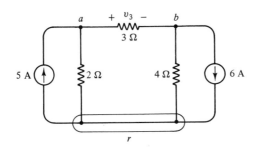

Figure 2.28 Using v_a and v_b to determine v_3.

The second line of Eq. (2.20) was converted to a form where r was the second subscript and then r was dropped, as is our custom when writing node voltages. Thus v_3 with the polarity marking of Fig. 2.28 is $2.44 - (-8.89)$ or 11.3 V. Had we marked the + at node b, we would have followed a similar procedure. You can confirm for yourself that this would have reversed the right sides of Eq. (2.20) and resulted in v_3' being -11.3 V, where the primed v_3 has the + at node b.

Another Example.   Before considering refinements of the node voltage method, let us practice on another circuit. Figure 2.29 shows a circuit similar to the one we just solved, except that it contains an additional current source. We wish to find the voltage across the 6-Ω resistor with the + polarity symbol at the left. We have marked one node as the reference node and labeled the other nodes c and d. Using the standard patterns, we write KCL at node c:

$$\frac{v_c - (0)}{6} + \frac{v_c - v_d}{7} = +(-9) + (+10)$$

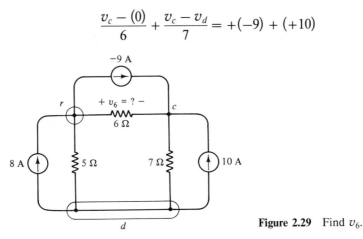

Figure 2.29   Find v_6.

Similarly, KCL for node d is the following:

$$\frac{v_d - v_c}{7} + \frac{v_d - (0)}{5} = -(+10) - (+8)$$

The solution is routine. We find v_c to be -26.0 V and v_d to be -63.3 V. These are our secondary unknowns; our primary unknown was v_6, which is v_{rc} ($-v_c$ or $+26.0$ V).

2.4.3 Some Refinements

How to Handle Voltage Sources.   The method thus far has dealt only with current sources. The circuit of Fig. 2.30, for example, is currently beyond the reach of the method. In Fig. 2.30 we have identified a reference node and labeled the other two nodes a and b. Look at the circuit carefully and you will likely see what to do with node a. If the reference node is considered to be zero volts, it constrains v_a to be $+10$ V. To assure you, we will write KVL from r to a and back to r through the source:

$$v_{ra} + 10 = 0 \Rightarrow v_{ar} = v_a = +10$$

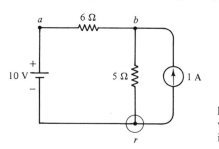

Figure 2.30 Node voltage problem with a voltage source. The unknown is v_6.

Thus when we write KCL for node b, we treat v_a as known rather than unknown:

$$\frac{v_b - (+10)}{6} + \frac{v_b - (0)}{5} = +(+1)$$

The solution for v_b is 7.27 V, which you can confirm using superposition.

In Fig. 2.31 we show another circuit with a constrained node. Here we have one node constrained to another independent node rather than to the reference node. We recognize that although the voltages at nodes a and b are unknown, they are not independent. If we knew the voltage at either of them, we could determine the voltage at the other; thus it would be inappropriate to treat them as independent unknowns. We can determine the relationship between them by writing KVL from r to a to b and back to r:

$$v_{ra} + 6 + v_{br} = 0 \Rightarrow -v_a + 6 + v_b = 0 \tag{2.21}$$

$$v_b = v_a - 6$$

In the second line of Eq. (2.21) we have written the equation as if v_a were our unknown

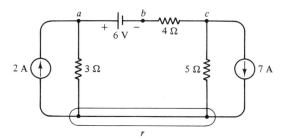

Figure 2.31   Nodes a and b are constrained by the 6-V source.

and v_b were to be expressed from v_a. Now let us write KCL for node c:

$$\frac{v_c - (0)}{5} + \frac{v_c - (v_a - 6)}{4} = -7$$

In the second term on the left side, we write $(v_a - 6)$ for v_b in expressing the current referenced out of node c in the 4-Ω resistor. When we write KCL for node a, we find that our familiar pattern breaks down when we try to write the current departing node a for b because we would have zero in numerator and denominator. This does not mean that the current is indefinite; it means that the method does not apply. In order to express the current departing node a for b, we must recognize that this current is identical to the current departing node b for c, which we can express

$$\frac{v_a - (0)}{3} + \frac{(v_a - 6) - v_c}{4} = +2$$

We now have two equations in two unknowns, v_a and v_c, which solve to -2.75 V and -20.4 V, respectively. From these we could calculate any primary unknown that was required.

Summary. We recognize that a voltage source will establish a constraint between two nodes. We can express the node voltage at one end of a voltage source in terms of the node voltage at the other end plus or minus the source value, depending on the polarity of the source. To determine the sign, we may have to write KVL around a loop containing the reference node and the two constrained nodes.

Independent Nodes. The preceding paragraph implies that a first count of the nodes of a circuit could overestimate the number of unknown node voltages to be determined. If there are voltage sources, some nodes will be constrained and the number of unknowns (and equations to be solved) will be reduced. We will designate an *independent node* as a node whose voltage cannot be derived from the voltage of another node. When we analyze a circuit using the method of node voltages, we will have as many unknowns (and equations to solve) as we have independent nodes. Here is a rule for counting independent nodes: Turn off all sources and count the nodes that remain. The number of independent nodes is one less than the number of remaining nodes. Of course, you understand what we mean by "turning off" all sources—voltage sources are replaced by short circuits and current sources are replaced by open circuits.

This concept of an independent node leads us back to the second step in our method for node voltage analysis. Step Two is to turn off all sources and determine the number of independent nodes. The third step is to label only the independent nodes, because these are the only secondary unknowns. (Of course, you may temporarily have to label the constrained nodes to facilitate writing KVL to establish how their voltages depend on the independent nodes, but this is not crucial to the method. Before long you will get the knack of how the signs go and dispense with that exercise.)

Practice for Step 2.   Let us practice by counting the independent nodes for the circuit in Fig. 2.32. When we turn off the sources, all the nodes are tied to the reference node with short circuits except that node where the 5-Ω and 6-Ω resistors connect. Thus we have only two nodes, one of which is the reference node. We thus have only one unknown node voltage and one equation to write, KCL at the independent node. See if you can correctly solve for the unknown node voltage (14.6 V).

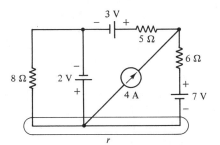

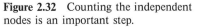

Figure 2.32   Counting the independent nodes is an important step.

But what happened to the 8-Ω resistor? It got shorted out when we turned off the 2-V source. This means that the voltage across the 8-Ω resistor is fixed by the 2-V source, independent of what goes on in the rest of the circuit. We can thus compute its voltage, current, and power independent of the remainder of the circuit. The 8-Ω resistor merely *looks* like part of the same circuit. The 2-V source, being an ideal voltage source, acts independently on the 8-Ω resistor (and any number of parallel resistors) and the rest of the circuit.

2.4.4 Critique

A Good Method.   Nodal analysis is our first really systematic method for analyzing circuits. It can be implemented somewhat routinely and always works. It is probably the favorite method of the electrical engineer for analyzing a circuit, all other factors being equal. Loop currents, the next method we present, also qualifies as a popular and powerful method.

What About KVL?   We might pause to ask: How can we solve a circuit without consideration of KVL? This is certainly what we appear to do when we use the method of node voltages. Kirchhoff's voltage law refers to branch voltages around a loop. Node voltages, on the other hand, are not branch voltages but are all referred to the same point. Hence KVL does not apply directly. If we were to use the node voltages (secondary unknowns) to compute the branch voltages (primary unknowns) around a loop and if we then add up the derived voltages, we would find that they added to zero as required by KVL.

The situation is analogous to surveying elevations with a transit level. If you level around a loop, you have to worry about closure as you return to your starting point. But if you keep the transit level at the same point and shoot all elevations from that point, there is no possible problem with closure, is there? In like manner, node voltages never violate KVL because we are measuring all voltages in the circuit relative to the same point, the reference node.

The Reference Node Versus "Ground."   When the node voltage method is presented in books and used in practice, the reference node is often called the *ground node* and given the symbol (⏚). Strictly speaking, the ground in an electrical circuit usually identifies the point which is physically connected via a thick wire to the moist earth, usually for safety purposes. We will discuss grounding in Chapter 5; here we only comment on the relationship between the reference node of nodal analysis and the physical ground of an electrical system. The grounded portion of an electrical circuit usually has many wires connecting to it. That is the whole idea—to ground much of the circuit to reduce the danger of electrical shock. Because many wires connect to the ground, the electrical ground is often designated the reference node in a nodal analysis. But this is mere coincidence: the reference node and the ground are totally different concepts. We have avoided referring to the reference node as the ground to establish the concept of the reference node without confusion with the concept of electrical grounding. You should be aware, however, that many people use the terms interchangeably when discussing nodal analysis.

2.5 LOOP CURRENT ANALYSIS

2.5.1 The Simple Method of Loop Current Analysis

Consider the circuit shown in Fig. 2.33; we are to solve for the voltage across the 2-Ω resistor. We will solve this circuit using loop currents. As we did with the node voltage method, we will give step-by-step instructions.

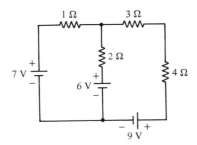

Figure 2.33 The voltage across the 2-Ω resistor is to be determined.

1. Label the independent loops 1, 2, 3, . . . and define loop currents i_1, i_2, i_3, . . . going clockwise (or counterclockwise, just be consistent) around the loops. An independent loop is a loop that does not pass through a current source. In this circuit both loops are independent loops and we get the picture of Fig. 2.34.

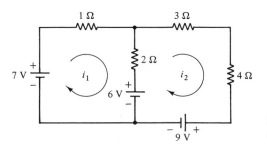

Figure 2.34 Same circuit with loop currents i_1 and i_2 defined.

2. Write KVL going clockwise (or counterclockwise) around all loops in a special form:

$$\sum \text{voltages across } R\text{'s} = \sum (+ \text{ or } -) \text{ voltage sources} \qquad (2.22)$$

On the right-hand side of Eq. (2.22) we use + if the voltage aids the loop current for that loop and − if the voltage source opposes the loop current for that loop. That is, if the voltage source were very large, would it force the loop current in that loop to be large and positive or to be large and negative relative to the clockwise reference direction? If the source forces its loop current to be positive, use a plus for that source term; if negative, use a minus.

3. Solve for the loop currents and from them compute the branch currents and voltages as required. Note that because we write one KVL equation for each loop and we have one loop current for each loop, we always get the same number of equations as unknowns.

We will now follow this plan for the circuit in Fig. 2.33. The loop currents have already been defined in Fig. 2.34. The KVL equation for loop 1 is

$$i_1(1\ \Omega) + (i_1 - i_2)(2\ \Omega) = +7 - (+6) \qquad (2.23)$$

The first term in Eq. (2.23) is the voltage across the 1-Ω resistor. We are going clockwise and we are using a load set for the voltage across the resistor; thus we automatically get a + sign for that voltage. The second term is the voltage across the 2-Ω resistor. The current in that resistor is $i_1 - i_2$, the difference between the two loop currents. Because i_1 is referenced down and i_2 is referenced up, the current in the resistor is their difference. We are going clockwise (with i_1), so we write the voltage across that resistor as $+i \times R$, where i is the current referenced in the direction we are going, $i_1 - i_2$. Thus the + sign in the equation is automatic, just like the + sign on the first term. The +7 on the right side is due to the 7-V source. It has a + sign because that source tends to force the loop current in that loop (i_1) in its positive direction. The $-(+6)$ is due to the 6-V source. It has the minus sign outside the parentheses because it opposes the positive flow of i_1.

Of course, there are not two physical currents in the 2-Ω resistor, one going up and the other going down. The loop "currents" are mathematical variables which may or may not be identical to a physical current somewhere in the circuit. In this case i_1 is the physical current in the 1-Ω resistor, i_2 is the physical current in the 3-Ω and 4-Ω resistors, and $i_1 - i_2$ is the physical current referenced downward in the 2-Ω resistor.

The KVL equation for the second loop is

$$i_2(3\ \Omega) + i_2(4\ \Omega) + (i_2 - i_1)(2\ \Omega) = +6 - (+9) \tag{2.24}$$

We are now going clockwise around loop 2. The first two terms in Eq. (2.24) are due to i_2 going through the 3-Ω and 4-Ω resistors. We automatically get + signs because we are going the same direction as the loop current. When we get to the 2-Ω resistor, we write the current as $i_2 - i_1$ because now we are going up, in the same direction as i_2. This term in Eq. (2.24) is the negative of the corresponding term in Eq. (2.23). The sign is changed because in both cases we are writing voltages across resistors going clockwise: in loop 1 this requires going through the 2-Ω resistor going down and in loop 2 this requires going up.

We now have two equations in two unknowns:

$$(1 + 2)i_1 - (2)i_2 = +1$$
$$-(2)i_1 + (3 + 4 + 2)i_2 = -3$$

Therefore, i_1 is 0.130 A and i_2 is −0.304 A. This result is not the end of the problem, however, for we set out to calculate the voltage across the 2-Ω resistor. We were deliberately silent about the reference direction on this voltage because we wanted to consider both possibilities. If we put the + polarity symbol at the top of the 2-Ω resistor, the voltage would be $+(i_1 - i_2)(2\ \Omega)$ or $[0.130 - (-0.304)](2) = +0.870$ V. If we put the + polarity symbol at the bottom of the 2-Ω resistor, the voltage would be $(i_2 - i_1)(2\ \Omega)$ or $[(-0.304) - (+0.130)](2) = -0.870$ V. That is, to avoid the minus sign in Ohm's law, we must use the current referenced into the + polarity symbol, which is either $i_1 - i_2$ or $i_2 - i_1$, depending on which polarity marking we require.

Another Example. As a second example, we invite you to write the equations for the three-loop circuit shown in Fig. 2.35. We have already labeled the loop currents for you so that you can check your results with those below. Turn back to the step-by-step instructions on page 47 and follow the directions. The KVL loop equations are

$$2i_1 + 8(i_1 - i_2) + 3(i_1 - i_3) = +10 - 9$$
$$4i_2 + 5(i_2 - i_3) + 8(i_2 - i_1) = -11 \tag{2.25}$$
$$(6 + 7)i_3 + 3(i_3 - i_1) + 5(i_3 - i_2) = +12 + 9$$

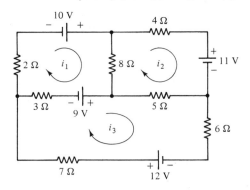

Figure 2.35 Circuit with three loops.

Note the form of the equations. On the left-hand side all the signs outside the current differences are positive. That occurs because you are writing KVL *going with the loop current* in that loop and because we use load sets for resistors. The only minus signs occur in front of the loop currents in the loops adjacent to the one you are going around. These minus signs occur because all the loop currents are referenced clockwise and thus go opposite directions in the resistors that are common to two loops. Consequently, the signs on the left-hand side exhibit a simple pattern. On the right-hand side of the equations the signs result from the direction of the sources: Do they aid or oppose the loop current of that loop?

Solving three equations in three unknowns is no fun, but you may have a fancy calculator which solves the problem automatically. If so, it ought to give these results: $i_1 = 0.0892$ A, $i_2 = -0.330$ A, and i_3 is 0.934 A.

2.5.2 Some Extensions and Fine Points

How to Handle Current Sources.   Perhaps you have noticed that, as it now stands, our loop current method cannot handle current sources. Current sources require a modest extension of the standard procedure. Consider the circuit in Fig. 2.36. The circuit has a current source; but, not knowing any better, we have defined and labeled our loop currents nevertheless. The effect of the current source would be to constrain the two loop currents passing through it; for by definition of a current source we must require that

$$i_2 - i_1 = 2 \text{ A} \tag{2.26}$$

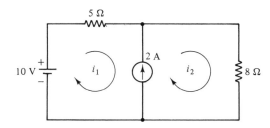

Figure 2.36 The current source constrains i_1 and i_2.

Having two unknowns, we require another equation. The second equation comes from KVL around the outer loop:

$$5i_1 + 8i_2 - 10 = 0 \tag{2.27}$$

In writing Eq. (2.27) we have departed from the standard procedure in two ways: we went around the outer loop, which does not strictly follow any single loop current; and we wrote KVL with all terms on the left-hand side of the equation, which is the way we originally learned to write KVL equations. We went around the outer loop because we do not know how to handle the current source in a KVL equation. Furthermore, we have already accounted for its effect in Eq. (2.26). Equations (2.26) and (2.27) solve to $i_1 = -0.462$ A and $i_2 = 1.538$ A. We note that i_1 is the current in the voltage source and the 5-Ω resistor and i_2 is the current in the 8-Ω resistor. Note that $i_2 - i_1$ is 2 A, as required by the current source.

The solution presented above gives the correct answer but is not the recommended way to handle current sources. We now present two additional solutions which use the concept of a "constrained loop." Figure 2.37 shows the same circuit with one unknown loop current defined, i_1', and one known loop current of 2 A defined to flow around the right-hand loop. "Wait," you should say, "that current will divide at the top and go both ways." Yes, that is true, but bear with us until we finish the two examples and then you will understand this approach. Because we now have only one unknown, we need only one equation, which comes from KVL around the loop of i_1':

$$5i_1' + 8(i_1' + 2) - 10 = 0 \qquad (2.28)$$

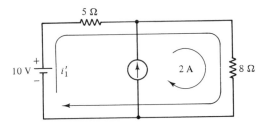

Figure 2.37 The loop current i_1' is routed to avoid the current source.

In Eq. (2.28) the term for the 8-Ω resistor includes the effect of both the unknown and the constrained loop currents. They add because both are referenced in the same direction through that resistor. The solution to Eq. (2.28) is routine: $i_1' = -0.462$. Using this and the constrained loop current, we calculate the current in the 8-Ω resistor to be $i_1' + 2 = 1.538$ A. Thus by using the constrained loop concept to handle the current source, we get the same results for the branch currents (and voltages, if we cared to compute them) as we obtained before with the first, more straightforward method. Hence we get the same results for the branch currents from this solution.

The third solution is similar to the second, except now we shall constrain the 2-A loop current from the source to flow through the 5-Ω resistor and voltage source, as shown in Fig. 2.38. (We realize that we have not yet answered your objection to our authority over the 2-A current. Please bear with us a little longer and we will clarify.) The KVL equation around the loop of i_1'' is

$$5(i_1'' - 2) + 8i_1'' - 10 = 0 \qquad (2.29)$$

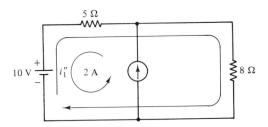

Figure 2.38 Again the loop current misses the current source. The constrained loop current is rerouted.

Solution of Eq. (2.29) is routine: $i_1'' = 1.538$ A, and the current referenced left to right in the 5-Ω resistor is $i_1'' - 2 = -0.462$ A. Again we have the same answers for the branch currents.

We have shown how the constrained loop concept works, but why does it work? Who are we to make the current from the current source go wherever we wish? The answer is: We are the ones defining variables in the problem. We are defining variables by numbering loops and by drawing currents going clockwise around those loops; and when there are current sources we are defining variables by assigning the paths by which those currents are defined to flow. Look at Figs. 2.37 and 2.38. The unknown loop current appears to be the same in both but in fact the *meaning* of that unknown differs in the two cases. By changing the path of the 2-A current, we changed the meaning of the unknown loop current. But when we computed the branch current, we got the same answer. Thus we can handle current sources by defining their currents to flow in certain paths, modifying our interpretation of the unknown loop currents accordingly.

Counting Independent Loops.   The first step in our standard procedure was to number the independent loops. We identify independent loops by turning off all sources, as we did when we wished to identify independent nodes in the method of node voltages. For example, when we turn off both sources in Fig. 2.36, the current source becomes an open circuit and the voltage source becomes a short circuit. We are left with one loop containing two resistors. Thus we have one independent loop, requiring one unknown and one KVL equation.

Loop Currents and Mesh Currents.   Strictly speaking, what we have been using up to now are mesh currents, a special class of loop currents. In circuit terminology, a loop is any closed path. A *mesh* is a special loop, namely, the smallest loop one can have. A mesh is thus a loop that contains no other loops. In the fuller sense of the loop current method, we can define the loop currents with great freedom, allowing them to go wherever we wish within certain guidelines. For our relatively simple circuits, the guidelines are that we must define the correct number of currents and that we must go through each resistor with at least one loop current.

We will rework the circuit in Fig. 2.33 as an example of this more general loop current method. The circuit is redrawn in Fig. 2.39 with true loop currents drawn as shown. Notice that we define one current clockwise and the other counterclockwise. We write KVL around the loops of the unknown currents with the following results:

$$-7 + 1(i_1' - i_2') + 2i_1' + 6 = 0$$
$$+7 - 9 + (3 + 4)i_2' + 1(i_2' - i_1') = 0$$

Therefore, $i_1' = 0.435$ A and $i_2' = 0.304$ A. Because i_1' is the only current through the

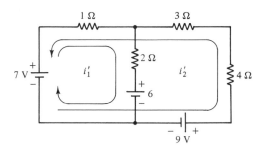

Figure 2.39 These are true loop currents, not mesh currents.

2-Ω resistor, we could have solved only for it in order to compute the voltage across that resistor, the original unknown.

The more general loop current method is useful when we wish to determine only one branch current or voltage, because we can define all the unknowns to avoid that branch except one loop current. The resulting equations can then be solved for that one loop current, which will be the desired branch current. The trouble with the generalized loop current method is that all the nice symmetries and automatic sign patterns of the mesh method vanish, and once again we are required to pay careful attention to the signs. Usually, the mesh method works best. As you will soon see, the mesh method leads to such routine patterns that we usually can write the equations by inspection.

How We Can Ignore KCL.   With the loop current method we solve for the currents of a circuit by writing only KVL equations. How can we ignore KCL? We can ignore KCL because we defined the current to flow in complete loops rather than from one point to another in the circuit. In Fig. 2.40 we show two loop currents flowing through a node. If we wrote KCL for such a node, each loop current would contribute two equal and opposite terms: each loop current must add to zero at every node. Thus KCL is satisfied automatically when we use loop currents to describe the circuit; we have only to satisfy KVL to find the solution.

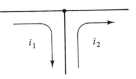

Figure 2.40   Node with two loop currents.

2.5.3 Standard Forms for the Mesh Currents and Node Voltage Methods

The Standard Form for Mesh Current Analysis.   The mesh current equations for the circuit in Fig. 2.41 are

$$R_1 i_1 + R_3(i_1 - i_2) + R_2 i_1 = +V_1 - V_2$$
$$R_4 i_2 + R_5 i_2 + R_3(i_2 - i_1) = +V_2 - V_3$$

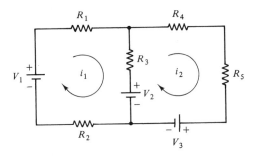

Figure 2.41   The loop current equations can be written by inspection.

When we collect terms and rewrite the equations, we see an obvious pattern:

$$(R_1 + R_2 + R_3)i_1 - (R_3)i_2 = +V_1 - V_2$$
$$-(R_3)i_1 + (R_3 + R_4 + R_5)i_2 = +V_2 - V_3 \tag{2.30}$$

Look first at the signs. The sign of the i_1 term is positive in the equation for loop 1, and the sign of the i_2 term is positive in the equation for loop 2. This results from our writing KVL equations going in the same direction as the loop currents. Similarly, the other terms on the left-hand side of each equation are negative. This also results from the way we write the equations, for the loop currents of adjacent loops always subtract in the shared resistor. Thus, if we wrote Eqs. (2.30) in matrix (or determinant) form, the diagonal terms would be positive and the off-diagonal terms would be negative. This pattern of signs will hold for all circuits solved by the mesh current method, no matter how many meshes the circuit contains.

Now look at the resistor values in Eqs. (2.30). The coefficient of i_1 in the KVL equation for loop 1 is the sum of the resistors around loop 1. This happens because we get one term for each resistor around the loop, always with a positive sign. Similarly, the coefficient of i_2 in the second equation is the sum of the resistors around loop 2. The off-diagonal coefficients are the resistances shared by the two loops. Thus the resistance in common between loops 1 and 2 will appear twice with a negative sign, as the coefficient of i_2 in the KVL equation of loop 1 and as the coefficient of i_1 in the KVL equation of loop 2.

The pattern on the right side of the equations follows from the special form of KVL: We get a term for each voltage source in the loop, with the sign determined according to whether the source aids or opposes the loop current for that loop. Thus we might summarize these patterns as follows. When the coefficients of the mesh equations are written in a matrix or determinant form, the diagonal terms are the positive sums of the resistances in the various loops. The off-diagonal terms are the negative of the resistances shared by the various loops. For example, the term in the third row and second column would be the negative of the resistance in common between loops 2 and 3. The terms on the right side of the equations are the sums of the voltage sources in the loops, the value of the source being subtracted if it opposes the current flow in that loop.

With this pattern in mind, turn back to Fig. 2.35 and write the equations for the three loop currents by inspection. Then reduce Eqs. (2.25) to the standard form and confirm that you get the correct equations. The writing of circuit equations by inspection is attractive—we have come a long way from the nonmethod with which we began this chapter—but it does require a special type of circuit. At our present level of understanding, we can use inspection provided that we have only voltage sources in a circuit and we wish to analyze by the method of mesh currents.

The Standard Form for Node Voltage Analysis.   A similar pattern emerges when we put node voltage equations into a standard form. For example, turn back to page 41 and reexamine the analysis of the circuit in Fig. 2.27. Rewrite Eqs. (2.18) and (2.19) into the following standard form:

$$(\qquad)v_a - (\qquad)v_b = (\qquad)$$
$$-(\qquad)v_a + (\qquad)v_b = (\qquad)$$

Notice that the pattern of signs is identical to what we obtained in the standard form for mesh equations. Also, the coefficients follow the same pattern, except that instead of resistances we use reciprocal resistances, or conductances. We can say, for example, that the term multiplying v_a in the nodal equation for node a is the sum of the conductances connected directly to node a, and that the coefficient for v_b in the nodal equation for node a is the negative of the conductance connected directly between nodes a and b. We can write the standard nodal equations by inspection, provided that we have only current sources in the circuit.

2.5.4 Summary of the Methods Presented So Far

When to Set up the Equations by Inspection. We have just presented methods which permit the writing of mesh and node equations in standard form by inspection. Both the signs and the magnitudes of the coefficients of the unknowns can readily be determined. Solution of the resulting equations is routine, but not trivial if more than two unknowns are to be found. However, the methods do not work when the sources are of the "wrong" kind. Later in this chapter we will show how to overcome this limitation.

Which Method to Use. We now have three methods for solving circuit problems: the method of current and voltage dividers, the method of node voltages, and the method of loop currents. How can we select the best one to use, say, on a test question? Here are some of the factors to consider:

1. One important consideration is: Are you trying to analyze or design the circuit? If the circuit is completely specified and we are trying to determine some aspect of its response, say, the power out of a source or the voltage across some resistor, the nodal or loop methods are indicated. These are general methods of analysis, which solve for the entire circuit response all at once. Special attention is not given toward the effect of a single resistor or source; rather, everything is incorporated into the equations at the beginning.

   On the other hand, if you are designing the circuit in some regard, determining a resistor here or there to yield a desired result, the method of voltage and current dividers is indicated. This method focuses on the effects of individual resistors and sources at specific points in the circuit. Design must, of course, deal in such details and this method is well suited for allowing the designer to control the way voltage and current distribute throughout a circuit.

2. Another important consideration is: How many equations must be solved? It is possible for a circuit to have more independent nodes than loops, or vice versa. The circuit shown in Fig. 2.42, for example, has two independent loops but only one independent node. Thus you would favor nodal analysis for simpler mathematics. The solution, by the way, is 1.73 A.

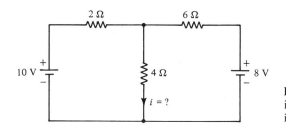

Figure 2.42 The circuit has two independent loops but only one independent node.

3. Finally, there are a host of minor considerations which would suggest a method if those discussed above fail to dictate a choice. If there are many current sources, nodal analysis is indicated, but if many voltage sources, loops might be easier. If the unknown is a voltage, nodes might be best, but if the unknown is a current, loops might be more efficient. If there is only one source and the circuit is not too complicated, the method of voltage and current dividers is indicated. These decisions come easily as a result of much experience in circuit analysis. As a beginner, you will have to practice the various methods on a number of circuits before you develop the ability to choose the most efficient method.

2.6 THÉVENIN'S AND NORTON'S EQUIVALENT CIRCUITS

2.6.1 An Example to Justify the Concept

We shall analyze the circuit shown in Fig. 2.43 by a method which you will surely think strange. We are to determine the current in the variable resistor R as a function of that resistance. This method was first proposed by a French telegrapher named Thévenin and is important for two reasons. The first reason is that it is useful as a method for circuit analysis, but the second reason is equally important. Thévenin's equivalent circuit leads to one of the most useful ideas of electrical engineering, namely, the idea of the "output resistance" of a circuit. This concept influences the thinking of electrical engineers in much of the work they do. We will now justify the method as we analyze the circuit in Fig. 2.43. Later we will distill the approach to a simpler form.

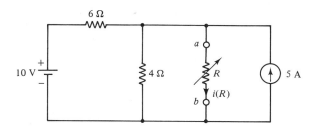

Figure 2.43 Solve for the current in R as a function of R, $i(R)$.

1. Remove the resistor. Yes, the first step is to dismantle the circuit you are trying to solve. In the lab, you could literally cut out the resistor; on paper, you merely erase it or redraw the circuit without it. The resulting circuit is shown in Fig. 2.44.

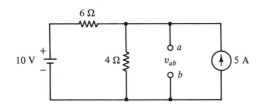

Figure 2.44 The open-circuit voltage is $v_{ab} = 16$ V.

2. Measure (in the lab) or calculate (on paper) the open circuit voltage between a and b, v_{ab}. We call v_{ab} the *open-circuit voltage* because this voltage appears between a and b with the original resistor gone, that is, with the circuit "open" between a and b. This is a good opportunity for you to calculate v_{ab} by a method of your choosing. If you get 16 V, you are correct.

3. Connect to b, but not to a, a voltage source equal to the open-circuit voltage (a 16-V battery in this case), as shown in Fig. 2.45. The voltage across the gap, $v_{aa'}$, is now zero. If you do not believe that, write KVL around a to b to a' back to a and you ought to get zero for $v_{aa'}$.

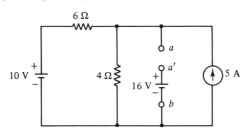

Figure 2.45 With the 16-V source inserted, $v_{aa'} = 0$ V.

4. Now connect the original resistor between a and a'. Because there is no voltage across that gap, replacing the resistor will disturb nothing and the voltage across the resistor will remain at zero. Hence *no current will exist in the resistor*. This is an important conclusion and deserves careful attention. By replacing the resistor we have restored our original circuit, except that now we have inserted a voltage source in series with the resistor. That voltage source has a polarity to oppose the flow of current through the resistor and has the exact magnitude to prevent any current from flowing. We might say that it "bucks out" the current in the resistor. In hydraulics, it would be like inserting a pump to stop fluid flow.

Now consider superposition. Normally, we would calculate the total current in the modified circuit as the combined results of all three sources, but this time we will distinguish between the original sources and the inserted source, as indicated in the equation

$$i_{\text{total}} = i_{\text{original}} - i_{\text{inserted}} = 0 \qquad (2.31)$$

In Eq. (2.31) i_{original} stands for the current through R from the original sources in the circuit, the 10-V battery and the 5-A current source; and i_{inserted} stands for the current due to the open-circuit voltage source which we inserted, a 16-V battery in this case. The minus sign comes from the polarity with which we inserted the battery, namely, so as to reduce the current to zero.

The two components must be equal. (This is where this method begins to make sense.) The component due to the original sources is what we set out to calculate in the first place. Because this is equal to the current due to the inserted source, we reason that we can calculate the current due to one source instead of calculating the current due to multiple sources (two in this case).

When we use superposition, we turn all the sources "off" except the source we are considering at the moment. Hence we calculate the effect of the inserted source by turning off the original sources. This produces the circuit shown in Fig. 2.46. Now R is seen to be connected in series with an equivalent resistance of $4 \| 6 = 2.4 \ \Omega$ and the resulting current is easily seen to be

$$i(R) = \frac{v_{\text{open circuit}}}{R + R_{\text{eq}}} = \frac{16}{R + 2.4} \tag{2.32}$$

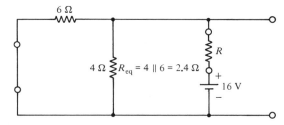

Figure 2.46 With the original sources off, the circuit is reduced to an equivalent resistance.

Equation (2.32) gives the current in R as a function of R, which was what we set out to find. Because this was an easy problem to begin with, you may wonder why we solved it by this roundabout method. The point is this: We would have derived the same simple equation as Eq. (2.32) no matter how complicated the original circuit. There could have been hundreds of sources and thousands of resistors in the original circuit and we still would have reduced the circuit to two parameters: an open-circuit voltage and an equivalent resistance. We have illustrated this method with a simple circuit to avoid distracting you with complexity. But the method itself has powerful consequences.

2.6.2 Thévenin's Equivalent Circuit

The Basic Concept.   Equation (2.32) suggests the simple equivalent circuit shown in Fig. 2.47. With this equivalent circuit, we can forget about inserting sources and putting resistors in and out of the circuit. That was merely a way to justify the equivalent circuit, which is the really important result of our derivation. The Thévenin equiva-

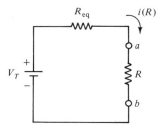

Figure 2.47 Equivalent circuit suggested by Eq. (2.32).

lent circuit consists of a voltage source, V_T, in series with an equivalent resistance. (We use the uppercase symbol for V_T to go with the battery symbol in the circuit. Actually, the technique can easily be extended to circuits with time-variable sources.) The magnitude and polarity of V_T are identical to the open-circuit voltage at *a-b*, the terminals of the resistor of interest. The equivalent resistor (R_{eq}) is computed with all sources in the circuit turned off.

Another Example.   Let us solve another problem using a Thévenin's equivalent circuit, this time skipping the justifying steps. The circuit in Fig. 2.48 is drawn with everything in a box except the resistor of special interest, which is usually called the "load." We wish to calculate the voltage across that load resistor using the method of Thévenin equivalent circuits. That is, we will replace the circuit in the box with the simpler circuit shown in Fig. 2.49. First we calculate the Thévenin voltage source (V_T), the open-circuit voltage between terminals *x* and *y* with R removed. Because the circuit has only one loop with R gone, we can employ the method of loop currents to find the current in the 20-Ω resistor and hence the required voltage. This analysis shows V_T to be 15 V with the + polarity symbol at terminal *x*. The polarity is important because we require that the two circuits in the boxes in Figs. 2.48 and 2.49 be fully equivalent to the load. This requires that the polarity in the Thévenin equivalent circuit be identical to that in the original circuit. If, for example, the open-circuit voltage had come out −15 V with

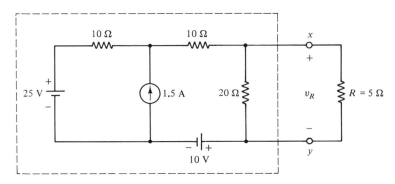

Figure 2.48   Replace the circuit in the box by a Thévenin equivalent circuit.

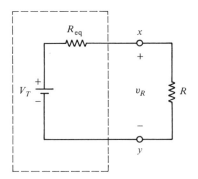

Figure 2.49   The Thévenin equivalent circuit consists of the open-circuit voltage in series with the equivalent resistance.

the + polarity symbol at terminal x, we would have turned V_T around in Fig. 2.49 or else put x at the bottom and y at the top.

The other step is to compute R_{eq}, the equivalent resistance. To make this computation, we turn off all three sources within the box. This leaves the two 10-Ω resistors in series with each other and together in parallel with the 20-Ω resistor. Thus we compute

$$R_{eq} = (10 + 10) \parallel (20) = 10 \ \Omega$$

We now have the Thévenin equivalent circuit shown in Fig. 2.50. The solution to our original problem becomes obvious: v_R is +5 V. Of course, with modest effort we could have computed this directly from the original circuit in Fig. 2.48. With this method, however, we gain the freedom to ask many additional questions such as: What value of R makes the voltage 20 V? or What value of R withdraws the most power from the circuit? These questions lead to mathematical complexities with the original circuit but can be handled simply with the equivalent circuit. The answer to the first question is that no value of R will give 20 V. The investigation of the second question leads to an interesting and important result, to which we will soon turn.

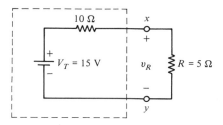

Figure 2.50 Thévenin equivalent circuit for the circuit in Fig. 2.48.

First, let us consider further what we mean by an equivalent circuit. Thévenin's equivalent circuit replaces the circuit within the box only for effects external to the box. We can no longer ask questions about what goes on in the box after we have replaced it by an equivalent circuit. For example, if we are interested in the current in the 20-Ω resistor or the total power consumed by the resistors in the box, the equivalent circuit is useless. It is useless because it is equivalent only for the external effects, not for the internal effects.

Maximum Power Transfer.   Let us now investigate the question of maximizing the power in R. This is a straightforward problem in differential calculus. The power in R as a function of R, $P(R)$, is

$$P(R) = i^2 R = \frac{V_T^2 R}{(R + R_{eq})^2} = \frac{(15)^2 R}{(R + 10)^2} \tag{2.33}$$

To maximize $P(R)$ as given in Eq. (2.33), we take the derivative with respect to R and set it equal to zero. You can confirm that the maximum (or minimum) occurs at R equals 10 Ω. To confirm that we have a maximum and not a minimum, a second derivative can be taken, but a better way in this case is to make a simple sketch of the function to show that we have found a maximum. Figure 2.51 shows such a sketch reflecting the result of

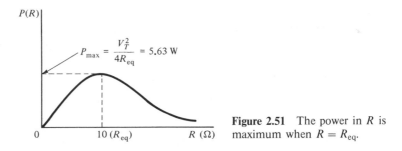

Figure 2.51 The power in R is maximum when $R = R_{eq}$.

our computation. Note that the maximum power is $V_T^2/4R_{eq}$, which is the value of $P(R_{eq})$ in Eq. (2.33).

The Importance of Maximum Power Transfer.   Why are we interested in the maximum power out of a circuit? This is an important result because often we deal in electronics with small amounts of power and wish to make full use of the power which is available. On a TV set, for example, we pull out the "rabbit ears" antenna to receive power from radio waves originating at a transmitter many miles away. The antenna on your set does not collect much power, so the TV receiver is designed to make maximum use of the power provided by the antenna. Although our results were derived for a simple battery and resistor, they can be applied to a TV antenna. Our results show that we should design the receiver input circuit, represented here by a load resistor, to have a special value for withdrawing the maximum power from the antenna. In general, the equivalent resistance of a source of power is called its *output resistance* because it is the resistance the source presents to a load. Back on page 31, for example, we represented the output resistance of a 12-V car battery by a 0.04-Ω resistor.

We might point out in closing this discussion of the Thévenin equivalent circuit that the maximum power which can be extracted from a circuit is given by $V_T^2/4R_{eq}$, as shown in Fig. 2.51. This is known as the available power from the source, the "source" being in this case the entire circuit represented by the equivalent circuit. If the source has a low output resistance, as in the case of the car battery mentioned above, it can supply much power to an external load. You might wish to compute for your own information how much power the car battery can provide, based on the output resistance given.

2.6.3 Norton's Equivalent Circuit

Not to be outdone by the French, an American engineer named Norton came up with a different equivalent circuit. Norton's equivalent circuit consists of a current source connected in parallel with an equivalent resistance, as shown in Fig. 2.52. The derivation of the values for the resistance and the current source, I_N, is similar to that for Thévenin's circuit and will not be repeated here. Indeed, the equivalent resistance is the same as before, namely, the resistance of the circuit presented to the load after all internal sources are turned off. The Norton current source has a magnitude identical to the current which would flow in a short circuit of the output terminals.

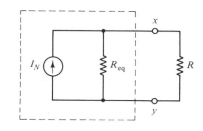

Figure 2.52 The Norton equivalent circuit appears in the box.

As an example, we will find the Norton equivalent circuit for the circuit in Fig. 2.48. The value of R_{eq} is the same as before, 10 Ω. The value of the Norton current source, I_N, can be determined by replacing the load with a short circuit. This gives the circuit in Fig. 2.53. The short circuit effectively removes the 20-Ω resistor from the circuit because it forces its voltage, and hence its current, to be zero. The current flowing through the short circuit is easily calculated by loops. The result is 1.5 A in the short circuit; hence the Norton equivalent circuit is as shown in Fig. 2.54. From this simple circuit, v_R is seen to be $1.5 \times 10 \| 5$ or +5 V, as before. Notice that the polarity of the equivalent source must again be such as to produce the current in the proper direction through the load.

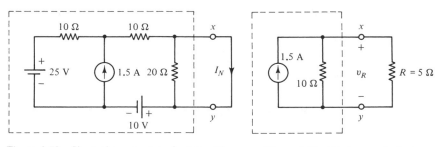

Figure 2.53 Short the output to find I_N. The 20-Ω resistor is effectively removed.

Figure 2.54 Norton equivalent circuit.

2.6.4 Relationship Between Thévenin's and Norton's Equivalent Circuits

If two circuits are equivalent to the same circuit, they must be equivalent to each other. Hence Thévenin's and Norton's circuits must be related. If the Norton's circuit in Fig. 2.54 is open-circuited, the voltage must be the same as the voltage source in the Thévenin circuit of Fig. 2.50. You can see that this proves true in this case, as it must be in all. In general, it must be true that

$$V_T = I_N R_{eq} \tag{2.34}$$

Equation (2.34) is sometimes useful in theoretical work, but it also can be applied in the laboratory to find the output resistance of a source. That is, we may be unable to get inside and turn off the internal sources, particularly in an electronic circuit, but we can measure the open-circuit output voltage and the current that flows when the output terminals are shorted. From the measured voltage and current, we can compute the output resistance from the equation

$$R_{eq} = \frac{V_T}{I_N} \tag{2.35}$$

2.6.5 Source Transformations

The transformations between Thévenin's and Norton's equivalent circuits can be used to handle "wrong" kinds of sources in nodal and loop analysis. In the nodal analysis of the circuit in Fig. 2.55a, for example, the branch from a to c can be converted to a Norton circuit to convert the 6-V source to an equivalent current source, as shown in Fig. 2.55b. Notice that the circuits in the boxes are equivalent. With the transformed circuit, the node voltage equations can be written by inspection.

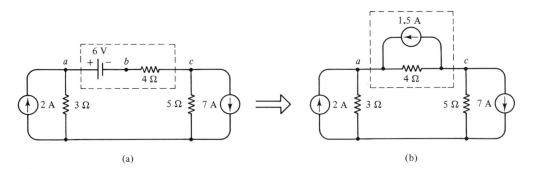

(a)   (b)

Figure 2.55   (a) The voltage source is inconvenient for nodal analysis (Fig. 2.31 repeated). (b) After the branch is transformed to a Norton equivalent circuit, the standard form may be used.

In closing this section, we wish to compare the method of equivalent circuits with our first method of voltage and current dividers. With voltage and current dividers, the strategy was to combine resistors until we have represented the entire circuit to the source as a single equivalent resistance. We then restore the original circuit, dividing voltage and current as we go. Eventually, we can calculate the current or voltage across a particular resistor of interest. If there are multiple sources, we must repeat this process for each source.

The method of equivalent circuits works in the opposite direction. Here we represent the entire circuit, including multiple sources, to the load of interest. We bring everything, as it were, to a particular point in the circuit that has special importance. For this reason, the method of equivalent circuits often plays an important role in both design and analysis of electrical circuits.

2.7 TIME-VARYING AND DEPENDENT SOURCES

2.7.1 Time-Varying Sources

Many of the sources in electrical circuits vary with time. Typically, constant sources (dc) find application in electronic circuits and automotive electrical systems; but most power circuits use alternating current (ac), and information signals must be time varying to carry information. Because Kirchhoff's laws and Ohm's law are independent of time, all equations and techniques presented in this chapter can be used for resistive circuits with time-varying sources. This statement includes the power relationships, provided that these are understood as instantaneous power as a function of time, not time-average power.

In Chapters 3, 4, and 5 we introduce circuit elements whose definitions, unlike resistors, contain time derivatives. These call for new concepts and techniques of analysis; nevertheless, many of the concepts of this chapter remain valid and relevant, for example, superposition, nodal and loop analysis, and Thévenin equivalent circuits.

2.7.2 Dependent Sources

Frequently, in modeling a physical device with a circuit, we must evoke a voltage or current source whose strength is mysteriously controlled by a voltage or current somewhere in the circuit distant from the source. Figure 2.56 shows such a circuit (actually, two connected circuits, for the single wire can carry no current and permits no voltage difference). The current source in the right-hand part of the circuit is shown to be controlled by the current in the left-hand part. For example, if $i_1 = 1$ mA, then $i_2 = -10$ mA. The current source is a dependent source, specifically, a current-controlled current source.

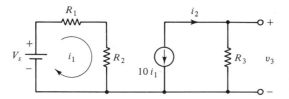

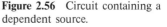

Figure 2.56 Circuit containing a dependent source.

Controlled sources appear mainly in electronics. When we listen to a telephone, for example, we rightly suppose that the current in the handset is controlled by a distant voice. "Well," you may object, "the telephone wires coming to the telephone are part of a circuit—there is no mystery here." True, but this "circuit" may include a telephone in London, a microwave repeater satellite midway over the Atlantic, and an interstate transmission system. For the purpose of circuit analysis, such a "circuit" is best modeled with controlled sources.

The four possible types of controlled sources are summarized in the table in Fig. 2.57. For the voltage source we use the general symbol for a voltage source rather than

Controlling quantity	Controlled source	Equation	Units of K	Symbol
Current, i_c	Current, i_s	$i_s = K_1 i_c$	None	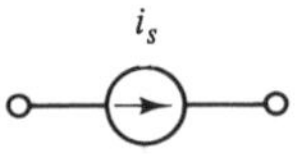
Current, i_c	Voltage, v_s	$v_s = K_2 i_c$	Ohms	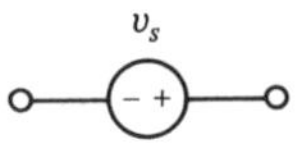
Voltage, v_c	Current, i_s	$i_s = K_3 v_c$	Siemens	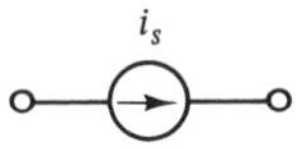
Voltage, v_c	Voltage, v_s	$v_s = K_4 v_c$	None	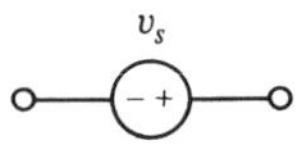

Figure 2.57 Defining equations and symbols for the four possible types of dependent sources.

the battery symbol. The units of the K's are useful in checking dimensional consistency of results. In Fig. 2.56, for example, $K_1 = 10$ and is dimensionless. The circuit in Fig. 2.56 is typical of many in electronics because it can be analyzed in stages: First the left-hand part is solved and then the right-hand part. The analysis is easy because the right-hand part has no influence on the left-hand part. More difficult are problems like that shown in Fig. 2.58, where interaction occurs between the dependent sources and their controlling variables. This circuit has one independent source (the battery), and two dependent sources. The current source depends on the voltage across R_1, and the voltage source is controlled by the current in R_2. We are required to solve for i_1. In Section 2.7.3 we apply to this problem the various methods presented in this chapter. The discussion focuses on the modifications required by the presence of the dependent sources.

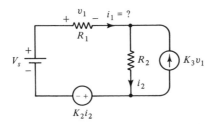

Figure 2.58 Circuit with two dependent sources. The interaction between sources and variables makes this circuit difficult.

2.7.3 Analyzing Circuits with Dependent Sources

The Nonmethod. Straightforward application of KVL, KCL, and Ohm's law must be successful on such circuits. There are four unknowns, and one can write two equations for the resistors, one KVL equation for the loop, and one KCL equation for either of the nodes. Solution of the resulting equations would be straightforward.

Voltage and Current Dividers; Series and Parallel Combinations.   This method is of limited value in problems involving controlled sources. Certainly, one can combine series or parallel resistors, provided that *none of the controlling variables disappear* in the process. But probably the circuit cannot in this way be much simplified. Furthermore, superposition can be applied to the *independent* sources but cannot be applied to the controlled sources. The controlled sources have the character of sources, but they also in part exhibit the character of resistors, as suggested by K_2 in Fig. 2.57 having the dimensions of ohms. In short, this method offers many disadvantages and no advantages when the circuit contains controlled sources.

Node Voltages and Loop Currents.   Used with care, nodal and loop analysis can be applied to circuits with dependent sources. However, the symmetries and patterns that led to the standard forms on page 52ff no longer apply. Hence we must avoid short-cuts and depend on the basic methods. As an example, we shall analyze the circuit in Fig. 2.58 by using node voltages.

The first step is to determine the number of independent nodes. For this purpose, *all* sources are turned off. For our circuit, turning off all three sources yields a circuit with one loop and two nodes, one of which is an independent node. Thus we have one unknown. In Fig. 2.59 we choose a reference node and define v_a as our secondary unknown. (Recall that we are to solve for i_1.) We shall now write KCL at node a. To write the current departing node a through the current-controlled voltage source, we must use Ohm's law on R_1. For this purpose we have identified node b, which is constrained to node a by two sources. The constraint is given by the equation

$$v_b = v_a - K_2 i_2 + V_s \tag{2.36}$$

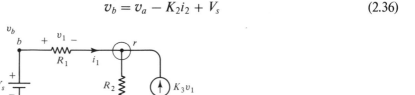

Figure 2.59   Analysis by the nodal technique.

To eliminate i_2, we note that $i_2 = -v_a/R_2$. Thus the current departing node a and passing through R_1 is

$$
\begin{aligned}
i_1 &= \frac{v_a - K_2(-v_a/R_2) + V_s - (0)}{R_1} \\
&= \frac{v_a(1 + K_2/R_2) + V_s}{R_1}
\end{aligned}
\tag{2.37}
$$

We now can write KCL at node a:

$$\frac{v_a - (0)}{R_2} + \frac{v_a(1 + K_2/R_2) + V_s}{R_1} = -K_3\left[v_a\left(1 + \frac{K_2}{R_2}\right) + V_s\right] \qquad (2.38)$$

Notice that we had to substitute the simplified version of Eq. (2.36) for $v_1(= v_b)$ as the controlling variable for the current source. The solution of Eq. (2.38) follows from

$$v_a\left[\frac{1}{R_2} + \frac{1 + K_2/R_2}{R_1} + K_3\left(1 + \frac{K_2}{R_2}\right)\right] = -\frac{V_s}{R_1} - K_3V_s \qquad (2.39)$$

Now that we have solved for our secondary unknown, we must further manipulate the equations to solve for i_1, our primary unknown. We leave this for a homework problem. At the risk of understatement, we note that one must in such problems proceed with care because the dependent sources introduce unexpected interactions between circuit variables.

For this circuit, loop current analysis works out better than nodal analysis because the unknown is a current. We leave this for a homework problem.

Thévenin and Norton Equivalent Circuits. Equivalent circuits can be used to analyze circuits with dependent sources, again with new cautions and complexities. In Fig. 2.60 we show the circuit of Fig. 2.58 with R_1 removed for the open-circuit calculation. We can determine V_T $(= v_1)$ by writing KVL around the opened loop

$$-V_s + V_T + R_2(K_3V_T) + K_2(K_3V_T) = 0 \qquad (2.40)$$

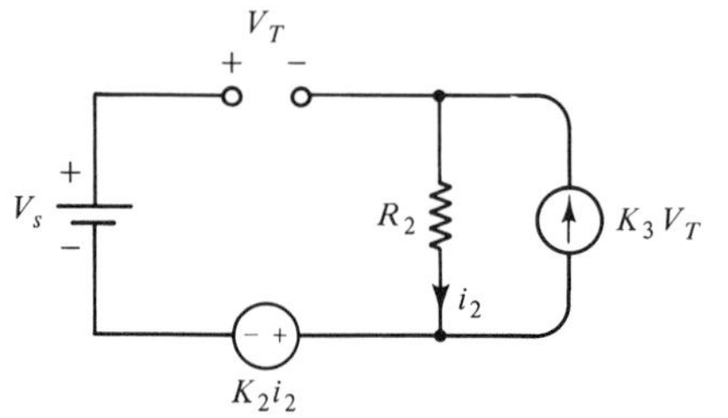

Figure 2.60 Open-circuit voltage calculation.

Note that we can write i_2 as K_3V_T because, with R_1 removed, the current from the dependent current source must pass totally through R_2. Equation (2.40) can be solved for V_T.

$$V_T = \frac{V_s}{1 + R_2K_3 + K_2K_3} \qquad (2.41)$$

We cannot combine resistors in series or parallel to determine the output resistance, R_{eq}. This approach fails because we must leave the dependent sources turned on, since they affect the output resistance. Here we have two possible approaches. One is to leave all sources on, short the output to calculate I_N, and compute the output resistance from Eq. 2.35. The other approach is to turn off the independent sources in the circuit (V_s in this case) while leaving the dependent sources operating, excite the circuit at its output with an external source (say, connect a 1-V voltage source or a 1-A current source), and compute the circuit response. If, for example, we excite the circuit with a

1-A current source and the resulting voltage is 12 V, the output resistance of the circuit would be 12 Ω.

We shall use the first approach and leave the second for a homework problem. Figure 2.61 shows the circuit with the output shorted. Note that this eliminates the dependent current source. Thus I_N is identical to i_2, and can be determined by the application of KVL around the loop:

$$-V_s + R_2 I_N + K_2 I_N = 0$$

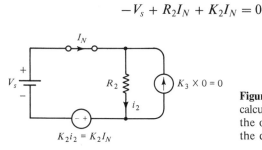

Figure 2.61   Short-circuit current calculation. Note that the shorting of the output effectively removes one of the dependent sources.

Therefore, the Norton current is

$$I_N = \frac{V_s}{R_2 + K_2} \tag{2.42}$$

Finally, we may compute the equivalent output resistance from the ratio, $R_{eq} = V_T / I_N$:

$$R_{eq} = \frac{R_2 + K_2}{1 + R_2 K_3 + K_2 K_3} \tag{2.43}$$

We leave for a homework problem the verification of the dimensional correctness of this result.

From the values of V_T and R_{eq}, we derive the Thévenin equivalent circuit in Fig. 2.62, from which the calculation of i_1 is straightforward.

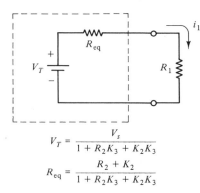

$$V_T = \frac{V_s}{1 + R_2 K_3 + K_2 K_3}$$

$$R_{eq} = \frac{R_2 + K_2}{1 + R_2 K_3 + K_2 K_3}$$

Figure 2.62   Thévenin equivalent circuit. We find R_{eq} from $R_{eq} = V_N / I_N$.

Summary.   We may use Thévenin and Norton equivalent circuit concepts for circuits with dependent sources. However, we must leave on the dependent sources for the calculation of the equivalent-circuit parameters. Such calculations may require careful consideration, but can be simpler than straightforward analysis.

PROBLEMS

Practice Problems

SECTION 2.1.1

P2.1. A resistor capable of handling safely 2 W is to be placed across the terminals of a 12-V battery. What range of resistances will not exceed the 2-W limit?

Ans: R must be greater than or equal to 72 Ω

SECTIONS 2.1.2–2.1.3

P2.2. For the circuit shown in Fig. P2.2, use the nonmethod shown in Section 2.1.3 to solve for v_5, i_{10}, and i_{20}. Compute the power out of each source and into each resistor, and show that power is conserved.

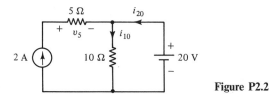

Figure P2.2

P2.3. Using the nonmethod described in Section 2.1.3, solve for all voltages and currents in the circuit in Fig. P2.3. The polarity directions have been assigned arbitrarily. Bear in mind as you proceed that this method is being displayed as a standard to make attractive the more efficient methods of the remainder of this chapter.

Ans: (currents only) $i_1 = -4$ A, $i_2 = 0.8$ A, $i_3 = 0.8$ A, and $i_4 = 4.8$ A

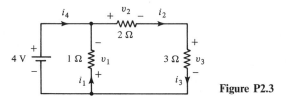

Figure P2.3

SECTION 2.2.1

P2.4. **(a)** Three resistors connected in series have resistance values in the ratio $1:2:3$ and combine to an equivalent resistance of 120 Ω. What is the smallest resistor?

   (b) The three resistors, still in series, are placed across a 300-V dc voltage source. What is the voltage that will appear across the largest resistor?

   (c) Two series resistors are to work as a voltage divider, with the smaller getting 30% of the total voltage. What are the resistors, given that their combined resistance is to be 200 Ω?

Ans: (a) 20 Ω; (b) 150 V; (c) 60 Ω, 140 Ω

P2.5. For the circuit shown in Fig. P2.5, the resistor R is variable; hence the voltage across R, $v(R)$, will depend on the value of R, as the notation indicates. Determine $v(R)$ and, from that, the value of R to make the voltage 5.5 V.

Ans: $R = 8.46$ Ω

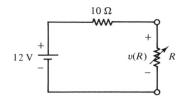

Figure P2.5 The arrow means that R can vary.

SECTION 2.2.2

P2.6. (a) Find the equivalent resistance of the parallel combination shown in Fig. P2.6.
(b) If 10 A enters the parallel combination, referenced in at a and out at b, what is the current referenced downward in the 5-Ω resistor? What is the current referenced upward in the 5-Ω resistor? What is the voltage, v_{ab}?

Ans: (a) 2.73 Ω; (b) 5.45 A, -5.45 A, 27.3 V

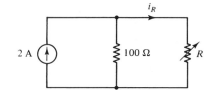

Figure P2.6

P2.7. For the circuit shown in Fig. P2.7:
(a) Find R such that $i_R = 0.5$ A.
(b) Find R (a different R) such that the power in R is 50 W.

Ans: 300 Ω, 17.2 Ω or 583 Ω

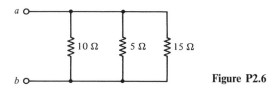

Figure P2.7 The arrow means that R can vary.

P2.8. Design a current divider that has an equivalent resistance of 50 Ω and divides the current in the ratio of 2:1. This problem is summarized in Fig. P2.8.

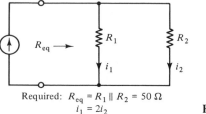

Required: $R_{eq} = R_1 \| R_2 = 50\ \Omega$
$i_1 = 2i_2$

Figure P2.8

P2.9. (a) Evaluate $R_{eq} = 7 \| (5 + 6 \| 8 + 1 \| 2)$.
(b) Draw the circuit corresponding to this expression.
(c) What resistance in parallel with 90 Ω reduces the parallel resistance to 70 Ω?

Ans: (a) 3.96 Ω; (c) 315 Ω

P2.10. For the circuits shown in Fig. 2.10, solve for the indicated unknown using voltage-
and current-divider techniques.

   Ans: (a) $i = 5.38$ A, $v = 46.2$ V; (b) $i = 0.182$ A, $v = 3.15$ V; (c) $i = -1.38$ A;
   (d) $P_{\text{out}} = -20$ W, $v_{10} = -2.22$ V

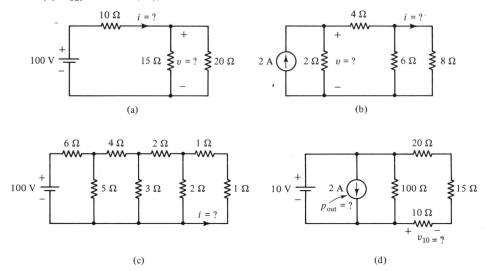

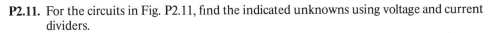

Figure P2.10

P2.11. For the circuits in Fig. P2.11, find the indicated unknowns using voltage and current
dividers.

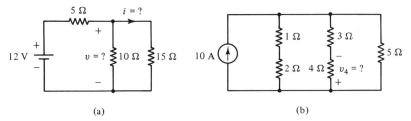

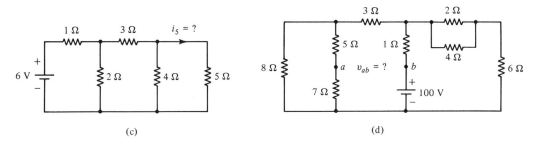

Figure P2.11

SECTIONS 2.3.1–2.3.3

P2.12. Find the current in the 100-Ω resistor in Fig. P2.12 using superposition.

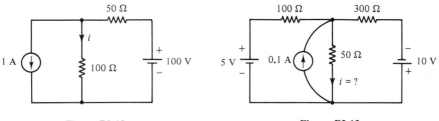

Figure P2.12   **Figure P2.13**

P2.13. In Fig. P2.13, find i using the principle of superposition.

P2.14. For the circuit in Fig. P2.14, determine the current in the 10-Ω resistor with the reference direction shown using the principle of superposition.

Ans: +0.67 A

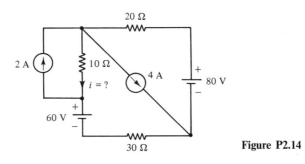

Figure P2.14

SECTIONS 2.4.1–2.4.4

P2.15. For the circuit shown in Fig. P2.15, solve for v_a and v_b using node voltage techniques. Current dividers would provide an easy check on your results.

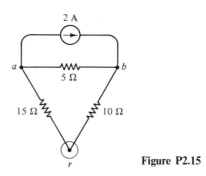

Figure P2.15

P2.16. For the circuit shown in Fig. P2.16, $v_{ar} = v_a = +6$ V due to the battery. Write the KCL equation for node b using the node voltage patterns and solve for $v_{br} = v_b$. Check your solution using the voltage-divider method.

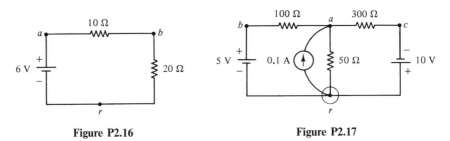

Figure P2.16   **Figure P2.17**

P2.17. For the circuit in Fig. P2.17:
(a) Write KVL to show $v_{br} = v_b = +5$ and $v_{cr} = v_c = -10$ V.
(b) Find the current downward in the 50-Ω resistor by first solving for v_a using nodal analysis.
Hint: Nodes b and c are constrained to the reference node.

P2.18. Using node voltage analysis, solve for the indicated unknowns in Fig. P2.18. *Hints:* In part (a), note that kΩ, mA, and volts make a consistent set of units; in part (c), find v_a, then v_{10} from the voltage divider.
Ans: (a) 3.33 V; (b) $v_{20} = +22.5$ V; (c) $i = 1.40$ A, $v_{10} = 14.0$ V

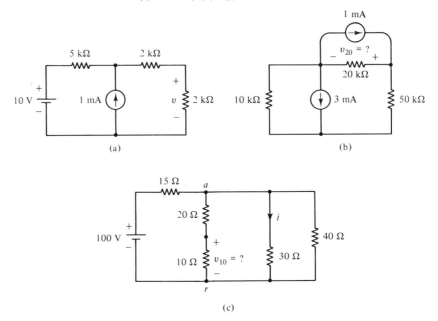

(a)   (b)

(c)

Figure P2.18

SECTION 2.5.1

P2.19. Solve for all currents in the circuit in Fig. P2.19.

Ans: $i_4 = +1$ A upward

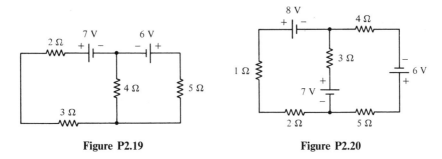

Figure P2.19   **Figure P2.20**

P2.20. Find the power out of the 7-V source in Fig. P2.20 using loop currents to analyze the circuit.

Ans: $P_{out} = 19.3$ W

SECTION 2.5.2

P2.21. Solve for the current referenced downward in the 100-Ω resistor in the circuit in Fig. P2.12 using the constrained loop concept to handle the current source.

P2.22. Solve for i in Fig. P2.22 using the loop current method.

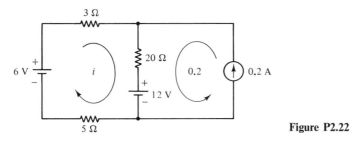

Figure P2.22

P2.23. For the circuit in Fig. P2.23, solve for i_4 using mesh current variables to analyze the circuit. Note that you may direct the constrained loops to miss the 4-Ω resistor. Be sure to count independent loops before definding variables.

Ans: $i_4 = -1.33$ A

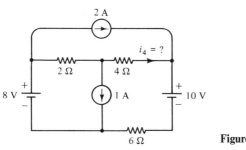

Figure P2.23

P2.24. Solve for the power out of the 80-V source in the circuit in Fig. P2.24, using loop current variables. Remember to count independent loops first.

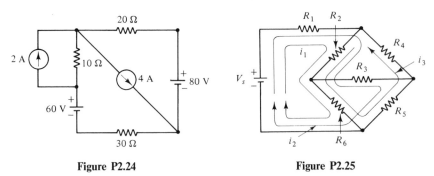

Figure P2.24   **Figure P2.25**

P2.25. True loop (not mesh) currents are defined in Fig. P2.25 satisfying the rules given in Section 2.5.2. Write the KVL equations for the circuit using these loop variables and put into standard form. You are not required to solve the resulting equations.

SECTION 2.5.3

P2.26. **(a)** By inspection write the equations for the mesh currents in the circuit shown in Fig. P2.10a.
   (b) Write by inspection the node voltage equations for the circuit in Fig. P2.18b. Put the reference node at the $-$ end of v_{20} and node a at the other end so that v_a will be v_{20}.

P2.27. For the circuit in Fig. P2.27, write node voltage equations in standard form. Use the inspection method first and then confirm by writing KCL equations.

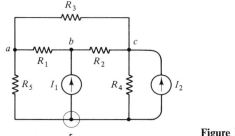

Figure P2.27

SECTIONS 2.6.1–2.6.2

P2.28. Place a connection (a short circuit, no resistor) between a and a' in Fig. 2.45. Solve for the current flowing down from a to a' by the most efficient method. The answer should be zero current, as argued in Section 2.6.1.

P2.29. Develop a Thévenin equivalent circuit for the part of the circuit shown in the box in Fig. P2.29. Use this equivalent circuit to solve for i, as shown. *Hint:* After you have removed the load, use the standard form for nodes with v_{ar} your only unknown, then determine v_{br} (which is V_T) with a voltage divider.

Ans: $V_T = 0.89$ V (+ at y), $R_{eq} = 2.22 \ \Omega$, $i = -0.123$ A

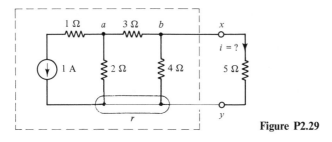

Figure P2.29

P2.30. For the circuit in Fig. P2.30:

(a) Replace the circuit in the box by a Thévenin equivalent circuit. Solve for V_T by the most efficient method.

(b) Find v_{ab} for $R = 3 \, \Omega$.

(c) What value of R receives maximum power from the circuit?

(d) What values of R will receive 20 W from the circuit?

Ans: (b) $+8$ V, (d) $R = 8.66 \, \Omega$ or $2.34 \, \Omega$ for 20 W

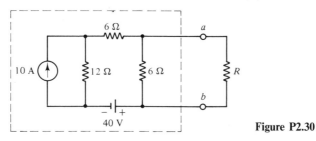

Figure P2.30

P2.31. A potentiometer or "pot" is a variable resistor connected so as to produce a voltage that is adjustable. This is the device commonly used, for example, as a volume control in a radio. The model of a pot is given in Fig. P2.31. Determine the Thévenin equivalent circuit and make a graph of V_T and R_{eq} versus x for $R = 10 \, k\Omega$ and $V_s = 100$ V.

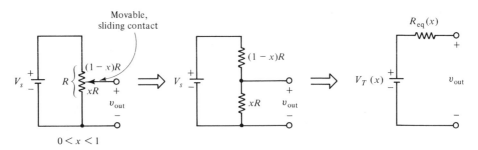

Figure P2.31 Thévenin equivalent circuit for a potentiometer. Note that the output voltage and output resistance depend upon the setting of the potentiometer, x.

P2.32. A black box with a circuit in it was connected to a variable resistor and the power in the resistor was measured as the resistance was varied. The results are shown in Fig. P2.32. From this graph, determine the Thévenin equivalent circuit for the circuit in the black box.

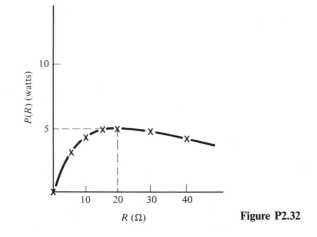

Figure P2.32

SECTIONS 2.6.3–2.6.5

P2.33. Rework Problem P2.29, this time using a Norton equivalent circuit for the portion of the circuit in the box.

P2.34. A mysterious black box is found in the electrical engineering lab. A curious student measured the output voltage to be 12.6 V. Then he shorted the output through an ammeter (consider the ammeter to have zero resistance) and read a current of 56 A. Give the Norton equivalent circuit for the box. How much power can be gotten out of the box if a variable resistor is connected to its terminals and adjusted for maximum power?

Ans: $R_{eq} = 0.225\ \Omega$, $P_{available} = 176\ \text{W}$

SECTION 2.7.2

P2.35. **(a)** Using the results in Eq. 2.39, solve for i_1 in Fig. 2.59.
(b) Show that the result in part (a) is dimensionally correct.

P2.36. Using loop current analysis, solve for i_1 in Fig. 2.58.

P2.37. Connect a 1-A current source to the output in Fig. 2.60 (V_s off) and compute the resulting voltage. This would be the value of R_{eq} and hence should agree with Eq. 2.43.

P2.38. Show that Eq. 2.43 is dimensionally correct.

Application Problems

A2.1. A car radio designed to operate from a 6-V system uses 2 A of current. What resistance should be placed in series with this radio if it is to be used in a 12-V system? What should be the power rating of this resistor?

A2.2. A 16-Ω loudspeaker would draw maximum power from the output of its amplifier, which is capable of producing 25 W in the speaker. What would be the power produced in an 8-Ω speaker? Represent the loudspeakers as resistors of 16 Ω and 8 Ω, respectively.

A2.3. An electric stove (dc or ac, it does not matter, because the same power formulas apply) requires 230 V for the line voltage. The stove uses two heater elements which can be switched in one at a time or placed in series or parallel, making four heating temperatures. If the highest setting requires 2000 W and the lowest 444 W, what are the powers for the two intermediate settings?

A2.4. A black box has two terminals, to which a variable resistor is attached. The current and voltage are measured, with the results shown in the table in Fig. A2.4a.

   (a) Plot i versus v and show that the data form a straight line. Write the equation of the line in slope-intercept form.

   (b) According to Eq. (2.32), these data must fit an equation of the form $i = V_T/(R + R_{eq})$. Treat V_T and R_{eq} as unknowns and determine them both, using any two of the data points and simultaneous equations.

   (c) Show that for this circuit, the voltage intercept of the line in part (a) is V_T, the current axis intercept is I_N, and the slope is $-1/R_{eq}$.

   (d) Show that the relations of part (c) are generally true by deriving the equation of the output current for a generalized Thévenin equivalent circuit. The circuit in Fig. A2.4b may prove helpful.

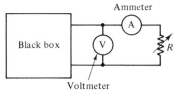

Voltage (V)	Current (A)	R (Ω)
10	1	10
8	2	4
6	3	2
4	4	1

(a)

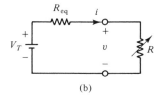

(b)

Figure A2.4

A2.5. For the circuit in Fig. A2.5, show that the resistance between any two nodes is 0.5 Ω (all R's are 1 Ω). *Hint:* Use symmetry—picture the circuit as a tetrahedron. Note that the resistor opposite the two terminals of entry has no current and may be replaced by either a short or an open circuit.

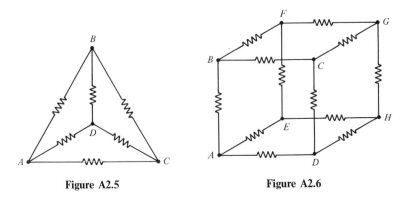

Figure A2.5   Figure A2.6

A2.6. For the circuit shown in Fig. A2.6, determine the resistance between the opposite corners of the cube of resistors (all R's are 1 Ω). *Hint:* Utilize the symmetry of the circuit. Attach a 1-A current source between opposite corners and reason that the three resistors connecting to these corners have the same current and thus the corners to which they connect have no voltage between them. Hence these may be shorted together; in other words, they are connected in parallel in spite of appearances.

A2.7. An attenuator is a circuit that reduces the signal level. The circuit shown in Fig. A2.7 will operate as an attenuator. Design the attenuator circuit (specify R_s and R_p) such that the attenuator would reduce the voltage at the load to a factor of 10 less than the load would have if connected directly to the source. Also require that the output resistance seen by the load be 50 Ω with the attenuator in place.

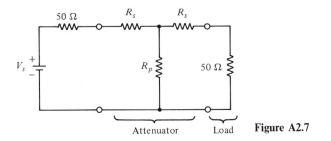

Attenuator   Load   **Figure A2.7**

A2.8. A voltmeter must draw some current from the circuit it is measuring in order to operate (see Fig. A2.8). The amount of current it draws depends on the meter scale and is specified in terms of the "ohms per volt" of the meter. For example, a 10-kΩ/V meter would have a resistance of 10 kΩ on the 1-V scale, 100 kΩ on the 10-V scale, and so on. For this problem, assume that we are measuring with a 10-kΩ/V meter.
   (a) If we measure 5 V on the 10-V scale, what current does the meter draw from the circuit?
   (b) If the output resistance of the circuit is 600 Ω and we measure 5 V on the 10-V

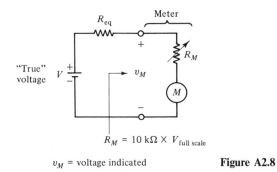

$R_M = 10 \text{ k}\Omega \times V_{\text{full scale}}$

v_M = voltage indicated   **Figure A2.8**

scale, what is the true open-circuit voltage of the source, that is, what would be measured by an ideal meter which drew no current from the circuit?

(c) With our 10 kΩ/V meter, on an unknown circuit we measure 24 V on the 30-V scale but 30 V on the 100-V scale. Explain the reason for this apparent discrepancy and determine from these measurements the Thévenin equivalent circuit for the source we are measuring.

THE DYNAMICS
OF CIRCUITS

3.1 THE THEORY OF INDUCTORS AND CAPACITORS

3.1.1 Statics and Dynamics

Energy and Time. In mechanics, dynamics is usually taught after statics. Statics deals with the distribution of forces in a structure; time is not a factor. Dynamics deals with the exchange of energy between components in a mechanical system, and time is an important factor because energy cannot be exchanged except as a time process.

In our study of electrical circuits, we have not yet considered time as an important quantity. Because our dc circuits involve only KVL, KCL, and Ohm's law, and because none of these equations have time as a factor, we have not worried about time processes. Indeed, even if we had allowed one of our voltage or current sources to have an output that varied with time, the solution would not become more complicated. It would be like letting the force in a statics problem vary slowly with time: the method of solution would remain valid provided energy exchanges between components of the system are small. In true dynamics problem, the rate of energy transfer between components must be considered.

With this chapter we begin the study of electrical circuits in which rates of energy exchanged between circuit components are an important factor. We begin by introducing two circuit components which store energy in electric circuits. The presence of inductors or capacitors in an electric circuit suggests a true dynamics problem. We first will identify the two types of energy that may be stored in a circuit.

Electric Energy and Magnetic Energy Storage. To understand what we mean by magnetic energy and electric energy, let us reconsider Eq. (1.2), which describes the form of the vector force between two charges:

$$\vec{F}_2 = \vec{F}_e + \vec{F}_m$$

Equation (1.2) describes the force Q_2 experiences in the presence of Q_1, the charges being a distance R apart and having velocities $\vec{u}_1$ and $\vec{u}_2$. When we first presented Eq. (1.2) we called attention to the two types of force, one positional and the other motional. Here we wish to emphasize energy. From mechanics, you recall that when a displacement is made against a force, work is done (mechanical energy is exchanged). Similarly, if we displace a charge in the presence of a motional (magnetic) force $(\vec{F}_m)$, magnetic energy is exchanged. To store up magnetic energy, we must get many moving charges close together. This is what an inductor does. On the other hand, when we move charges in the presence of positional (electrostatic) forces $(\vec{F}_e)$, electric energy is exchanged. To store up electric energy, we must separate charges yet have them close together. This is what a capacitor does.

The Analogy Between Mechanical and Electrical Energy.   Magnetic and electric energy are the two forms of electrical energy, arising from the two types of electrical force. In a mechanical system we also have two types of force and two types of energy. Forces that depend on position, as in a spring or a gravitational field, store potential energy. Forces that depend on changes in velocity lead to kinetic energy. Potential energy and kinetic energy in a mechanical system are analogous to electric energy and magnetic energy in an electrical system. We must emphasize, however, that this is only an analogy. Magnetic energy is not kinetic energy; it is merely analogous to kinetic energy. We shall continue to point out analogies between mechanical and electrical systems because your understanding of mechanics can help you understand the behavior of electrical systems.

3.1.2 Inductor Basics

A Physical Inductor.   In Fig. 3.1 we show a coil of wire, which will act as an inductor. When current flows in the wire, moving charges are close together and hence magnetic energy is stored.

     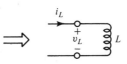

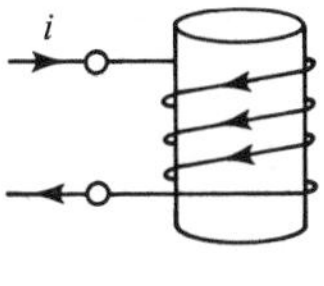

Figure 3.1 A coil of wire gets moving charges close together and acts as an inductor.

Figure 3.2 Circuit model for an inductor.

The Circuit Theory Definition of an Inductor.   Figure 3.2 shows the circuit symbol for an inductor. The equations describing the voltage–current characteristics of an inductor are based on Faraday's law of induction and are given in

$$v_L(t) = +\frac{d}{dt}[Li_L(t)] \quad \text{and} \quad i_L(t) = i_L(0) + \frac{1}{L}\int_0^t v_L(t')\,dt' \tag{3.1}$$

The circuit symbol and the accompanying equations together define a circuit theory model for an inductor. Normally, the inductance is considered a constant and brought outside the derivative. Note that v_L and i_L form a load set.

The Analogy with Newton's Law.   The differential form of Eq. (3.1) is analogous to Newton's second law. Just as changes in the velocity of a mass require a force, or vice versa, changes in the current through an inductor produce a voltage, or vice versa. If the current is increased (positive di/dt), the inductor voltage opposes the change by being numerically positive with the polarity shown. If the current is decreased (negative di/dt), the inductor acts momentarily as a source trying to keep the current going. Thus the physical polarity of the inductor voltage will tend to keep the current constant, just as a mass will tend to maintain constant velocity. This expresses Lenz's law for inductors.

The integral form of Eq. (3.1) is analogous to determining the velocity by integrating the acceleration, which is proportional to an exciting force. We note that only changes in current can be determined from the voltage across the inductor, just as only changes in velocity can be computed from the force acting on a mass. To determine the current, we need to know the inductance, the voltage as a function of time, and the initial current, $i_L(0)$.

Stored Magnetic Energy in an Inductor.   We pointed out that a load set of v–i variables is used to define the equation of an inductor. In Eq. (3.2) we compute the energy stored in the inductor by integrating the input power, $+v_L i_L$.

$$P_L = \frac{dW_m}{dt} = +v_L i_L = \left(L\,\frac{di_L}{dt}\right)i_L = \frac{d}{dt}\left(\tfrac{1}{2}Li_L^2\right) \tag{3.2}$$

$$W_m = \int P_L\,dt = \int \frac{d}{dt}(\tfrac{1}{2}Li_L^2)\,dt = \tfrac{1}{2}Li_L^2 \tag{3.3}$$

The constant of integration is set to zero because there is no stored energy when the current is zero. The result in Eq. (3.3) is analogous to the kinetic energy in a moving mass, with inductance playing the role of mass and current playing the role of velocity.

The Units of Inductance.   The inductance, L, depends on coil size and the number of turns of wire in the coil. The inductance also depends on the material (if any) located near the coil. In particular, the magnetic properties of iron increase greatly the inductance of a coil wound on an iron core. The units of inductance are volt-seconds per ampere, but to honor Joseph Henry (1797–1878) we use the name "henry" (H) for this unit; millihenries (10^{-3} H or mH) and microhenries (10^{-6} H or μH) are also in common use.

An Example.   We shall use the definition of an inductor to compute the voltage across, the power into, and the energy stored in a 2-H inductor with current shown in Fig. 3.3. Because the current function is piecewise continuous, we cannot represent it by a single mathematical formula but rather must represent it separately in its several regions. During interval I, $0 < t < 0.1$, the slope is constant at 20 A/s; thus during this period the inductor voltage will be 2 H $\times$ +20 A/s or +40 V. During interval II,

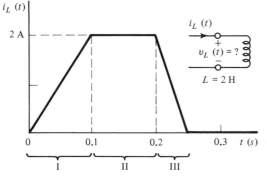

Figure 3.3   A voltage will result from the changing current.

$0.1 < t < 0.2$, the slope is zero and hence the voltage will be zero also. In interval III, $0.2 < t < 0.25$, the slope is -40 A/s and the inductor voltage will be -80 V.

When we compute the power into the inductor as the product of v_L and i_L, we note that the power is positive during interval I, zero during II, and negative during III. The magnetic energy stored in the inductor can be computed by integrating the input power, but the easier way uses Eq. (3.3), with the results shown in Fig. 3.4. During interval I, when the current is increasing in magnitude, we must supply power to the inductor to increase the stored magnetic energy. During interval II, the current is constant and hence no energy is exchanged between the inductor and source: the system is "coasting" like a mass in constant motion. During interval III, the inductor acts as a source, giving energy back to the circuit. The net energy exchanged is zero, for the ideal inductor is lossless, that is, does not convert electrical energy to nonelectrical form. The wire in a physical inductor would have resistance and would not give back all the energy delivered to it. Some of the energy would heat up the wire and be lost to the electrical circuit. Of course, this energy is not lost to the entire physical system because it appears as thermal energy. The circuit model for a real coil of wire, shown in Fig. 3.5, is more complicated

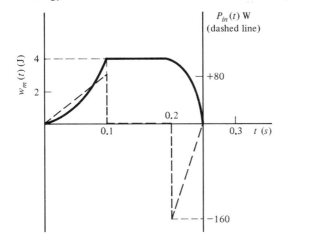

Figure 3.4   The power curve (dashed line) is the slope of the stored energy.

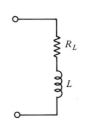

Figure 3.5   The circuit model for a coil of wire must contain a resistance to account for loss.

than a pure inductance because the model must account for loss as well as the storage of magnetic energy.

3.1.3 Capacitor Basics

A Physical Capacitor.   Electric (electrostatic) forces arise from interactions between separated charges, and the associated energy is called electric energy. To store electric energy, we must separate charges, as in Fig. 3.6.

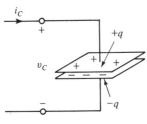

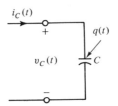

Figure 3.6 Structure having capacitance.

Figure 3.7 Circuit symbol for capacitance.

The Circuit Theory Definition of a Capacitor.   Capacitance is defined as the constant relating charge and voltage in a structure that supports a charge separation. If q is the charge on the + side of the capacitor and the voltage is v_C, as shown in Fig. 3.7, the capacitance C is defined as in

$$q = Cv_C \tag{3.4}$$

For every charge arriving at the + side of the capacitor, a charge of like sign will depart from the − side and the structure as a whole will remain charge neutral. Thus KCL will be obeyed because the charge flowing into the + terminal side must flow out of the − terminal, as for a resistor. Thus the current, but not the charge, passes through the capacitor. If current is positive into the + terminal, charge will accumulate there and will be increasing in proportion to the current. From the definition of current we can relate the charge in Eq. (3.4) and current as follows:

$$i_C = \frac{dq(t)}{dt} \tag{3.5}$$

We may thus define the relationship between current and voltage as

$$i_C = C\frac{dv_C}{dt} \tag{3.6}$$

Notice that we have again used a load set for the voltage and current variables.

The Units of Capacitance.   The capacitance, C, appears in Eq. (3.6) as a constant relating the current to the derivative of the voltage across the capacitor. The units of capacitance can be derived from fundamentals, but to honor Michael Faraday (1791–

1867) we use the name "farad." Realistic capacitor values come small and usually are specified in microfarads (10^{-6} F or μF), nanofarads (10^{-9} F or nF), or even picofarads (10^{-12} F or pF). When a capacitor is constructed from parallel plates, as in Fig. 3.6, the capacitance depends on the area, separation, and material (if any) lying between the plates.

   The Mechanical Analog of Capacitance.   The mechanical analog of a capacitor is a spring, as shown in Fig. 3.8. The analog of velocity is current and thus displacement of the spring is analogous to charge. The voltage across a capacitor is analogous to the force produced by the spring. Comparison of the equation in Fig. 3.8 with Eq. (3.6) shows that capacitance corresponds to the inverse of the stiffness constant, K, and is thus analogous to the compliance of a spring.

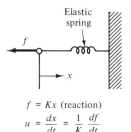

Elastic
spring

$f = Kx$ (reaction)

$$u = \frac{dx}{dt} = \frac{1}{K}\frac{df}{dt}$$

Figure 3.8 Mechanical analog for capacitance.

   The Integral i–υ Equation.   Equation (3.6) is useful in determining the current through a capacitor, given the voltage as a function of time. If we know the current and wish to determine the voltage, we must integrate the equation. Let us consider that we know the capacitor voltage at some time, say, $t = 0$, and wish to determine the voltage at a later time, t. We can integrate Eq. (3.6) from 0 to t, with the result

$$v_C(t) = v_C(0) + \int_0^t \frac{i_C(t')}{C}\,dt' \tag{3.7}$$

Note that we have used t' as the dummy variable of the integration process because t is one limit of the integral.

   An Example.   As an example of the use of Eq. (3.7), consider a capacitor with current through it as shown in Fig. 3.9. Here we know the voltage at the beginning time, $v_C(0) = +5$ V, and the current through the capacitor. We wish to compute the voltage for all $t > 0$. We will integrate, noting that we must divide the integration into several regions due to the piecewise nature of the current. During interval I, $0 < t < 1$ ms, the current is constant at -75 mA. Because the initial voltage is $+5$ V, the voltage is

$$v_C(t) = +5 + \frac{1}{10\,\mu\text{F}}\int_0^t -75 \times 10^{-3}\,dt' = +5 - 7500t \qquad \text{I: } 0 < t < 1\text{ ms}$$

During interval I, the voltage starts at $+5$ V and decreases linearly, reaching -2.5 V at

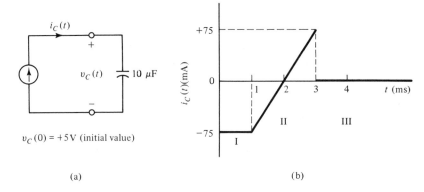

(a)

(b)

Figure 3.9 (a) The capacitor has an initial voltage; (b) current into the capacitor for positive time.

$t = 1$ ms, the end of interval I. At this time we must change formulas for i_C and start integrating again. Our initial value is now -2.5 V and the current is

$$i_C(t) = -0.150 + 75t \text{ A} \qquad \text{II: } 1 < t < 3 \text{ ms}$$

Thus during interval II, the voltage will be

$$v_C(t) = -2.5 + \frac{1}{10\,\mu\text{F}} \int_{1\text{ms}}^{t} (-0.150 + 75t')\, dt' \qquad \text{II: } 1 < t < 3 \text{ ms}$$

As shown in Fig. 3.10, the voltage continues in the negative direction until it reaches -6.25 V at $t = 2$ ms, and then it increases until it reaches -2.5 V again at the end of interval II.

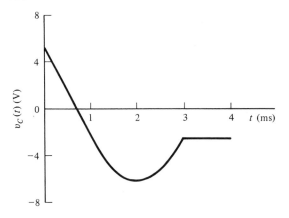

Figure 3.10 Capacitor voltage for positive time.

During interval III, which starts at $t = 3$ ms and continues indefinitely, the current is zero, indicating that the charge on the capacitor does not vary. Hence the voltage will remain at -2.5 V. This constant voltage applies for an ideal capacitor; a physical capacitor would leak and hence discharge over a period of time.

The voltage curve in Fig. 3.10 can be related to the physical processes in the capacitor. The negative current represents a discharging of the capacitor and the voltage decreases accordingly. At the moment when the voltage is zero, the capacitor is totally discharged, but the negative current continues and an excess of negative charge begins to accumulate on the $+$ side (excess positive on the $-$ side). This continues into interval II, although at a slower rate, until the time in the middle of interval II when the current is zero. At this time the $+$ side has its maximum negative charge and the charge is temporarily constant because the current is zero. During the second half of interval II, the current is positive and the negative charge on the $+$ side begins to diminish, resulting in a decrease of the negative voltage between the $+$ side and the $-$ side. At the end of interval II, the current stops and the voltage remains unchanged thereafter.

The Mechanical Analogy.   The process of integrating current through a capacitor to determine voltage is analogous to the process of integrating velocity to determine displacement in a mechanics problem. The only difference is that the value of the capacitance scales the integral of the current, whereas the scale factor is unity in the mechanical analog. Capacitors are in fact used to integrate signals in solving dynamics problems with analog computers, with capacitor voltage representing displacement in the mechanical system.

Stored Energy in a Capacitor.   The charge separation in a capacitor stores electric energy. This energy is analogous to potential energy stored in a stressed spring. We may derive the stored electric energy in a capacitor by integrating the power into the capacitor, as shown in

$$W_e = \int P_C \, dt = \int v_C \times C \frac{dv_C}{dt} \, dt = \int \frac{d}{dt} \left(\tfrac{1}{2} C v_C^2 \right) dt = \tfrac{1}{2} C v_C^2 \tag{3.8}$$

The constant of integration must be zero because the uncharged capacitor stores no energy. We see from Eq. (3.8) that the stored energy depends uniquely on the voltage (or the charge) and the capacitance. For example, if we take a 10-μF capacitor and connect it briefly to a 12-V battery, the capacitor will receive $\tfrac{1}{2} \times 10^{-5}(12)^2 = 7.2 \times 10^{-4}$ J from the battery, which it will store until it is discharged or perhaps until the charge leaks off over a period of time. Although this is not much energy, it would take only about 1 μs to charge the capacitor. Hence the rate of energy flow would be rather high, about 720 W.

The Mechanical Analog for Resistance.   A mechanical analog for a resistance is a frictional loss of a special type. Voltage is analogous to force, and current analogous to velocity. Voltage/current, or resistance, would thus imply a force which is proportional to velocity. We experience such a force when we try to move underwater. A common mechanical component having this property would be an automotive shock absorber. Table 3.1 summarizes the analogies between mechanical and electrical quantities.

TABLE 3.1   SUMMARY OF MECHANICAL
AND ELECTRICAL ANALOGS

Mechanical	Electrical
Force	Voltage
Velocity	Current
Displacement	Charge
Mass	Inductance
Spring compliance	Capacitance
Shock absorber	Resistance

3.2 FIRST-ORDER TRANSIENT RESPONSE OF RL AND RC CIRCUITS

3.2.1 The Type of Problem We Will Solve

In this section we will show how to solve an important class of problems. Typical problems of this class are shown in Fig. 3.11. These circuits are characterized by having a single energy storage element, one capacitor or inductor. They have dc sources and a switch that either opens or closes at a known time, normally defined to be the time origin, $t = 0$. The circuit will have one dc state before the switch action and another dc state long after the switch action. Consequently, the state of the circuit goes through a transition. This transition is transient in that it lasts for a brief period of time; thus these problems are often called *transient problems*.

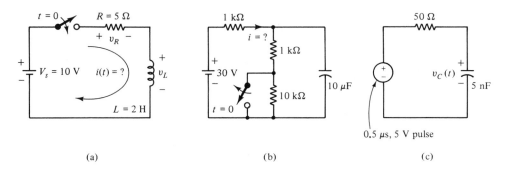

Figure 3.11   Each circuit has one energy storage element, a constant source, and a sudden change.

This is an important class of problems. It represents switching something on or off, which is often a critical period in the operation of a device. When, for example, does a light bulb normally burn out? Also included in this class of problems are many digital signals, such as a computer uses in processing information. Such digital signals can often be represented as a dc source being turned on and off.

3.2.2 The Classical Differential Equation Solution

Deriving the Differential Equation.   Our approach will be first to analyze a representative problem using the techniques of differential equations (DEs). Then we will show the pattern of the solution and develop a much simpler method of solution. Our simple method can also be applied to mechanical, thermal, or chemical problems that fall into the class described in Section 3.2.1.

We will analyze the circuit in Fig. 3.11(a). With the switch open, the full 10 V appears across the switch because there can be no current to cause a voltage across the resistor or the inductor. After the instant at which the switch is closed, KVL for the circuit is

$$-V_s + v_R(t) + v_L(t) = 0 \tag{3.9}$$

Equation (3.9) becomes a DE when we express v_R and v_L in terms of the unknown current.

$$L\frac{di(t)}{dt} + Ri(t) = V_s \qquad t > 0 \tag{3.10}$$

The Form of the Solution.   We now must solve this DE. Equation (3.10) is a linear DE with constant coefficients and a constant forcing term on the right side. The solution of a DE of this type usually proceeds by separating the unknown solution into two parts: the homogeneous part and the particular integral, also called the *forced response* or the *steady-state response*. These names are given because this portion of the solution is caused by the forcing function on the right side. In this problem, and all problems we will solve in this chapter, this response must be constant because the forcing function is constant in time.

The form of the homogeneous response is determined by the left-hand side of the equation. This part of the solution, also called the *natural response*, satisfies the equation with the forcing function set to zero. A linear DE with constant coefficients is always satisfied by a function of the form $e^{-t/\tau}$, where τ (Greek lowercase tau) is an unknown constant with the dimensions of time. This function works because the derivatives of the exponential function are proportional to the function itself and thus the terms on the left-hand side are capable of self-cancellation. The solution must, therefore, be of the form

$$i(t) = A + Be^{-t/\tau} \tag{3.11}$$

where A, B, and τ are unknown constants.

Determining the Unknown Constants.   We can determine A and τ by substituting Eq. (3.11) back into Eq. (3.10). The coefficient of the exponential term must vanish if the equation is valid for all times. This represents the self-cancellation required for the homogeneous part of the solution. Hence we determine τ and A to be

$$LB\left(\frac{-1}{\tau}\right)e^{-t/\tau} + R(A + Be^{-t/\tau}) = V_s$$

$$B\left(\frac{-L}{\tau} + R\right)e^{-t/\tau} + AR = V_s \Rightarrow \tau = \frac{L}{R} = 0.4 \text{ s}$$

Setting the coefficient of the exponential term to zero leads also to the value of A, the steady-state response.

$$AR = V_s \Rightarrow A = \frac{V_s}{R} = 2A$$

To determine B we must consider the initial condition. The initial condition for this, and for all such systems, arises from consideration of energy. Time is required for energy to be exchanged and hence processes involving energy carry the system from one state to another, particularly when, as here, a sudden change occurs. The closing of the switch will eventually allow the inductor to store energy, but at the instant after the switch is closed, the stored energy must still be zero. Zero energy implies zero current because the stored energy in an inductor [Eq. (3.3)] is $\frac{1}{2}Li^2$; hence, $i(0^+)$ is zero, where 0^+ denotes the instant after the switch is closed. This condition leads to the value for B:

$$0 = A + Be^{-0/\tau} \Rightarrow B = -A = -2 \text{ A}$$

The final solution is given in Eq. (3.12) and plotted in Fig. 3.12.

$$i(t) = \frac{V_s}{R} - \frac{V_s}{R}e^{-t/\tau} = 2 - 2e^{-t/0.4} \text{ A} \qquad (3.12)$$

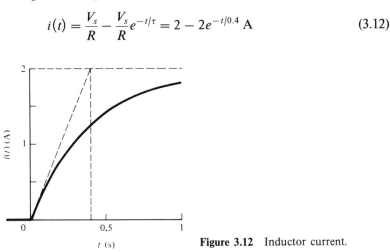

Figure 3.12   Inductor current.

The Physical Interpretation of the Solution.   The mathematical solution is complete, but we wish to examine carefully the physical interpretation of the response, for we shall develop our simpler method from a physical understanding of this type of problem. The current is zero before the switch is closed and approaches V_s/R as a final value. This final value does not depend on the value of the inductor, but would be the current if no inductor were in the circuit. The inductor cannot matter in the end because we have a dc excitation in the circuit and eventually the current must reach a constant value. The constant current renders the inductor to be invisible since an inductor will exert itself only when its current is changing.

Without the inductor, the current would change instantaneously from zero to V_s/R when the switch is closed. The effect of the inductor is to effect a smooth transition between the initial and final values of the current. The smooth change caused by the inductor takes place over a *characteristic time* of τ, which is also called the *time constant*. Note from Fig. 3.12 that the current initially increases with a rate as if to arrive at the final value in one time constant. This time constant for the *RL* circuit, L/R, can be interpreted from energy considerations. The inductor will require little energy if L is small or R is large; hence the transition will be rapid for small τ. But if the inductor is large or R is small, much energy will eventually be stored in the inductor; hence the transition period will be much longer for large τ.

Summary.   We see that the response consists of a transition between two constant states, an initial state and a final state. The effect of the energy storage element is to effect a smooth transition between these states. Because of the form of the DE for this class of problems, the transition between states is exponential and is characterized by a time constant. The stored energy in the inductor or capacitor carries the state of the circuit across the instant of sudden change and thus leads directly or indirectly to the initial condition.

3.2.3 A Simpler Method

We shall now rework this problem using a more efficient method. The goal is to write the circuit response directly based on physical understanding. There are four steps.

Find the Time Constant.   We already know the time constant for this type of problem, L/R, but in general we can determine the time constant for a first-order system by putting the DE into the special form of Eq. (3.13). We divide by the coefficient of the derivative term, and place the coefficient of the other term on the left side entirely in the denominator.

$$\frac{dx}{dt} + \frac{x}{\tau} = \text{constant} \tag{3.13}$$

In Eq. (3.13), x refers to the physical quantity being determined. It would represent a voltage or current in a circuit problem, but in other physical systems x might represent a temperature, a velocity, or something else. When we put the DE in this form, τ will always be the characteristic time, or time constant. To put Eq. (3.10) into this form, we have to divide by L and flip R into the denominator, but you can see τ works out in this case to L/R, as it must. Since the purpose of this method is to avoid DEs, we hesitate to suggest that you must write the DE to get started. Usually, you will know the equation of the time constant for the circuit or system from previous experience. For an *RL* circuit, for example, the time constant is always L/R. Because we solve circuits having only one energy storage element with this method, we have only one L. In Section 3.2.5 we will show you how to handle circuits with more than one resistor. Returning to the problem we are solving, we now know our time constant: L/R is 0.4 s for this circuit.

Find the Initial Condition. The initial condition (x_0 in general, i_0 in this case) always follows from consideration of the energy condition of the system at the beginning of the transient. In this case we argued earlier that because the inductor had no energy before the switch was closed, it must have no energy the instant after the switch is closed. Hence the initial current is zero (i.e., $i_0 = 0$). Later we will discuss generally how to determine initial values.

Find the Final Value. The final value (x_∞ in general, i_∞ in this case) follows from the steady-state solution of the system. Earlier we argued that the inductor eventually becomes invisible. This is true because the final state of the circuit is a dc state, and the inductor has no voltage across it for a constant current. Thus application of Ohm's law to the circuit in Fig. 3.11(a) shows that the final current must be 2 A.

Substitute the Time Constant and the Initial and Final Values into a Standard Form.

$$x(t) = x_\infty + (x_0 - x_\infty)e^{-t/\tau} \tag{3.14}$$

Equation (3.14) is the solution for all first-order systems having dc (constant) excitation. In our case, the x's are currents of known numerical value, so our solution is

$$i(t) = \frac{V_s}{R} + \left(0 - \frac{V_s}{R}\right)e^{-t/\tau}$$

$$i(t) = 2 + (0 - 2)e^{-t/0.4} \text{ A}$$

You can confirm that this is the same solution we obtained directly by solving the DE. This response is shown in Fig. 3.12.

Figure 3.13 shows a generalized plot of Eq. (3.14). The curve starts at x_0 and asymptotes to x_∞. Its initial rate is such as to move from x_0 to x_∞ in one time constant, but it only reaches $(1 - e^{-1})$ or 63% of the way during the first time constant. You might note that we have shown a discontinuity at the origin. This can happen in general, although it does not happen in our present example, as shown in Fig. 3.12.

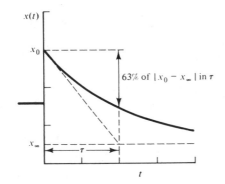

Figure 3.13 Generalized response in Eq. (3.14).

Summary.   Our efficient method consists of determining three constants and substituting these into a standard form. The time constant can be determined from the DE, but usually the formula for τ is known from prior experience. The other two constants are the initial value of the unknown, which is determined from energy considerations, and the final value, which is determined from the dc solution of the problem. The role of the energy storage element is to effect a smooth transition between the initial and final values and to influence the initial value through energy considerations.

Another Example.   As a second example of our simple method, we will solve the same circuit again, except that this time we will determine the voltage across the inductor. We already know the time constant. The initial condition follows indirectly from energy considerations. The initial energy in the inductor is zero, which implies zero current. Turn back to Fig. 3.11(a) and consider the state of the circuit at the instant after the switch is closed. Because of the inductor there can be no current at this initial instant; hence there can be no voltage across the resistor. If there is no voltage across the resistor and no voltage across the switch, the full 10 V of the battery must appear across the inductor. Hence the initial value of the inductor voltage is 10 V. The final value of the inductor voltage must be zero because we have a dc source and eventually the current becomes constant. Consequently, di/dt approaches zero, and the inductor voltage must also approach zero. We now know the time constant, the initial value, and the final value, and we can therefore write the full solution with our standard form,

$$v_L(t) = 0 + (10 - 0)e^{-t/0.4} = 10e^{-t/0.4} \text{ V} \tag{3.15}$$

The plot of Eq. (3.15) is shown in Fig. 3.14. Notice that we have shown a discontinuity in the inductor voltage at the origin. Before the switch closes, the inductor has no voltage. When the switch closes, the current begins to increase, but the inductor voltage instantly jumps to 10 V to oppose that increase. The initial rate of increase of the current is limited to

$$10 = 2 \left. \frac{di}{dt} \right|_{t=0} \Rightarrow \frac{di}{dt} = \frac{10}{2} = 5 \text{ A/s}$$

As the current increases, the voltage across the resistor increases accordingly, and the

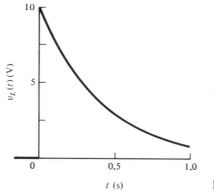

Figure 3.14   Inductor voltage.

voltage across the inductor must decrease. Finally, the current increases to a level where it is limited by the resistor. Thus the inductor dominates the beginning and the resistor dominates the end of the transient.

3.2.4 An RC Circuit

We will now apply our method to the RC circuit shown in Fig. 3.15. We will determine the voltage across the resistor, $v_R(t)$, assuming that the energy in the capacitor initially is zero.

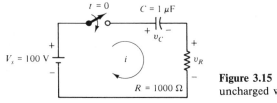

Figure 3.15   The capacitor is uncharged when the switch closes.

Write the DE.   We must write the DE because we have never analyzed an RC circuit before. After the switch is closed, KVL is

$$-V_s + v_C + v_R = 0 \tag{3.16}$$

To derive the DE for v_R, we must express v_C in terms of v_R. One approach is to differentiate Eq. (3.16) and then substitute v_R/RC for dv_c/dt. This substitution is a consequence of Ohm's law and the definition of a capacitor:

$$\frac{dv_C}{dt} = \frac{i}{C} = \frac{v_R/R}{C} = \frac{v_R}{RC} \tag{3.17}$$

The differentiation gets rid of the constant, V_s, and after the substitution from Eq. (3.17) we have

$$\frac{v_R}{RC} + \frac{dv_R}{dt} = 0 \Rightarrow \frac{dv_R}{dt} + \frac{v_R}{RC} = 0 \tag{3.18}$$

Comparison of Eq. (3.18) with Eq. (3.13) shows that the time constant for an RC circuit is RC. So τ is $1000 \times 10^{-6} = 1$ ms in this circuit. This is the only time we will write the DE for an RC circuit; henceforth we know that the time constant is RC.

Find the Initial Condition.   The initial condition emerges from consideration of energy. The capacitor is unenergized at $t = 0^-$, the instant before the switch is closed. Hence it remains unenergized at $t = 0^+$, the instant after the switch is closed, because delay is required for energy (and hence charge) to be stored in the capacitor. The electric energy stored in a capacitor is given in Eq. (3.8) as $\frac{1}{2}Cv_C^2$; thus the voltage across the capacitor will be zero before and after the closing of the switch. At $t = 0^+$, therefore, the full voltage of the battery appears across the resistor, and the initial value for v_R must be 100 V.

Find the Final Value.   The final value of the resistor voltage follows from the requirement that the voltage and current eventually become constant. This means that dv_C/dt is zero, i is zero, and v_R must therefore be zero.

Substitute the Time Constant and the Initial and Final Values into a Standard Form.   Now that we know the time constant, the initial value, and the final value for v_R, we substitute into the standard form of Eq. (3.14).

$$v_R(t) = 0 + (V_s - 0)e^{-t/\tau}$$
$$= 100e^{-t/1\,\text{ms}} \tag{3.19}$$

Equation (3.19) is plotted in Fig. 3.16. As the current flows, as evidenced by the voltage across the resistor, charge accumulates and builds up voltage across the capacitor. The voltage across the resistor must diminish with time and eventually vanish. Eventually, the full voltage of the battery appears across the capacitor and the current stops. We would say that the capacitor has been charged by the battery.

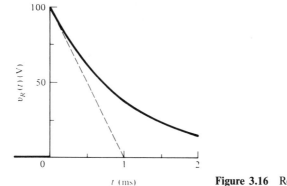

Figure 3.16   Resistor voltage.

3.2.5 Circuits with Multiple Resistors

The Thévenin Equivalent Circuit.   Circuits with more than one resistor can be analyzed with the concept of equivalent circuits. The circuit in Fig. 3.17a, for example, can be reduced to the equivalent circuit in Fig. 3.17b. From this equivalent circuit, it is

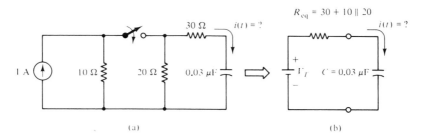

(a)   (b)

Figure 3.17   (a) The three resistors can be combined using a Thévenin equivalent circuit. (b) The time constant is easily identified in this equivalent circuit.

clear that the time constant is $R_{eq}C$, where R_{eq} is the equivalent resistance of the circuit with all sources turned off and the switch closed. Thus R_{eq} is the output resistance of the circuit to the capacitor as a load. In this case, R_{eq} is $30 + 20 \,\|\, 10 = 36.7\ \Omega$, and thus τ is $36.7 \times 0.03 \times 10^{-6}$ or $1.1\ \mu s$.

We may or may not be interested in the Thévenin equivalent circuit as a means for calculating the initial or final values of the unknown. That is a separate problem to be approached by the most efficient method. It is the *concept* of the Thévenin circuit, specifically the concept of the output resistance, which leads to the time constant for a transient circuit with multiple resistors. In general, the relevant resistance for the time constant is the output resistance presented to the energy storage element as a load.

Another Example.   The circuit in Fig. 3.11(b) can be approached with this technique. Here we clearly should not replace the circuit external to the capacitor with a Thévenin equivalent circuit, for in that case we would eliminate the unknown of interest, $i(t)$. We evoke the concept of the equivalent circuit only to calculate the equivalent resistance seen by the capacitor with the switch open, $(1\ k\Omega + 10\ k\Omega) \,\|\, 1\ k\Omega$ or $917\ \Omega$. We continue this example below.

3.2.6 The Initial and Final Conditions

What We Mean by "Initial" and "Final."   We now look generally at how to calculate the initial and final values. By "initial" we mean the instant after a change occurs in the circuit, usually a switch closing or opening. This does not have to be the time origin, although often we define $t = 0$ as the time of switch action. By "final" we mean the asymptotic state of the circuit, its state after a large amount of time. The final state may be hypothetical because the circuit may never reach that state due to subsequent switch action. That is, we may first close a switch and then open it before the circuit reaches the final state. Of course, the circuit cannot anticipate the second switch action and hence reacts to the first switch action as if it would reach steady state. In this section we develop guidelines for determining the final and initial states of the circuit.

Determining Final Values.   Final values arise out of the eventual steady state of the circuit. This implies that all time derivatives must eventually vanish. Consequently, the current through capacitors and the voltage across inductors must approach zero, as suggested in

$$v_C \longrightarrow \text{constant} \Rightarrow i_C = C\,\frac{dv_C}{dt} \longrightarrow 0$$

$$i_L \longrightarrow \text{constant} \Rightarrow v_L = L\,\frac{di_L}{dt} \longrightarrow 0$$

(3.20)

This means that capacitors act as open circuits and inductors act as short circuits in establishing final values, as shown in Figs. 3.18 and 3.19, respectively.

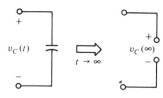

Figure 3.18 The capacitor acts as an open circuit as time becomes large.

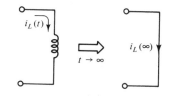

Figure 3.19 The inductor acts as a short circuit as time becomes large.

Determining Initial Values.   Initial values always follow from energy considerations. The fundamental principle is that the stored electric energy in a capacitor must be a continuous function of time and the stored magnetic energy in an inductor must be a continuous function of time. From continuity of energy we conclude that the voltage across a capacitor and the current through an inductor must be continuous functions of time. Thus we always calculate capacitor voltage or inductor current before the switch is thrown and then carry this value over to the moment after the switch is thrown. From these known values of the capacitor voltage or inductor current, other circuit variables can be calculated.

A model for an energized capacitor at the instant after the switch action is shown in Fig. 3.20. For that first instant, the charged capacitor acts like a voltage source because the voltage across the capacitor cannot change instantaneously. As current flows through the capacitor, its voltage will change, and hence the capacitor acts as a voltage source only at that first instant. Similarly, a current source models an energized inductor at the instant after switch action, as shown in Fig. 3.21. Although these models are valid only for the first instant of the new regime, they suffice for the calculation of the initial values of the circuit variables of interest.

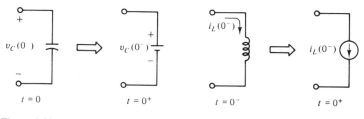

Figure 3.20 A battery models the initial voltage of the capacitor.

Figure 3.21 A current source models the initial current of the inductor.

An Example.   We shall illustrate this principle in the circuit of Fig. 3.22. The switch is opened after having been closed for a time long enough to establish steady state throughout the circuit. We are interested in the voltage across the inductor, specifically, its initial value after the switch is opened. Following the suggestions made above, we first

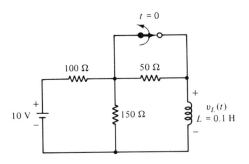

Figure 3.22 After being closed a long time, the switch is opened.

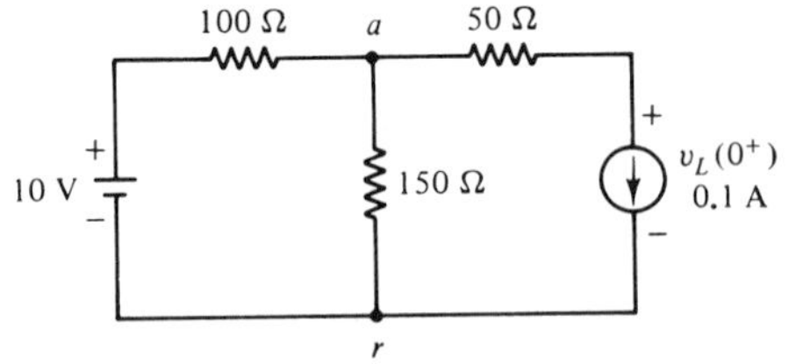

Figure 3.23 Equivalent circuit at $t = 0^+$.

calculate the inductor current at the instant before the switch is opened. We note that with the switch closed and with the circuit in steady state, the 50-Ω resistor is short-circuited by the switch and the 150-Ω resistor is shorted by the switch and the dormant inductor. Thus the current from the battery is 10 V/100 Ω or 0.1 A and this current flows through the switch and inductor. This information does not give us the initial value of $v_L(t)$ directly, but it leads indirectly to the initial value through the analysis of the circuit in Fig. 3.23. We have replaced the inductor by a current source and eliminated the switch because it is now an open circuit. As you can see, we have notated the circuit to suggest solution by nodal analysis. You may confirm that V_a is zero at $t = 0^+$. From this information we can calculate v_L from KVL around the right-hand loop, with the result that $v_L(0^+) = -5$ V. The inductor acts initially as a source to keep the current going at the same rate, even though it must now flow through the 50-Ω resistor.

Having made all these preparations, let us finish the problem. The time constant is L/R_{eq}, where R_{eq} is the equivalent resistance seen by the inductor with the switch open. Thus R_{eq} is $50 + 150 \| 100$ or 110 Ω, and the time constant is 0.1/110 or 909 μs. Since the final value of the inductor voltage is clearly zero, the full solution is

$$v_L(t) = 0 + (-5 - 0)e^{-t/909\mu s} = -5e^{-t/909\mu s}$$

Summary. An energized capacitor acts as a voltage source in the calculation of the initial values. A special case is an unenergized capacitor, which acts as a short circuit. In the calculation of the final value, a capacitor acts as an open circuit. An energized inductor acts as a current source in the calculation of the initial values. A special case is an unenergized inductor, which acts as an open circuit. In the final calculation, an inductor acts as a short circuit. We can solve for initial and final values by replacing capacitors and inductors by these models. The initial and final values thus are derived from the analysis of a circuit containing only resistors and sources.

Another Example. Consider the circuit in Fig. 3.11(b). We have already determined the equivalent resistance to be 917 Ω, so the time constant is $\tau = RC = 9.17$ ms. To find the initial value, we must determine the voltage across the capacitor at $t = 0^+$. At the instant before the opening of the switch, the circuit will be in steady state; indeed, this is the final state from previous actions which established the circuit. Thus the capacitor will act as an open circuit, as shown in Fig. 3.24. The voltage across C is 15 V because the 30 V of the source divides equally between the two 1-kΩ resistors (the 10-kΩ resistor being shorted by the switch.)

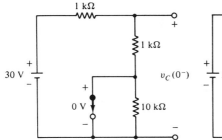
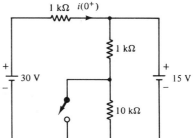

Figure 3.24 The capacitor acts as an open circuit in the steady state before the switch is opened.

Figure 3.25 The capacitor now acts as a battery in the initial-value calculation.

Consequently, the circuit which must be solved for the initial value of $i(t)$ is shown in Fig. 3.25. Notice that $i(0^+)$ can be determined by writing KVL around the outer loop containing the two sources, with the result, $i(0^+) = 15$ mA.

The final value of $i(t)$ can be established from the equivalent circuit in Fig. 3.26. Note that the capacitor is treated as an open circuit in this calculation of the final value. The value of i_∞ is easily derived from this simple series circuit: $i_\infty = 30$ V$/(12$ kΩ$)$ $= 2.5$ mA.

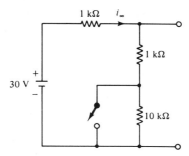

Figure 3.26 Equivalent circuit for final-value calculation.

We now have determined the time constant, the initial value, and the final value; hence the solution follows from Eq. (3.14).

3.2.7 The Pulse Problem

Pulse Circuits.   Finally, we consider the circuit in Fig. 3.11(c), as detailed in Fig. 3.27. Although this problem appears to differ significantly from the others, we can analyze it with the same techniques.

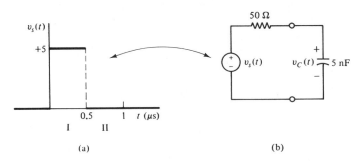

Figure 3.27 (a) Voltage pulse; (b) the capacitor is uncharged when the pulse begins.

Charging Transient.   We assume that the capacitor is initially unenergized. When the source voltage jumps to +5 V, current will flow and the capacitor will begin charging toward +5 V. Bear in mind that the circuit cannot anticipate the end of the pulse. The circuit will respond as if a 5-V battery has been permanently attached. Thus the transient proceeds toward a final value, even though that value will never be attained. During interval I, the time constant will be $RC = 0.25$ μs, the initial value will be zero and the final value will be +5 V. Thus the capacitor voltage during interval I will be

$$v_C(t) = 5 + (0 - 5)e^{-t/0.25\mu s} = 5(1 - e^{-t/0.25\mu s}) \text{ V}$$

Discharging Transient.   When the trailing edge (the end) of the pulse comes along 0.5 μs after the leading edge, the capacitor now has a voltage across it. Since the time constant is half the pulse width, the value of the capacitor voltage at the beginning of the second transient would be $5(1 - e^{-2}) = 4.32$ V. This becomes the initial value for the transient during interval II. During this interval, the voltage source is off and thus is equivalent to a short circuit. The situation in interval II is therefore equivalent to that shown in Fig. 3.28. We know the initial voltage, the final value is clearly zero, and the time constant is unchanged. Hence the capacitor voltage during interval II is

$$v_C(t) = 0 + (4.32 - 0)e^{-t'/0.25\mu s}$$

$$= 4.32 \exp\left(-\frac{t - 0.5\ \mu s}{0.25\ \mu s}\right) \text{V} \qquad t > 0.5\ \mu s$$

50 Ω

v_C 5 nF

"Initial" voltage = 4.32 V

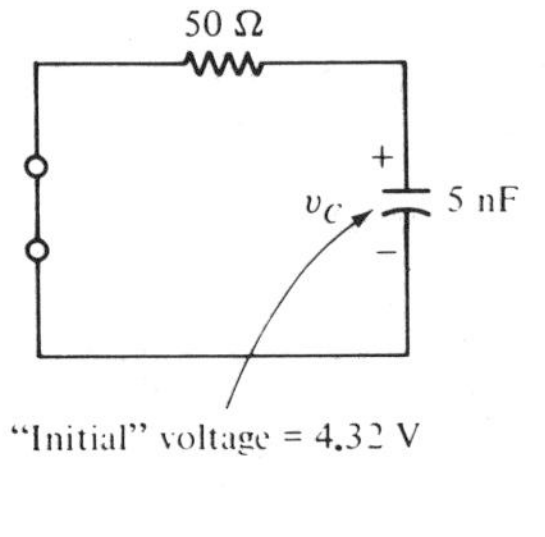

Figure 3.28 During interval II the "initial" voltage is 4.32 V.

For example, at $t = 1 \, \mu$s we calculate the voltage to be

$$v_C(1 \, \mu\text{s}) = 4.32 e^{-(1.0-0.5)/0.25} = 0.585 \text{ V}$$

Putting the two parts of the transient together, we plot the results in Fig. 3.29.

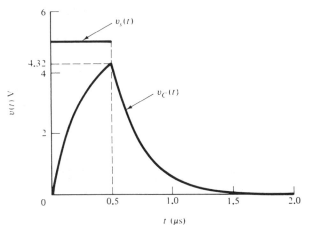

Figure 3.29 The capacitor voltage is a distorted version of the input pulse.

Comparing a Pulse with a Battery–Switch. It is interesting to contrast the pulse problem with that shown in Fig. 3.30. Here we simulate the pulse with a battery and a switch that closes for 0.5 μs. The charging part of the transient will be the same as for the pulse problem, but when the switch is opened, there is no discharge path for the capacitor. Hence the capacitor will charge up to 4.32 V in both cases, but with the battery–switch no discharge will occur. An ideal capacitor would retain its charge forever.

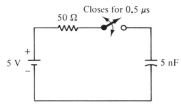

Figure 3.30 The battery–switch has the same Thévenin voltage but different output resistance from the pulse source.

We can understand the difference between the pulse source and the battery–switch source by considering the Thévenin equivalent circuits for each. The Thévenin voltage source is the same for both, but the output resistance differs. With the switch closed, the output resistance is 50 Ω for both cases, but with the switch open the output resistance becomes infinite (an open circuit) for the battery–switch source. Thus the Thévenin representations differ significantly.

A Puzzle. While we are considering battery–switch sources, let us consider what would happen in the circuit in Fig. 3.11(a) if we opened the switch after the current is established. Energy considerations require that the current in the inductor continue after the switch is opened. But no current can flow through an open switch. What will happen?

This question can be answered on two levels. At the theoretical level, we must outlaw this situation. We have created our dilemma by violating the definitions of the circuit elements we are using. Strictly speaking, we can no more open the switch on an inductor than we can short circuit an ideal voltage source—the definitions of these elements are contradictory. On the theoretical level, opening the switch on an inductor is like setting $1 = 0$ in mathematics—it is nonsense.

But this answer does not fully satisfy us, does it? There are, after all, real inductors and real switches. What happens when we perform the experiment? If you try it, you will witness a spark when you open a switch connected in series with an inductor. The voltage across the inductor, and hence the voltage across the switch, rises instantaneously to a high value, such that the air between the switch contacts becomes ionized. The ionized air provided a resistive path for deenergizing the inductor, a mini-lightning bolt, so to speak. Indeed, this is the principle behind the conventional automotive ignition system: the coil is the inductor, the points are the switch, and you know where the spark occurs.

PROBLEMS

Practice Problems

SECTIONS 3.1.1–3.1.2

P3.1. As shown in Fig. P3.1, a 50-mH inductor has an ac current of

$$i_{ac} = 10 \cos (500t) \qquad \text{amperes}$$

Calculate the voltage across the inductor with the polarity shown in Fig. P3.1.

Ans: $v_{ac}(t) = -250 \sin (500t) = 250 \cos (500t + 90°)$ V

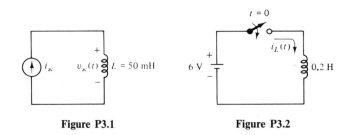

| Figure P3.1 | Figure P3.2 |

P3.2. A 6-V battery is connected to an ideal 0.2 H inductor, as shown in Fig. P3.2.
  (a) Calculate the current for positive time, that is, after the switch is closed.
  (b) At what time does the stored energy in the inductor reach 10 J? Verify the stored energy by integrating the input power (product of v and i) from $t = 0$ to the time you calculate.

Ans: $t = 0.333$ s

P3.3. Prove from the defining equation of inductance that two inductors (L_1 and L_2) in series combine like resistors in series, that is, $L_{eq} = L_1 + L_2$. Show also that two inductors in parallel combine like resistors in parallel, that is,

$$L_{eq} = L_1 \| L_2 = \cfrac{1}{\cfrac{1}{L_1} + \cfrac{1}{L_2}}$$

SECTION 3.1.3

P3.4. How much charge and how much energy are stored in a 1000-μF capacitor which has been charged to 200 V? If this energy were converted to kinetic energy and distributed to the electrons on the capacitor, what would be their velocity?

Ans: $v = 5.93 \times 10^6$ m/s

P3.5. A 0.1-μF capacitor is charged with a 1-μs pulse of current, as shown in Fig. P3.5. Find the voltage across the capacitor with the polarity shown, as a function of time.

Ans: Starts at 0, charges to 1 V in 1 μs, and stays at 1 V thereafter because with $i_s = 0$ the charge is trapped on the ideal capacitor.

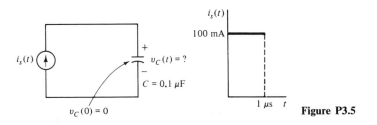

Figure P3.5

P3.6. In order to move the spot of a CRT across the screen, the voltage across a pair of deflection plates must be increased in a linear fashion, as shown in Fig. P3.6. If the capacitance of the plates is 0.001 μF, sketch the resulting current through the capacitor.

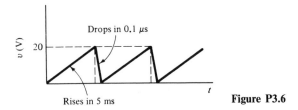

Figure P3.6

P3.7. Prove from the defining equation of capacitance that two capacitors (C_1 and C_2) connected in parallel are equivalent to a single capacitor of value $C_{eq} = C_1 + C_2$, but in series combine like resistors in parallel,

$$C_{eq} = C_1 \| C_2 = \cfrac{1}{\cfrac{1}{C_1} + \cfrac{1}{C_2}}$$

P3.8. An inductor and a capacitor are placed in series with a current source whose current increases with time as shown in Fig. P3.8.

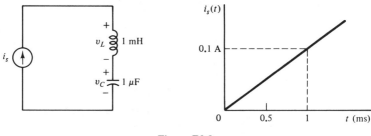

Figure P3.8

(a) Find the voltage across the inductor, $v_L(t)$.
(b) Assuming no initial charge on the capacitor, find its voltage, $v_C(t)$.
(c) Calculate the instant when the stored energy in the capacitor first exceeds that in the inductor.

SECTIONS 3.2.1–3.2.2

P3.9. The solution of a first-order DE with constant coefficients and a constant term on the right side is

$$x(t) = 5 - 5e^{-t/0.1}$$

(a) Plot this solution. What is the value at the origin and as time becomes large?
(b) What is the DE that this solution satisfies? Verify by substitution that your DE is correct.

SECTION 3.2.3

P3.10. Verify that the standard solution [Eq. (3.14)] satisfies Eq (3.13) and has the required values at $t = 0$ and t very large.

P3.11. A system is described by the DE

$$10\frac{dx}{dt} + 5x = 50$$

(a) What is the time constant for this system?
(b) What would be the final value, x_∞?
(c) If $x_0 = 12$, find and sketch $x(t)$ for $t > 0$.

P3.12. Figure P3.12 shows a switch which is changed from a to b at $t = 0$. Assume that the switch has been in position a for a long time before it is changed. Find and sketch $v_L(t)$, the voltage across the inductor with the polarity shown.

Ans: $v_L(t) = -10e^{-t/15 \text{ ms}}$ V

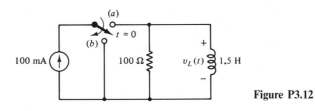

Figure P3.12

P3.13. A current source is placed in series with a resistor and inductor, as shown in Fig. P3.13. During this period the switch is open. Then the switch is closed, which action separates the circuit into two independent loops which share a common short circuit but do not interact.

    (a) Calculate the current in the right-hand loop after the switch closes at $t = 0$.

    (b) Plot the current in the switch, referenced downward.

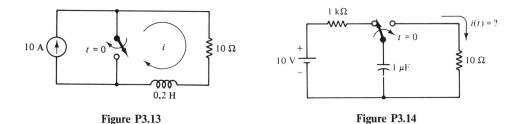

Figure P3.13   **Figure P3.14**

P3.14. For the circuit shown in Fig. P3.14, assume that the switch has been in the position shown for a long time and then is thrown to the right-hand position.

    (a) Find and sketch $i(t)$.

    (b) Calculate by integration the energy lost in the resistor during positive time. Confirm that all the energy stored initially in the capacitor is accounted for by the loss in the resistor.

Ans: (a) $i(t) = 1e^{-t/10\mu s}$ A

SECTIONS 3.2.4–3.2.6

P3.15. For the circuit shown in Fig. P3.15, the switch has been open a long time and is closed at $t = 0$.

    (a) What is the time constant of the circuit with the switch closed?

    (b) Determine the capacitor voltage with the polarity shown and sketch.

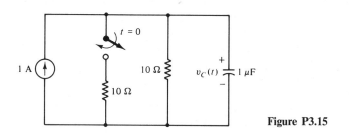

Figure P3.15

P3.16. Solve for and sketch $i(t)$ for the circuit in Fig. 3.11(b).

P3.17. For the circuit shown in Fig. P3.17, the switch is open a long time, closed for 5 ms, and then opened again. Find and sketch $i(t)$.

Ans: $i(t) = 2.5(1 - e^{-t/10\text{ms}})$ mA for $0 < t < 5$ ms, and
$i(t) = 0.98 \exp\left[-(t - 5\text{ ms})/20\text{ ms}\right]$ mA for $t > 5$ ms

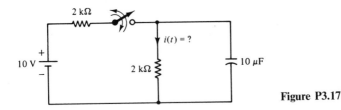

Figure P3.17

P3.18. For the circuit shown in Fig. P3.18, determine and plot $i_C(t)$, assuming that the capacitor has no initial energy.

Ans: $i_C(t) = 0.689 e^{-t/0.331\text{ms}}$ A

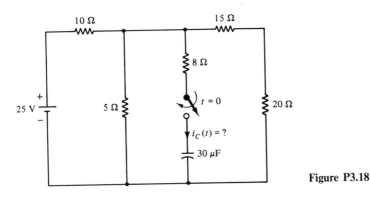

Figure P3.18

SECTION 3.2.7

P3.19. The parallel RL circuit shown in Fig. P3.19 is excited by the current pulse as shown. Calculate and sketch the resulting voltage, v_L. Assume that the inductor is initially unenergized.

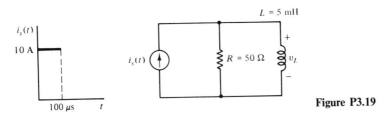

Figure P3.19

P3.20. A pulse can be modeled as two sources, one switching the voltage on and the other switching it off, as shown in Fig. P3.20. Note that the two sources add to zero except during the period $0 < t < t_1$. The voltage across the capacitor can be calculated by superposition. Use this model to rework the problem in Fig. 3.27. Specifically, solve for the voltage across the capacitor at $t = 1.0$ μs and verify the result given on page 100.

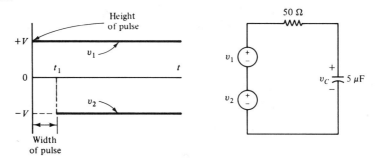

Figure P3.20   The pulse can be made by superposing two sources.

Application Problems

A3.1. In the process of charge separation in rain clouds, positive charges get left on the larger droplets and fall to earth. The cloud–earth system acts as a capacitor and hence this charge separation causes a large voltage to develop between the earth and the cloud. Voltage builds up until an ionized column forms, lightning discharges the cloud, and the process begins again. Here are some details:

1. The capacitance between cloud and earth is 0.01 μF.
2. The charge carried by the rain is 3×10^{19} electronic charges/s.
3. The resistance of the ionized column is 10 kΩ.
4. The lightning starts when the voltage reaches 3×10^9 V and stops at 10% of this value.

The circuit model for this problem is given in Fig. A3.1. Here are the questions:
(a) What is the rate of voltage buildup developed between cloud and ground in V/s?
(b) How often does lightning occur with this model?
(c) How long does the lightning discharge last?
(d) What percent of the stored energy is released by the lightning bolt?

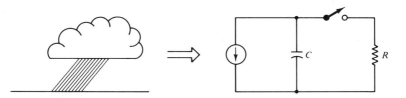

Figure A3.1   Circuit model for a lightning discharge.

A3.2. A computer puts out 5-V pulses which are 0.2 μs in duration. The output resistance is 600 Ω. The computer is connected to various pieces of equipment via coaxial cable which has 27 pF of capacitance per foot. The load resistance of the various pieces of equipment is 600 Ω. This information is summarized in Fig. A3.2. Sketch the output pulse for three cable lengths: 1 ft, 10 ft, and 100 ft. *Note:* These calculations suggest that pulses cannot be sent over long cables. This is a bit misleading. With proper output and load resistances, and with consideration of the cable inductance, one can show that pulses can be sent over long cables without significant distortion.

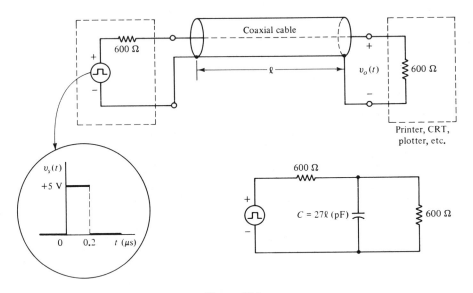

Figure A3.2

A3.3. When the switch in the *RC* circuit of Fig. A3.3 is closed, energy will for a period of time flow from the battery into the circuit. Once the current stops, the energy flow will cease.

    (a) Calculate by integration the total energy given to the circuit by the battery for $t > 0$. Show that this is Vq, where q is the charge on the capacitor.

    (b) Show by direct calculation that one-half this total energy is stored in the capacitor and the other half is lost to the circuit as losses in the resistor.

    (c) In view of the above, discuss the case where an ideal voltage source is connected to an ideal capacitance. Is energy conserved? Is this a legal circuit theory problem?

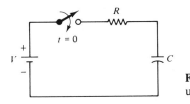

Figure A3.3 The capacitor initially is unenergized.

A3.4. The circuit in Fig. A3.4 will operate as a variable delay in operating the alarm. The alarm operates when its input current exceeds 100 μA. Find the range of R such that the delay is between 0.1 and 1 s.

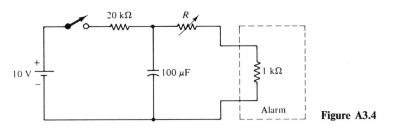

20 kΩ R

10 V   100 μF   1 kΩ

Alarm   **Figure A3.4**

A3.5. A microwave oven will boil water in 3 min, starting with water at 75°F. If the initial heating rate is 55°F/min, estimate the temperature to which the water would heat were it not for the phase change at 212°F.

A3.6. The circuit in Fig. A3.6 contains a capacitor, initially charged to a voltage V_0, a switch, and an inductor. What is the maximum value of the current which will flow after the switch is closed. *Hint:* Review the mechanical analogies in Table 3.1 and draw a mechanical analog for the circuit. Consider the behavior of the mechanical system and apply to the circuit. Consider conservation of energy.

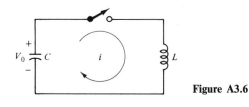

V_0 C i L

Figure A3.6

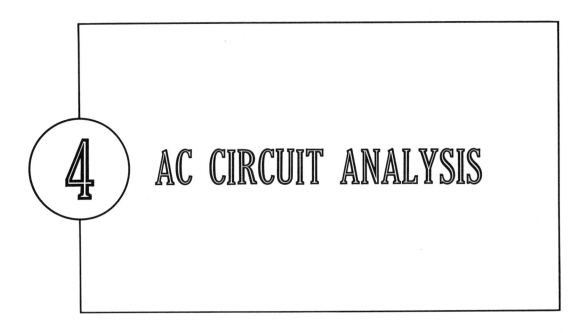

AC CIRCUIT ANALYSIS

4.1 INTRODUCTION TO ALTERNATING CURRENT

4.1.1 The Importance of AC

Thomas A. Edison was a clever, determined inventor whose activities excited the public imagination toward the practical uses of electricity. He pioneered, among other things, in the generation and distribution of electrical power for lighting. But Edison was committed to direct current. The power plants built by the Edison Electric Lighting Company produced dc.

Edison had a lot of young inventors and scientists working for him. His was, in fact, the first industrial research laboratory. One of these underlings was Nikola Tesla, a brilliant European engineer. Tesla appears to have been the first person to recognize the possibilities of ac and he is credited with inventing the ac induction motor. But his efforts to convince Boss Edison of the benefits of ac were in vain, and Tesla eventually quit.

Tesla went to a rival company and battle was pitched: Was it to be dc or ac? Nasty ads were placed in the newspapers by Edison's group, claiming that ac was unsafe. From our perspective, these warnings of the dangers of ac seem ridiculous, but at the time they created a serious debate.

Needless to say, Edison was wrong and Tesla was right. Today the vast majority of all electrical power is generated, distributed, and consumed in the form of ac power. Before World War II, to major in electrical engineering meant to study ac generators, motors, transformers, transmission lines, and the like. Currently, the ac power industry employs a mature technology and is a vital part of modern civilization. Count, for example, the number of electric motors in your dwelling. Of course you will think first of the air conditioner or heater blower and the refrigerator and other large appliances, but do not overiook the hair dryer, the electric clocks, the timer in the dishwasher, the phonograph turntable.

In this chapter you will learn how to analyze ac circuits and, more broadly, how electrical engineers think about ac waveforms. This chapter may tax your patience and your intellect, for the methods by which electrical engineers attack ac problems initially appear abstract and roundabout. We begin by deriving a simple DE, only to lead you into a review of complex numbers. But these methods, once mastered, become powerful, both in solving problems and in stimulating insight. What first appears as opaque becomes a window to a new way of viewing the physical world: the frequency-domain concept. We begin by acquainting you with the central figure in our development, the sinusoidal waveform.

4.1.2 Sinusoids

A Physical Model for a Sinusoid.   We use the word "sinusoid" to indicate a class of waveforms. Most of us were introduced to sines and cosines through the study of triangles. Later we learned that circular motion leads to sine and cosine functions. Figure 4.1 shows a crank. The horizontal projection of the crank is the length times the cosine of the angle ϕ. If the rotation speed is uniform, the horizontal projection becomes a sinusoidal function of time. We may describe this waveform mathematically as a sine function or a cosine function, but we will simply call it a sinusoid, or sinusoidal waveform.

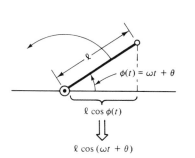

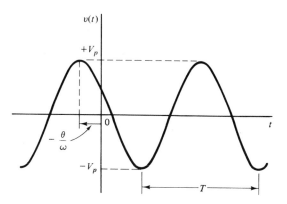

Figure 4.1 The projection of the rotating lever is a sinusoid.

Figure 4.2   The sinusoidal function is defined by its peak value, its phase, and its period (or frequency).

The Mathematical Form for a Sinusoid.   Electrical engineers have adopted the cosine function as the standard mathematical form for sinusoidal waveforms. In Fig. 4.2 we show the peak value, period, and phase of a sinusoidal waveform. The corresponding mathematical form of such a waveform is

$$v(t) = V_p \cos(\omega t + \theta) \tag{4.1}$$

The peak value of the voltage is V_p. The period, T, is related to the frequency of the sinusoid. The "event" frequency, namely, the number of cycles during a period of time, is the reciprocal of the period:

$$f = \frac{1}{T} \tag{4.2}$$

For example, if the period were 0.02 s, the event frequency would be 1/0.02 or 50 cycles per second. Because the unit "cycles per second" is awkward to say, most people tend to shorten it to "cycles," which is misleading to the novice and offensive to the purist. To stop this dimensional abuse, and to honor Heinrich Hertz (1857–1894), the unit hertz, abbreviated Hz, has replaced cycles per second as the common unit for event frequency. So 50 hertz means 50 cycles/second.

When most people talk about a frequency, they mean the event frequency. When we write mathematical expressions for sinusoidal waveforms, however, we more often deal with the proper mathematical measure, the angular frequency, ω (the Greek lowercase omega), in radians per second. Because there are 2π radians in a full circle (a cycle), the relationship between event frequency, f, and radian frequency is

$$\omega = 2\pi f = \frac{2\pi}{T}$$

The scientific dimensions of frequency are reciprocal seconds. The numerator (radians or cycles) represents an angle and is therefore dimensionless. Just as we wisely retain radians or degrees to remind us which measure of an angle we are using, so we need to state the units for frequency to make explicit which frequency we mean, f or ω.

The Units of Phase.   The phase, θ (the Greek lowercase theta), is what permits the waveform in Fig. 4.2 to represent a general sinusoid. We have drawn the curve for a phase of $+60°$, but we can shift the position of the sinusoid by varying the phase. The phase is related to the time origin when we use a mathematical description, but in an ac problem significance belongs only on the relative phases of the various sinusoidal voltages and currents. Here you must tolerate one of the traditional inconsistencies of electrical engineers. The proper mathematical unit for ωt is radians and hence the correct units for phase should also be radians. For example, we may wish to compute the time when the voltage reaches its positive peak. The cosine is maximum for zero angle, so the peak occurs when the total angle is zero:

$$\omega t_{\text{peak}} + \theta = 0 \Rightarrow t_{\text{peak}} = -\frac{\theta}{\omega} = -\frac{\theta}{2\pi} T \tag{4.3}$$

When we solve an equation like Eq. (4.3), we must use radian measure for the phase, as the last form of Eq. (4.3) suggests. But electrical engineers usually speak of phase in degrees, as we did above $(+60°)$. This inconsistency is tolerable because only relative phase is significant in most situations. Electrical engineers, like most people, still are more comfortable thinking about and sketching angles in degree measure. Probably you also think best in degree measure, so we will continue to express phase in degrees unless the mathematics demands radians.

Some Familiar Frequencies.   Frequency is a familiar concept. When we speak of a motor speed as 4000 rpm, for example, we are indicating a frequency of 66.7 Hz, and each spark plug would be firing with a frequency of 33.3 Hz. The power system frequency is 60 Hz in this country, although 50 Hz is used in some other parts of the world and 400 Hz is used in airborne and naval applications. When the radio announcer tells you that you are "tuned to 1200 on your radio dial," he is giving his station frequency, 1200 kHz (kilohertz) or 1.2×10^6 Hz. The FM stations broadcast at frequencies of about 100 MHz (megahertz) or about 10^8 Hz. The UHF TV band extends to about 800 MHz and communication satellites relay signals at about 5 GHz (gigahertz) or 5×10^9 Hz. The highest frequencies currently used for radio signals are about 300 GHz, 3×10^{11} Hz. We find infrared, optical, and x-ray radiation at even higher frequencies.

Later in this book we will explore more fully the importance of frequency in electrical engineering, particularly in communication systems. For now, we wish merely to introduce the concept of frequency and to show how ac waveforms are represented for ac circuit analysis.

4.1.3 An AC Circuit Problem

Figure 4.3 shows the ac circuit we will analyze first. The ac source has a frequency of 60 Hz or 120π, about 377, rad/s and is connected at $t = 0$. We wish to solve for the current, $i(t)$. After the switch is closed, we write KVL as

$$-v_s(t) + v_R(t) + v_L(t) = 0$$

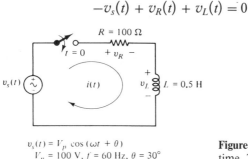

$$v_s(t) = V_p \cos(\omega t + \theta)$$
$$V_p = 100 \text{ V}, f = 60 \text{ Hz}, \theta = 30°$$

Figure 4.3 Solve for $i(t)$ for positive time. The switch closes at $t = 0$.

We may introduce $i(t)$ through Ohm's law and the definition of inductance, with the results

$$L\frac{di}{dt} + Ri = V_p \cos(\omega t + \theta) \qquad (4.4)$$

Equation (4.4) is a linear first-order DE, but the technique we used in Chapter 3 does not apply because we do not have a dc term on the right-hand side. The character of the response, however, is similar to our earlier results. Here also there will be a transition from one state of the circuit, before the switch was closed, to another state corresponding to the closed switch. Here also the transition period is expressed in terms of the L/R time constant of the circuit. In fact, the form of the solution is

$$i(t) = Ae^{-t/\tau} + i_p(t) \qquad \tau = \frac{L}{R} \tag{4.5}$$

where A is an unknown constant, τ the time constant, and $i_p(t)$ the particular integral, or steady-state response. The final state, $i_p(t)$, is no longer a constant but results from a dynamic equilibrium between source, resistor, and inductor. Methods for determining the steady-state response are the focus of this chapter. Equation (4.5) can be solved with several methods. The equation can be integrated directly with the aid of an integrating factor. This approach fails, however, when we try it on more complicated circuits and, besides, our goal is to learn how electrical engineers analyze ac problems. No electrical engineer would integrate this equation directly to find the steady-state solution. To learn the best procedure, you have to proceed through a long chain of reasoning. We will lead you down the traditional path—and please be patient. Once we arrive you will be amazed how easily we can solve ac circuit problems.

4.2 REPRESENTING SINUSOIDS WITH PHASORS

4.2.1 Sinusoids and Linear Systems

Sinusoid In, Sinusoid Out.   An important idea is suggested in Fig. 4.4. Think of the circuit as a linear system, linear because the equations of R and L are linear equations and a "system" because the circuit is an interconnection of such elements. (We could also have capacitors in our circuit, but that would complicate this first effort.)

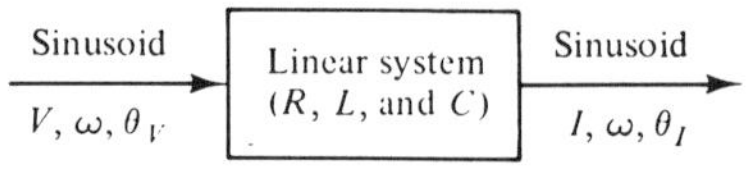

Figure 4.4   A linear system responds at the frequency of excitation.

Think of the voltage source as an input to this system and the current $i(t)$ as an output of the system. The important idea is the following: If the input is a sinusoid, the output is also a sinusoid at the same frequency. This assertion can be justified through examination of Eq. (4.4) and reflection on the properties of sinusoids. The sinusoidal steady-state solution of Eq. (4.4) must be a function which, when differentiated and added to itself, will result in a sinusoid of frequency ω. The only mathematical function that qualifies is a sinusoid of the same frequency because the "shape" of the sinusoidal function is invariant to linear operations such as addition, differentiation, and integration. To get a feeling for this, think about a system of springs and masses. If you shake such a system with a certain frequency (the ac input is an electrical shaking), the entire system will shake at the same frequency. In other words, no new frequencies are generated in the system; the only frequency that exists is the one you applied externally. So if we stimulate the circuit at 60 Hz, the current will respond at 60 Hz.

New Unknowns.   The input, being a sinusoid, is completely described by three numbers: the amplitude (V), the frequency (ω), and the phase (θ_V). The output must also be a sinusoid, and hence can also be described by an amplitude (I), frequency (ω), and phase (θ_I). Because the output frequency is known, only the amplitude and phase need to be derived to solve the problem, that is, to find the current in Fig. 4.3. Our object, therefore, is to develop an efficient method for finding the amplitude and phase of the output, which thus become our new unknowns. We will now develop a mathematical model suited to finding the unknown amplitude and phase. The first step is to represent the amplitude and phase of a sinusoid by a complex number.

A Mechanical Picture.   The rotating handle in Fig. 4.1 gives a picture of the amplitude and phase of a sinusoid. If the total angle, $\phi(t)$, were $\omega t + \theta$, the x-axis projection of the handle would be a sinusoid of amplitude equal to the length of the handle, ℓ, of the proper frequency, and having the phase θ. The side view of the crank even gives us a satisfying geometric picture of amplitude and phase. The problem with this model is that we cannot substitute a rotating crank into a DE, that is, that the mathematics of this model become awkward. To take derivatives, we have to introduce vector mathematics, and no one has yet to offer a satisfying definition of the reciprocal (for example) of a rotating crank.

A Mathematical Model.   The scheme adopted by electrical engineers for modeling sinusoids resembles the rotating crank but has outstanding mathematical credentials. To lay all our cards on the table, we suggest that we can model a sinusoid as a rotating point in the complex plane. An initial difficulty is that many students are not up to speed in the complex number system without considerable review. So now we must follow a detour in this development to refresh your knowledge of the complex plane. If you do not need the detour, jump to Section 4.2.3.

4.2.2 The Mathematics of the Complex Plane

Most of us were introduced to complex numbers through the study of quadratic equations. We discovered that the solution to certain equations such as $x^2 = -4$ required the introduction of a new type of number:

$$x = \sqrt{-4} = i2$$

These new numbers are called *imaginary* and the symbol i is chosen to identify these new numbers, like "$-$" is used to identify negative numbers. The connotation of the word "imaginary" is unfortunate, because these new numbers are no less the product of mathematical imagination than "real" numbers. The solutions of other quadratic equations are combinations of the real and imaginary numbers, such as $2 \pm i\sqrt{2}$. These numbers are called *complex*, another unpleasant word. Complex numbers are perhaps well named because the rules for manipulating them are more complicated that those for real numbers, but on the whole the complex numbers represent a reasonable and useful extension of our number system.

We expect that you already have some skill in dealing with complex numbers. However, we will review the properties of complex numbers, because we will need many of these properties to understand fully why a rotating point in the complex plane is an ingenious method for analyzing ac circuits. We now list and illustrate some of the important properties of complex numbers.

The Complex Plane.   All numbers can be represented as points in a "complex" plane, such as we show in Fig. 4.5. The horizontal axis represents real numbers, and the vertical axis represents imaginary numbers. You will note that we have begun using j instead of i to indicate imaginary numbers. This is customary among electrical engineers to avoid confusion between imaginary numbers and currents, which have traditionally been symbolized by i. In Fig. 4.5 we show two complex numbers, z_1 and z_2, each having a real and imaginary component. In general we can state that a complex number has a real and imaginary part,

$$z = x + jy$$

where x and y stand for ordinary numbers. Below, we give the principal algebraic and geometric properties of complex numbers, for we must understand these properties to accomplish our goal of representing sinusoids as rotating points in the complex plane.

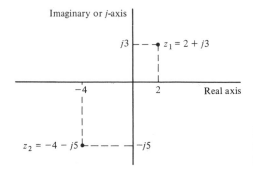

Figure 4.5 A complex number can be represented by a point in the complex plane.

Addition, Subtraction, Multiplication, and Equality of Complex Numbers.   A good first approximation to the algebra of complex numbers is to use ordinary algebra plus two rules: (1) keep real and imaginary parts separate and (2) treat j^2 as -1. For example, when we add complex numbers, we add the real parts and the imaginary parts separately:

$$z_1 + z_2 = (2 + j3) + (-4 - j5) = (2 - 4) + j(3 - 5) = -2 - j2$$

and similarly for subtraction. Thus complex numbers add and subtract like vectors in a plane. This is one of the few properties common between complex numbers and vectors.

As a second example, we shall find the square root of $-1 + j2$. We have two unknowns, the real and the imaginary parts of the root, so we can proceed as follows:

$$(x + jy)^2 = -1 + j2$$
$$x^2 + 2x(jy) + (jy)^2 = -1 + j2$$
$$x^2 - y^2 + j(2xy) = -1 + j2$$
$$x^2 - y^2 = -1 \quad \text{and} \quad 2xy = 2$$

Notice that we treated j^2 as -1 and we kept real and imaginary parts separate. Thus an equation involving complex numbers is equivalent to two ordinary equations, one for the real part and one for the imaginary part. We could continue the problem by solving simultaneously for x and y, but we will stop here. As we shall soon see, this is not the best way to extract the roots of complex numbers.

Division, Conjugation, and Absolute Values.   Division requires a trick to get the results into a standard form:

$$\frac{z_2}{z_1} = \frac{-4 - j5}{2 + j3} = \frac{-4 - j5}{2 + j3} \times \frac{2 - j3}{2 - j3}$$

$$= \frac{(-4)(2) + (-j5)(-j3) + (-4)(-j3) + (-j5)(2)}{(2)^2 - (j3)^2}$$

$$= \frac{-8 - 15 + j(+12 - 10)}{4 + 9} = -\frac{23}{13} + j\frac{2}{13}$$

The first form we wrote to the right of the first equal sign is considered nonstandard because it contains a complex number in the denominator. To force the denominator to be real, we multiply top and bottom by the *complex conjugate* of z, which is the same complex number except that the sign of the imaginary part is changed. This causes the cross term in the product in the denominator to drop out and thus forces the denominator to be real and positive. Meanwhile the numerator requires lots of careful work, but the rules are simple: Keep real and imaginary parts separate and let $j^2 = -1$. The final form is now considered standard because and we can at a glance identify the real and imaginary parts of the quotient.

If these tedious manipulations of complex numbers discourage you, take heart—there is a better way to multiply and divide complex numbers. Before we present this better way, however, we must look more closely at the "complex conjugate." In Fig. 4.6 we show the complex conjugate of z_1, denoted by z_1^*. We note that z_1^* is the mirror image of z_1. We have already shown that the product of a complex number with its conjugate is a real and positive number:

$$(x + jy)(x - jy) = x^2 - (jy)^2 = x^2 + y^2$$

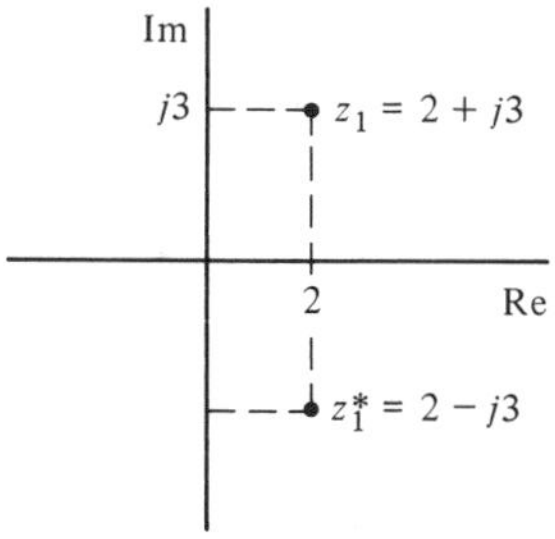

Figure 4.6   The complex conjugate of Z_1 is Z_1^*.

This suggests the definition of the absolute value (or the magnitude) of a complex number

$$|z| = \sqrt{zz^*} = \sqrt{x^2 + y^2} \tag{4.6}$$

The absolute value of a complex number thus is the Pythagorean sum of the real and imaginary parts. Geometrically, we identify this with the distance from the origin to the point representing the complex number, as shown in Fig. 4.7.

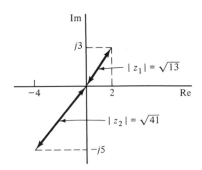

Figure 4.7 The magnitude of a complex number is its distance from the origin.

Trigonometric Form. Figure 4.8 shows a general complex number. We have shown that z, a point in the complex plane, can be located either by its x, y location or by a distance and an angle. We have used r for the distance; clearly, r is $|z|$, the absolute value of z. We call the x, y form of z the rectangular form and the r, ϕ form the polar form. We can symbolize the two forms as

$$z = x + jy = r\,\underline{/\phi} = |z|\,\underline{/\phi} \tag{4.7}$$

where the symbol $\underline{/\quad}$ is read "at an angle of." The right triangle yields simple transformations between rectangular and polar form:

$$x, y \Longleftrightarrow r, \phi$$

$$x = r\cos\phi \quad\underset{\Longleftarrow}{\Longrightarrow}\quad r = \sqrt{x^2 + y^2} \tag{4.8}$$

$$y = r\sin\phi \qquad \phi = \tan^{-1}\frac{y}{x}$$

Most engineering calculators have built-in programs to accomplish this transformation both ways. Presumably, you have used this feature in working with vectors.

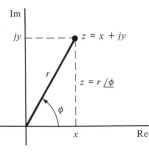

Figure 4.8 Complex number in rectangular and polar forms.

At this stage of our review, the polar form represents only a notation. If we wanted to multiply or take the square root of a complex number, we would have to do it with the rectangular form. But the polar form is closely related to the exponential form for a complex number and because of this relationship, the polar form proves to be extremely useful, not merely as a notation but also as a computational aid. All of this arises out of Euler's theorem, an amazing relationship between the exponential function and angles in the complex plane.

Exponential Form.   The Swiss mathematician Euler (pronounced "oiler," as in Houston Oilers) discovered an important property of complex numbers. He began with the series expansion for the function e^x,

$$e^x = 1 + x + \frac{x^2}{2!} + \frac{x^3}{3!} + \cdots$$

and made a series expansion for e^x when x is imaginary, say, $x = j\theta$.

$$e^{j\theta} = 1 + (j\theta) + \frac{(j\theta)^2}{2!} + \frac{(j\theta)^3}{3!} + \frac{(j\theta)^4}{4!} + \cdots$$

Euler simplified with $(j\theta)^2 = -\theta^2, (j\theta)^3 = -j\theta^3, \cdots$, and followed the rule of grouping together real and imaginary parts. The results were

$$e^{j\theta} = \left(1 - \frac{\theta^2}{2!} + \frac{\theta^4}{4!} - \cdots\right) + j\left(\theta - \frac{\theta^3}{3!} + \frac{\theta^5}{5!} + \cdots\right)$$

The series in the parentheses Euler identified as the expansions for cosine and sine. We would speculate that he wrote something like

$$e^{j\theta} = \cos \theta + j \sin \theta \qquad (?) \tag{4.9}$$

The question mark is not there because of some suspicion about the mathematics—the algebra is impeccable—but it is there because Euler must have wondered what this could mean. Look, for example, at the strange combination of mathematical symbols on the left-hand side of Eq. (4.9). A special number from differential calculus (e), the square root of -1, and the ratio of the arc to the radius of a circle (θ)—what can these have to do with each other? And on the right-hand side we find the ratios of the legs to the hypotenuse of a right triangle. What can these have to do with e and j? The right side of Eq. (4.9) is interpreted in Fig. 4.9. Since the Pythagorean sum of cosine and sine is unity, the right side of Eq. (4.9) must be a point in the complex plane one unit from the origin located at an angle θ counterclockwise from the positive real axis, as shown. Euler's theorem, if valid, requires that this point also be expressed by the function $e^{j\theta}$, as indicated by Eq. (4.9).

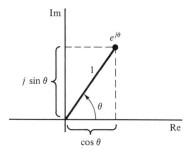

Figure 4.9 Implication of Eq. (4.9).

The best way to get acquainted with a new mathematical relationship often is to try it on some familiar specific cases to see how it works. Let us try Euler's formula on the identity $\sqrt{-1} = j1$. We will express the left-hand side of this equation with Euler's formula and follow the rules of exponents:

$$\sqrt{-1} = \sqrt{e^{j\pi}} = (e^{j\pi})^{1/2} = e^{j\pi/2} = \cos\frac{\pi}{2} + j\sin\frac{\pi}{2}$$

$$= 0 + j1$$

The first substitution follows from the fact that -1 in the complex plane is one unit from the origin at an angle π from the positive real axis. The second change replaces the square root symbol by the one-half power, and the third change follows an ordinary rule of exponents. The last form on the first line is a direct application of Euler's formula, and the second line gives the cosine and sine of $\pi/2$ radians. We conclude that Euler's formula leads to the correct answer for this special case. If we tried it on other powers and roots, it would work out there also.

Let us try Euler's formula on another, more demanding calculation. In Fig. 4.10 we show two complex numbers to be multiplied. The numbers are chosen to give common right triangles so that we can use familiar angles for the Euler formula. Note that these numbers are not on the unit circle, so we must multiply the complex exponential by a constant to move it out to the correct distance from the origin. In other words, because the exponential function represents only the angle of the complex number, we may multiply it by a magnitude to represent any complex number.

$$\sqrt{2}\,e^{j\pi/4} = 1 + j1$$

$$2e^{j2\pi/3} = -1 + j\sqrt{3}$$

$$(\sqrt{2}\,e^{j\pi/4})(2e^{j2\pi/3}) \overset{?}{=} (1+j1)(-1+j\sqrt{3})$$

$$2\sqrt{2}\,e^{j11\pi/12} = (1)(-1) + (j1)(j\sqrt{3}) + (j1)(-1) + (1)(j\sqrt{3})$$

$$2\sqrt{2}\left(\cos\frac{11\pi}{12} + j\sin\frac{11\pi}{12}\right) = (-1 - 1.732) + j(1.732 - 1)$$

$$-2.732 + j0.732 = -2.732 + j0.732$$

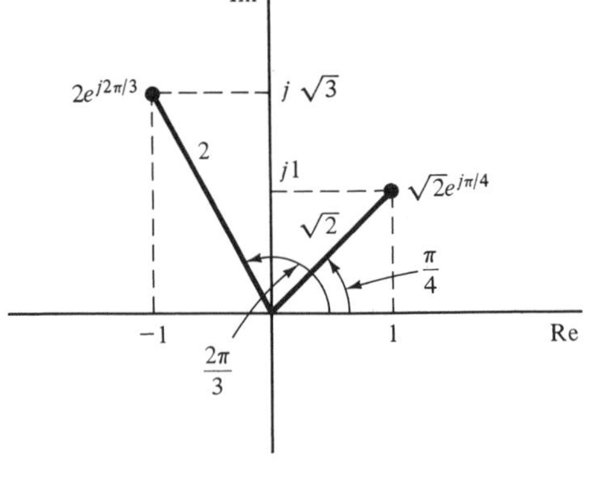

Figure 4.10 Two complex numbers in exponential form.

On the left-hand side we have combined the exponential forms according to the laws of exponents, and on the right-hand side we have followed the usual rules for complex numbers in rectangular form. We again conclude that Euler's formula works. Indeed, the strange formula in Eq. (4.9) is valid and provides the foundation for many important techniques in dealing with complex numbers. As you will see, our model for an ac waveform arises directly out of Euler's formula. Before returning to the main road from this detour, we wish first to show how the exponential form of complex numbers, based on Euler's formula, provides a simple way to multiply, divide, and take roots of complex numbers.

The relationship between our polar form and the exponential form follows from a comparison between Eqs. (4.8) and (4.9).

$$z = r \underline{/\theta} = re^{j\theta} \tag{4.10}$$

In Eq. (4.10) we normally express the angle of the polar form in degrees, whereas in the exponential form the angle *must* be expressed in radians for the Euler formula to be valid. This inconsistency arises from our traditions but does not diminish the fact that both forms give the same information, namely, that the complex number falls in the complex plane at a radius r and an angle θ counterclockwise from the positive real axis. We stress that the polar form is merely a compact notation, whereas the exponential form is a respectable mathematical function. The laws of exponents reveal that the magnitude of the product of two complex numbers is the product of the magnitudes of the numbers and the angle of the product is the sum of their angles.

$$(z_1)(z_2) = (r_1 e^{j\theta_1})(r_2 e^{j\theta_2}) = r_1 r_2 e^{j(\theta_1 + \theta_2)}$$

Although the proof rests on the exponential form, the results are more easily expressed in polar form:

$$(r_1 \underline{/\theta_1})(r_2 \underline{/\theta_2}) = r_1 r_2 \underline{/\theta_1 + \theta_2}$$

In a similar way, we can show that dividing two complex numbers requires dividing the magnitudes and subtracting the angles:

$$\frac{r_1 \underline{/\theta_1}}{r_2 \underline{/\theta_2}} = \frac{r_1}{r_2} \underline{/\theta_1 - \theta_2}$$

Summary of How to Calculate with Complex Numbers.   When we add and subtract complex numbers, the rectangular form must be used. This is true because we add and subtract real and imaginary parts separately. When we multiply (or divide) complex numbers, the polar form should be used because the magnitudes multiply (or divide) and the angles add (or subtract).

For the analysis of ac circuits, we must master the manipulation of complex numbers. Our techniques require frequent changes between rectangular and polar form. In the days of slide rules, these conversions were a pain, but with modern calculators they usually require a simple push of a special button on the calculator. If you currently do not know about that button, locate the manual on your calculator and look it up. This review of complex numbers is vital to your gaining skill in solving ac circuit problems. These operations, like the vector manipulations required in statics, are simple in principle, but care is requisite to reliable calculation.

Some Examples. Here are some additional examples of complex number manipulation, which give practice and show some new principles.

$$(2 - j6)(5\underline{/+30°})^* = (6.32\underline{/-71.6°})(5\underline{/-30°})$$
$$= 31.6\underline{/-101.6°}$$
$$= -6.34 - j31.0$$

Note that complex conjugation merely changes the sign of the angle. Find z, where $z^3 = -1 + j0.5$.

$$z^3 = -1 + j0.5 = 1.12\underline{/153.4°} = 1.12e^{j(153.4)(\pi)/180}$$
$$z = (1.12e^{j2.68})^{1/3} = \sqrt[3]{1.12}\ e^{j2.68/3}$$
$$= \text{also } \sqrt[3]{1.12}\ e^{j(2.68+2\pi)/3} \text{ and } \sqrt[3]{1.12}\ e^{j(2.68+4\pi)/3}$$
$$= 1.04\underline{/51.1°}, 1.04\underline{/171.1°}, 1.04\underline{/291.1°}$$

Note that to find all three cube roots, we must add 2π and 4π (one and two additional rotations) to the angle before dividing by 3.

4.2.3 The Phasor Idea

Expressing a Sinusoid with a Complex Number. Now that we have reviewed the mathematics of the complex plane, we will return to the main quest, that of finding the steady-state solution of the ac circuit problem described in Fig. 4.3 and Eq. (4.4). Specifically, we wish to find a way to represent sinusoidal sources, such as that appearing in Eq. (4.4), repeated below:

$$v_s(t) = V_p \cos(\omega t + \theta) \qquad V_p = 100, \quad \omega = 2\pi \times 60, \quad \theta = 30°$$

Euler's formula allows us to express a sinusoidal waveform with the function

$$V_p \cos(\omega t + \theta) = \text{Re}\{V_p e^{j(\omega t + \theta)}\} \tag{4.11}$$

where Re, the "real part of," is the complex-plane notation for the projection on the real axis. Equation (4.11) follows from Eq. (4.12) below, which expresses Euler's formula in appropriate terms:

$$V_p e^{j(\omega t + \theta)} = V_p[\cos(\omega t + \theta) + j\sin(\omega t + \theta)] \tag{4.12}$$

The Rotating Point in the Complex Plane. The left-hand side of Eq. (4.12) has great importance in this development and deserves more attention. We investigate it in the equation

$$V_p e^{j(\omega t + \theta)} = V_p(e^{j\omega t})(e^{j\theta}) = (V_p e^{j\theta})e^{j\omega t}$$
$$= (V_p\underline{/\theta})e^{j\omega t} = (100\underline{/30°})e^{j120\pi t} \tag{4.13}$$

The first line results from some juggling of terms and the rules of exponents. The second line emphasizes that this term consists of a complex constant, $100\underline{/30°}$ in this case, and a time function. The time function, $e^{j\omega t}$, describes the movement of a point in the complex plane. If we follow its progress as time increases, as shown in Fig. 4.11, we find it to be a point at unit distance from the origin rotating around the origin, starting at the real axis at $t = 0$ and passing through that point 60 times every second. The time function $e^{j\omega t}$ describes a point rotating in the complex plane at an angular frequency of ω radians per second. When we multiple this rotation function by the complex number $100\underline{/30°}$, we move the point out to a magnitude of 100 and counterclockwise by an angle of $30°$ at $t = 0$ as shown in Fig. 4.12 (point a). The rotation begins from that point and rotates around a circle of radius 100, 60 times a second. The projection on the real axis is shown below the complex plane (you have to turn the page on its side to look at the resulting time function). Clearly, we have the desired sinusoid—correct amplitude, phase, and frequency. Figure 4.12 is a graphical interpretation of Eq. (4.11), illustrated for the values appropriate for our specific case. We have now shown how to represent a sinusoid by a rotating point in the complex plane. To see why this is useful, we must return to the differential equation we are solving.

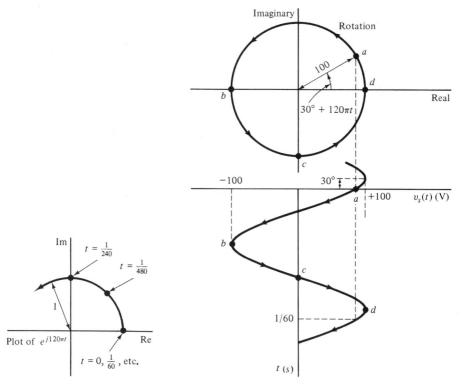

Figure 4.11 Rotating point in the complex plane.

Figure 4.12 The correspondence between the rotating point in the complex plane and the sinusoidal waveform.

Using Complex Numbers to Solve the DE:

$$L\frac{di}{dt} + Ri = V_p \cos(\omega t + \theta) = \text{Re}\left\{V_p \underline{\lfloor \theta}\, e^{j\omega t}\right\}$$

$$= \text{Re}\left\{\underline{\mathbf{V}}_s e^{j\omega t}\right\}$$

(4.14)

We have rewritten the DE in Eq. (4.14) and used the rotating point form on the right-hand side. The third form on the right-hand side uses the symbol $\underline{\mathbf{V}}_s$ to represent the complex number containing the source amplitude and phase, $V_p \underline{\lfloor \theta}$ in general, $100 \underline{\lfloor 30°}$ in this case. We use the uppercase bold, underlined symbol to indicate that it is a constant complex number.

Recall that we know something about the unknown current, $i(t)$. Earlier we argued that the output of the system (RL circuit in this case) had to be a sinusoid of the same frequency as the input. Only the amplitude and phase of that sinusoid must be determined. Using our scheme for representing sinusoids by complex numbers, we know that $i(t)$ may be represented in the form

$$i(t) = \text{Re}\left\{\underline{\mathbf{I}}e^{j\omega t}\right\} \qquad \text{where } \underline{\mathbf{I}} = I_p \underline{\lfloor \theta_I}$$

(4.15)

We have thus introduced $\underline{\mathbf{I}}$ as the unknown, everything else in the equation having been specified in the problem statement. We have numerical values for L and R, and of course ω is known from the input. We can now substitute Eq. (4.15) into Eq. (4.14):

$$L\frac{d}{dt}\text{Re}\left\{\underline{\mathbf{I}}e^{j\omega t}\right\} + R \times \text{Re}\left\{\underline{\mathbf{I}}e^{j\omega t}\right\} = \text{Re}\left\{\underline{\mathbf{V}}_s e^{j\omega t}\right\}$$

(4.16)

What to do about the derivative? We wish to avoid taking unnecessary excursions into mathematical proofs, so you will have to take our word for this assertion: The derivative can be taken inside the "real part" operation

$$\frac{d}{dt}\text{Re}\left\{\underline{\mathbf{I}}e^{j\omega t}\right\} = \text{Re}\left\{\frac{d}{dt}\underline{\mathbf{I}}e^{j\omega t}\right\} = \text{Re}\left\{\underline{\mathbf{I}}\frac{d}{dt}e^{j\omega t}\right\}$$

The exponential function with base e has the special property that its derivative is proportional to the function, as shown in the equation

$$\frac{d}{dx}e^{\alpha x} = e^{\alpha x}\frac{d}{dx}(\alpha x) = \alpha e^{\alpha x}$$

(4.17)

Applying this same procedure to the $e^{j\omega t}$ function, we obtain the result

$$\frac{d}{dt}e^{j\omega t} = e^{j\omega t}\frac{d}{dt}(j\omega t) = j\omega e^{j\omega t}$$

(4.18)

The Crank Again. The operation just completed is worthy of exploration. Recall our rotating crank in Fig. 4.1. If we express the tip of the crank as a rotating complex number, $\underline{\boldsymbol{\ell}}(t)$, we have

$$\underline{\boldsymbol{\ell}}(t) = \underline{\boldsymbol{\ell}}\,e^{j\omega t}$$

where $e^{j\omega t}$ represents the rotation in the complex plane and ℓ is the physical length of the crank. From mechanics, we know that the tip of the crank has a velocity $u = \omega\ell$, directed 90° from the position of the crank, as shown in Fig. 4.13. When we apply the derivative procedure of Eq. (4.18), we obtain

$$\frac{d}{dt}\,\underline{\boldsymbol{\ell}}(t) = j\omega\,\underline{\boldsymbol{\ell}}e^{j\omega t}$$

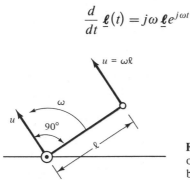

Figure 4.13 The position and velocity of the tip of the lever can be described by complex numbers.

We see that the magnitude comes out correct, but what about the j? We saw earlier that j represents an angle of 90°,

$$j1 = e^{j\pi/2} = 1\,\underline{/90°}$$

This 90° expresses the velocity in the correct direction in the complex plane. Thus we see that the mechanics of our crank comes out correct when we represent it as a complex number in the complex plane. We have thus been able to handle the "vectorness" of the problem without resorting to vectors—we simply took the derivative of a complex function. This simplicity of derivatives plays a role in the solution of DEs with this method.

__Back to the DE.__   We substitute the derivative back into Eq. (4.16). Because the "real part" operation distributes, we can collect terms:

$$\text{Re}\,\{L(j\omega\underline{\mathbf{I}}e^{j\omega t}) + R(\underline{\mathbf{I}}e^{j\omega t})\} = \text{Re}\,\{\underline{\mathbf{V}}_s e^{j\omega t}\} \tag{4.19}$$

$$\text{Re}\,\{(j\omega L\underline{\mathbf{I}} + R\underline{\mathbf{I}} - \underline{\mathbf{V}}_s)e^{j\omega t}\} = 0 \tag{4.20}$$

Equation (4.20) emerges from the manipulations; but what does it mean? Look first at the term, $j\omega L\underline{\mathbf{I}} + R\underline{\mathbf{I}} - \underline{\mathbf{V}}_s$. This represents a complex constant. Because it contains the complex number representing the unknown current amplitude and phase, $\underline{\mathbf{I}}$, this term in Eq. (4.20) is an unknown constant, a point somewhere in the complex plane. Of course, the $e^{j\omega t}$ term rotates this unknown constant, and the Re represents the projection of the rotating complex constant on the real axis. The right-hand side requires that this projection be zero at all times; hence the complex constant must itself be zero. If the point were not rotating, many complex constants could have zero projection (all points on the imaginary axis have this property), but the origin is the only point that continues to have zero projection when rotated about the origin. Equation (4.20) thus requires

$$j\omega L\underline{\mathbf{I}} + R\underline{\mathbf{I}} - \underline{\mathbf{V}}_s = 0 \tag{4.21}$$

Equation (4.21) involves only complex constants; time is no longer a factor. We know all the constants except $\underline{I}$, so we can solve for the unknown $\underline{I}$:

$$\underline{I} = \frac{\underline{V}_s}{R + j\omega L} = \frac{100\,\underline{/30°}}{100 + j(120\pi \times 0.5)}$$

$$= \frac{100\,\underline{/30°}}{213\,\underline{/62.1°}} = 0.469\,\underline{/-32.1°}$$

Recall that the magnitude of $\underline{I}$ represents the peak value of the sinusoidal steady-state current flowing in the RL circuit and the angle of $\underline{I}$ represents the phase angle of the current. Hence we may write

$$i_p(t) = 0.469 \cos(120\pi t - 32.1°) \tag{4.22}$$

This at last is the steady-state solution to the DE in Eq. (4.4) and hence this is the current that flows in the circuit shown in Fig. 4.3 after the transient period passes. Before finding the transient part of the solution, we will summarize the argument.

Summary of the Argument. Since the development of Eq. (4.22) has taken many pages and extensive math review, you may feel that you have been led through a complicated argument. Actually, there are but few steps, which we shall now review. First we wrote the DE and focused on the steady-state solution.

$$L\frac{di}{dt} + Ri = V_p \cos(\omega t + \theta)$$

We then represented the sinusoidal voltage and the unknown sinusoidal current by the real parts of rotating complex constants and took the derivative. The result was

$$\text{Re}\{j\omega L\underline{I}e^{j\omega t}\} + \text{Re}\{R\underline{I}e^{j\omega t}\} = \text{Re}\{\underline{V}_s e^{j\omega t}\} \tag{4.23}$$

We then collected terms and reasoned that the resulting complex constant had to vanish when the equation is put into the form of Eq. (4.21). This amounts to the same thing as dropping the Re and the $e^{j\omega t}$ in Eq. (4.23). Note that we did not *cancel* these factors—that would be mathematically incorrect—but the result is the same as if we had canceled them. The resulting equation is solved for $\underline{I}$, and the result is easily interpreted in terms of the amplitude and phase of the current in the circuit.

The Complete Solution. We may now determine the complete solution to Eq. (4.4) using principles developed in Chapter 3. The form of the solution is

$$i(t) = Ae^{-t/\tau} + 0.469 \cos(120\pi t - 32.1°)$$

where $\tau = L/R = 1/200$ s and A is an unknown constant to be determined from the initial condition. The initial current must be zero, $i(0^+) = 0$, due to the inductor; hence

$$0 = Ae^{0/\tau} + 0.469 \cos(-32.1°)$$

Hence $A = -0.397$. Thus the total solution is

$$i(t) = -0.397e^{-200t} + 0.496 \cos(120\pi t - 32.1°)$$

This response is shown in Fig. 4.14. Note that the current rapidly approaches the sinusoidal steady-state part of the solution.

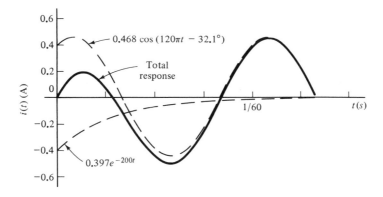

Figure 4.14   The current is the sum of the transient and steady-state responses.

Shortening the Procedure.   This method for determining the sinusoidal steady state admits to additional shortcuts. If we compare Eq. (4.21) with the original DE, we see a direct correlation between terms:

$$i(t) \Rightarrow \mathbf{I} \qquad v_s(t) \Rightarrow \mathbf{V}_s \qquad \frac{d}{dt} \Rightarrow j\omega \qquad (4.24)$$

We have shown that these are legitimate transformations. These suggest a shorter method:

1. Write the DE.
2. Perform the transformations shown in Eq. (4.24) on the known sinusoidal source, the unknown voltage or current, and the d/dt.
3. Solve the resulting complex equation for the unknown.
4. Interpret the results by substituting the amplitude and phase back into the standard form for a sinusoid.

Another Example, Solved with the Short Method.   Let us follow this procedure on another problem. Figure 4.15 shows an RC circuit driven by a sinusoidal current source. We wish to solve for the current in the capacitor. The relevant equations are KCL at the top node,

$$-i_s(t) + i_R(t) + i_C(t) = 0$$

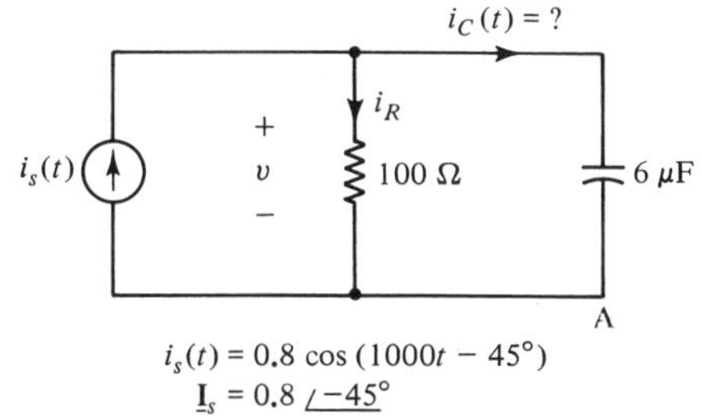

$i_C(t) = ?$

i_R

$i_s(t)$ $+$
v $100\ \Omega$ $6\ \mu\text{F}$
$-$

A

$i_s(t) = 0.8 \cos(1000t - 45°)$
$\mathbf{I}_s = 0.8 \angle{-45°}$

Figure 4.15   Solve for the steady-state current in the capacitor.

and the definitions of R and C,

$$i_R = \frac{v}{R} \quad \text{and} \quad i_C = C\frac{dv}{dt}$$

We can eliminate i_R from the KCL equation by taking the derivative of Ohm's law and substituting for dv/dt. The result is

$$\frac{di_R}{dt} = \frac{1}{R}\frac{dv}{dt} = \frac{1}{RC}i_C$$

Now we differentiate KCL, make this substitution, and rearrange terms into the usual form:

$$\frac{di_C}{dt} + \frac{1}{RC}i_C = \frac{d}{dt}i_s(t)$$

This completes the first step of our solution. The second step is to transform this DE into a complex equation with the changes suggested by Eq. (4.24).

$$i_s \Rightarrow \underline{I}_s \qquad i_C \Rightarrow \underline{I}_C \qquad \frac{d}{dt} \Rightarrow j\omega$$

The resulting complex equation is

$$j\omega\underline{I}_C + \frac{1}{RC}\underline{I}_C = j\omega\underline{I}_s$$

We can solve the equation for $\underline{I}_C$, which represents the amplitude and phase of the unknown current in the capacitor. The results are

$$\underline{I}_C = \frac{j\omega\underline{I}_s}{(1/RC) + j\omega} = \frac{j1000(0.8\underline{/-45°})}{1667 + j1000}$$

$$= \frac{800\underline{/90° - 45°}}{1944\underline{/31.0°}} = 0.412\underline{/14.0°}$$

The last step consists in writing the current in the standard form for sinusoids.

$$i_C(t) = 0.412\cos(1000t + 14.0°)\text{ A}$$

You must admit that apart from the manipulation of the complex numbers, which may still be unfamiliar to you, the derivation of the DE is now the hardest part of the solution. We will soon introduce additional shortcuts for simplifying that part of the problem. First we want to introduce some vocabulary to use in talking about these operations.

One important operation is representing a sinusoidal voltage or current by a complex number, as in Eq. (4.15). These complex representations are called *phasors*. A phasor voltage is the complex number which gives the time function sinusoid when rotated at the proper frequency and projected on the real axis. Because it is hard to see a single point, phasors are usually drawn as an arrow from the origin, as in Fig. 4.16. Because phasors add like vectors, they are often incorrectly referred to as vectors. The mathematics of phasors resembles the mathematics of vectors only in superficial ways.

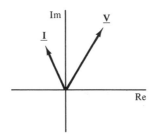

Figure 4.16 Voltage phasor and current phasor.

We use the expression *frequency domain* to refer to the form of the DE after it has been transformed into a complex equation. The DE is called the *time-domain* formulation of the problem because time is an important factor, but the transformed equation belongs to the frequency domain–time is no longer a factor. The phasor voltages and currents belong to the frequency domain. We move from the frequency domain to the time domain by rotating the phasors at the proper frequency and taking projections on the real axis. The factor $e^{j\omega t}$ represents the bridge between the time and frequency domains because it is used to move back and forth between the two domains. Now we move on to the final shortcut that electrical engineers use in analyzing ac circuits.

4.3 IMPEDANCE: REPRESENTING THE CIRCUIT IN THE FREQUENCY DOMAIN

4.3.1 The Basic Idea

Look at the Answer.   Let us look once more at the steady-state response of the *RL* circuit we solved, shown again in Fig. 4.17. This is a simple circuit with *R* and *L* in series; we know the sinusoidal driving voltage and we have solved for the current. When expressed as phasor voltages and currents, our results were

$$\underline{\mathbf{I}} = \frac{\mathbf{V}_s}{R + j\omega L} \tag{4.25}$$

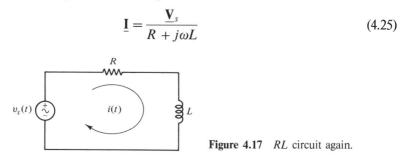

Figure 4.17 *RL* circuit again.

We interpret Eq. (4.25) as stating that the phasor current is found by dividing the phasor voltage by the sum of two terms, the resistance and $j\omega L$. Because Eq. (4.25) closely resembles Ohm's law, every part of this expression has meaning to us except the $j\omega L$ part. Clearly, this term represents the effect of the inductor in the circuit; the inductance appears in no other part of the answer. What can it mean?

Impedance.   This question, or one similar to it, sparked long ago an idea that leads to the final shortcut in analyzing ac circuits—the idea of impedance. This idea is the following: Since we are dealing only with sinusoidal voltages and currents, and since these can be represented by complex numbers, why not represent R's, L's, and C's by complex numbers also? This suggests transforming Ohm's law and the definitions of L and C into the frequency domain, that is, representing them by complex equations.

The Impedance of an Inductor.   Here is how it works out for the inductor: The defining equation for an inductor is

$$v_L = L\frac{di_L}{dt}$$

Because we are dealing solely with sinusoids, we can transform this equation into a frequency-domain equivalent with the changes

$$v_L(t) \Rightarrow \underline{\mathbf{V}}_L \qquad i_L(t) \Rightarrow \underline{\mathbf{I}}_L \qquad \frac{d}{dt} \Rightarrow j\omega$$

The result is

$$\underline{\mathbf{V}}_L = j\omega L\underline{\mathbf{I}}_L \tag{4.26}$$

The phasor interpretation of Eq. (4.26) is shown in Fig. 4.18. Equation (4.26) gives both the magnitude and phase relationships between the phasor voltage and phasor current for an inductor. The magnitude of the phasor voltage is ωL times the magnitude of the phasor current. Since $j1 = 1\underline{/90°}$, the phase of the phasor voltage leads the phase of the phasor current by 90°. We can verify this interpretation by taking the derivative in the time domain. If $i_L = I_p\cos(\omega t + \theta_I)$, then $v_L(t)$ is

$$v_L = L\frac{d}{dt}I_p\cos(\omega t + \theta_I) = -\omega L I_p\sin(\omega t + \theta_I)$$

$$= \omega L I_p\cos(\omega t + \theta_I + 90°)$$

where we have used the identity that $-\sin\theta = \cos(\theta + 90°)$. We use the word *impedance* to refer to the complex number relating the phasor voltage and phasor current of an element in an ac circuit, as shown in Fig. 4.19.

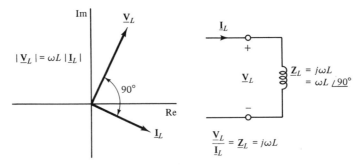

Figure 4.18   For an inductor, current lags voltage by 90°.

Figure 4.19   Representing an inductor by its impedance.

We define impedance as the phasor voltage divided by the phasor current, as in

$$\underline{Z} = \frac{\mathbf{V}}{\underline{\mathbf{I}}} \tag{4.27}$$

Applying this definition to Eq. (4.26), we obtain the result shown in Fig. 4.19. Hence $\underline{Z}_L = j\omega L$ is the impedance of an inductor at the frequency ω.

Kirchhoff's Laws in the Frequency Domain.   You will note that Fig. 4.19 also introduces the practice of putting the phasor voltage and current on the circuit diagram. This is legitimate and useful because, for sinusoidal voltages and currents, KVL and KCL can also be transformed directly into the frequency domain; for example,

$$-v_s(t) + v_R(t) + v_L(t) = 0 \Rightarrow -\underline{\mathbf{V}}_s + \underline{\mathbf{V}}_R + \underline{\mathbf{V}}_L = 0 \tag{4.28}$$

Let us review how the transformation in Eq. (4.28) is accomplished. Because all three voltages are sinusoids, they can be represented as the real part of complex numbers rotating in the complex plane. We can introduce the mathematical representations of such complex numbers and collect terms to obtain the equation

$$\mathrm{Re}\left\{(-\underline{\mathbf{V}}_s + \underline{\mathbf{V}}_R + \underline{\mathbf{V}}_L)e^{j\omega t}\right\} = 0$$

Only one point in the complex plane has the property that when rotated about the origin, its real part is always zero, and that point is the origin itself. Thus we can drop the real part and the $e^{j\omega t}$ part of the equation and obtain the phasor equation of Eq. (4.28). This is the frequency-domain version of KVL, as applied to the *RL* circuit. If there were nodes in the circuit, we could also write frequency-domain versions of KCL at these nodes.

Summary.   We can represent all voltages and currents in an ac circuit as phasors. These phasors are complex numbers representing the amplitudes and phases of the waveforms in the sinusoidal steady state. Phasor voltages and currents obey KVL and KCL. We can also represent *R* and *L* (and *C*, when we get around to it) as complex numbers, which are called *impedances*. Figure 4.20, gives the impedance of *R*, which is Ohm's law for phasors. The impedance of a resistor is real because its voltage and current are in phase.

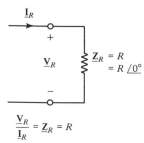

$$\frac{\underline{\mathbf{V}}_R}{\underline{\mathbf{I}}_R} = \underline{\mathbf{Z}}_R = R$$

Figure 4.20 The impedance of a resistor is real because no phase shift occurs.

Representing the Circuit in the Frequency Domain.   We can transform the entire circuit into the frequency domain, representing voltages and currents as phasors and representing the resistor and inductor as impedances. Because voltages and currents obey KVL and KCL, we can use the techniques of dc circuit theory to solve the circuit, only we must now work with impedances as if they were "complex resistors." This is the most efficient way to formulate the circuit equations (DEs in the time domain) and to determine the sinusoidal steady-state response.

The Benefits of the Impedance Concept.   We can now solve our original problem by an efficient method. Figure 4.21 shows the circuit transformed into the frequency domain. Because we have a simple series circuit, we can find $i(t)$ by the method of equivalent resistance (actually, equivalent impedance), combining the impedances of R and L in series. If we were faced with a more complicated circuit, we might analyze the circuit using node voltages, loop currents, or even a Thévenin equivalent circuit. Through the concept of impedance, all of the techniques of dc circuits can be applied to ac circuit problems.

By introducing the impedances we avoid the task of writing the DE. In fact, writing the DE corresponds to the "nonmethod" of solving dc circuits, because we introduce many variables and then seek sufficient equations to eliminate unwanted variables from the equations. The impedance concept eliminates that necessity.

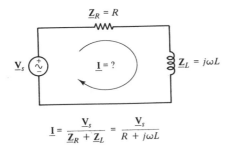

Figure 4.21   Frequency-domain version of the *RL* circuit.

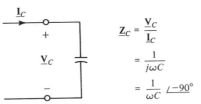

Figure 4.22   Impedance of a capacitor.

Let us rework the second example (Fig. 4.15) using frequency-domain techniques. Because the circuit contains a capacitor, we must derive the impedance of a capacitor, as symbolized in Fig. 4.22. This results from transforming the definition of a capacitor into the frequency domain, as shown in

$$i_C = C\frac{dv}{dt} \Rightarrow \underline{\mathbf{I}}_C = C(j\omega\underline{\mathbf{V}}_C) = j\omega C\underline{\mathbf{V}}_C$$

Thus

$$\underline{\mathbf{Z}}_C = \frac{1}{j\omega C} = \frac{j}{j^2\omega C} = j\frac{-1}{\omega C} \tag{4.29}$$

For a capacitor, the voltage lags the current by 90°, as shown in Fig. 4.23.

We now transform the circuit of Fig. 4.15 into the frequency domain using phasor currents and impedances, Fig. 4.24. Because R and C are connected in parallel, we may treat them like resistors in parallel and find $\underline{\mathbf{I}}_C$ with a current divider relationship. The results are the same as we obtained from the DE, once you do the complex algebra to show the equivalence.

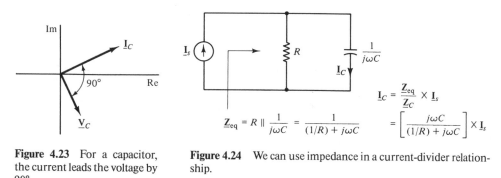

Figure 4.23 For a capacitor, the current leads the voltage by 90°.

Figure 4.24 We can use impedance in a current-divider relationship.

4.3.2 Phasor Diagrams for RL, RC, and RLC Circuits

You may wonder why we use the grandiose word "domain" for the techniques described above. You are correct in observing that we do not yet have much of a domain, only a technique for analyzing ac steady-state circuits at a single frequency. But there really is a domain; the exploration of this idea continues as an important theme of the remainder of this book. Extensive exploration of this concept is reserved for Chapter 8.

The viewpoint in this section is that we have available a sinusoidal source, the frequency of which we vary. Here we explore the effects of frequency changes on the response of series and parallel circuits in order to gain insight into the properties of inductors and capacitors in ac circuits.

RL Circuits

Series RL Circuit.   Figure 4.25 shows a series *RL* circuit. We shall investigate the effect of the inductor as frequency varies. If there were no inductor in the circuit, it would exhibit the same behavior at all frequencies; all voltages and currents would in phase with each other and in a fixed ratio. With an inductor in the circuit, however, the properties of the circuit change as frequency is varied.

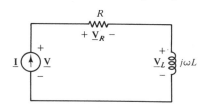

Figure 4.25 Series *RL* circuit.

The impedance of the series circuit is

$$\underline{\mathbf{Z}} = R + j\omega L$$

The relationship between the phasor voltage and current is thus

$$\underline{\mathbf{V}} = \underline{\mathbf{Z}}\underline{\mathbf{I}} = (R + j\omega L)\underline{\mathbf{I}}$$
$$= R\underline{\mathbf{I}} + j\omega L\underline{\mathbf{I}} = \underline{\mathbf{V}}_R + \underline{\mathbf{V}}_L \tag{4.30}$$

In Eq. (4.30) we have interpreted the two terms in the impedance as relating to the voltages across the resistor, $\underline{\mathbf{V}}_R$, and the inductor, $\underline{\mathbf{V}}_L$. Since the current is common to both R and L, we have drawn the phasor diagram in Fig. 4.26 with $\underline{\mathbf{I}}$ as the phase reference, that is, $\underline{\mathbf{I}} = I_p\underline{/0°}$. The phase of the voltage across the resistor, $\underline{\mathbf{V}}_R$, is the same as the phase of $\underline{\mathbf{I}}$, but the phase of the voltage across the inductor leads by 90°. The total voltage, $\underline{\mathbf{V}}$, the phasor sum of $\underline{\mathbf{V}}_R$ and $\underline{\mathbf{V}}_L$, thus leads the current by a phase angle somewhere between 0 and 90°. The phase angle by which the voltage leads the current is the geometric angle of $\underline{\mathbf{Z}}$ in the complex plane, θ, as given in Eq. (4.31) and Fig. 4.27.

$$\underline{\mathbf{Z}} = R + j\omega L = |\underline{\mathbf{Z}}|\underline{/\theta}$$

where

$$|\underline{\mathbf{Z}}| = \sqrt{R^2 + (\omega L)^2} \qquad \theta = \tan^{-1}\frac{\omega L}{R} \tag{4.31}$$

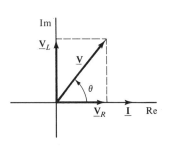

Figure 4.26 The inductor causes current to lag voltage.

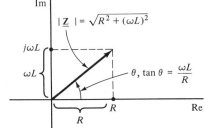

Figure 4.27 The angle of the impedance represents the phase shift.

Frequency Effects on the RL Series Circuit.   Now let us examine what happens to the phasor diagram when frequency is varied, other parameters being held constant. Consider first very low frequencies. By "low" we mean those frequencies where the imaginary part of $\underline{\mathbf{Z}}_s$ is small compared with the real part, that is, where $\omega L \ll R$. For low frequencies, the phase angle is small, meaning that $\underline{\mathbf{V}}$ and $\underline{\mathbf{I}}$ are almost in phase with each other, and the magnitude of the impedance is essentially equal to R. Thus the inductor has little effect at low frequencies. We saw in Chapter 3 that inductors become invisible at dc once their initial energy requirements are met. The foregoing discussion shows inductors to be virtually invisible at low ac frequencies. Or we can put it the other way around and treat dc behavior as a limiting case of ac as ω approaches zero. The mathematics supports this approach since $e^{j\omega t} \longrightarrow 1$ as $\omega \longrightarrow 0$. Whichever outlook we choose, Ohm's law dominates the low-frequency behavior of the circuit at low frequencies.

As frequency increases, the presence of the inductor is shown by an increase in the impedance of the inductor and hence an increase in the voltage across the inductor. This produces increased overall voltage and increased phase difference between the total voltage and the current. When $\omega L = R$, for example, the phase difference is 45° and the voltage is increased by $\sqrt{2}$ because the magnitude of the total impedance has increased by $\sqrt{2}$. As the frequency goes yet higher, the inductor comes to dominate the behavior of the circuit. The phase shift approaches 90° because the impedance approaches $j\omega L$. To get a feeling for how the impedance of the inductor gets large at high frequencies, imagine shaking a massive object in your hands. If you shake slowly (low frequencies), not much force is required, but as you attempt to shake faster, more force is required. Similarly, it takes more voltage to put the same current through an inductor as the frequency is increased.

The series RL circuit in Fig. 4.25 would be an appropriate model for an electric motor under steady load. The resistance would represent heat and friction losses in the motor, plus the energy converted to mechanical form and applied to a mechanical load. The inductance would represent magnetic energy storage in the motor structure. If we were to draw the phasor diagram for the motor, it would look like Fig. 4.26 except that the voltage would normally be used as the phase reference, as in Fig. 4.28.

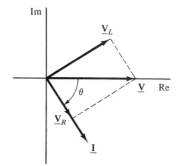

Figure 4.28 Phasor diagram redrawn with voltage as the phase reference.

Voltage is customarily used as the phase reference in ac power circuits because lights, motors, heaters, and so on, are parallel loads requiring a common voltage. Because voltage is the phase reference, current is said to *lag* for an inductive load and, as we shall soon see, *lead* for a capacitive load.

Resistive and Reactive Parts.   The foregoing discussion suggests that an impedance can represent more than a physical resistor and a physical inductor in an ac circuit. The real part of the impedance represents losses to the circuit, that is, energy leaving electrical form and converted to heat or mechanical work. The imaginary part of the impedance represents energy storage, in this instance magnetic energy storage in a motor. Because of these broader interpretations of impedance, which we shall explore more fully in the next chapter, technical names are given the real and imaginary parts of the impedance. The real part of $\underline{Z}$ is called the *resistive part* and the imaginary part is called the *reactive part*. The reactive part is given the symbol, X, as used in

$$\underline{Z} = R + jX = \text{(resistive)} + j\text{(reactive)} \tag{4.32}$$

The noun form of "reactive" is "reactance," such that it would be correct to say: "The reactance of an inductor is ωL." Reactance is a real number having the units ohms.

Why do we need these new words? Why can't we speak in terms of inductance? One reason we have already given—that the impedance represents more than simple R and L. The other reason becomes important when frequency is varied. The reactance of a true inductor varies linearly with frequency, but the reactance of more complicated circuits or devices does not vary linearly with frequency. Reactance is a more general concept than inductance in this broader context. Thus we need new words to extend the concept to cover new ranges of application of the ideas.

Parallel RL Load.   For the parallel RL load shown in Fig. 4.29, it is natural from both practical and mathematical considerations to use voltage as a phase reference. Since both R and L are connected in parallel with an ideal voltage source, we can determine their currents independently, as in

$$\mathbf{I}_R = \frac{\mathbf{V}}{\mathbf{Z}_R} = \frac{\mathbf{V}}{R} \qquad \mathbf{I}_L = \frac{\mathbf{V}}{\mathbf{Z}_L} = \frac{\mathbf{V}}{j\omega L} \qquad \mathbf{I} = \mathbf{I}_R + \mathbf{I}_L = \mathbf{V}\left(\frac{1}{R} + \frac{1}{j\omega L}\right) \qquad (4.33)$$

The impedance of the parallel load is calculated according to the generalized concept of impedance, $\mathbf{Z} = \mathbf{V}/\mathbf{I}$:

$$\mathbf{Z} = \frac{\mathbf{V}}{\mathbf{I}} = \frac{1}{1/R + 1/j\omega L} = R \,\|\, (j\omega L) = |\mathbf{Z}|\,\underline{/\theta} \qquad (4.34)$$

where $\theta = \tan^{-1}(R/\omega L)$.

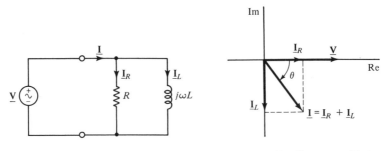

Figure 4.29   Parallel RL circuit.   Figure 4.30   Current still lags voltage due to the inductor.

Frequency Effects in RL Parallel Circuit.   The phasor diagram for the parallel RL circuit is shown in Fig. 4.30. Notice that the current lags the voltage. Let us now consider changes in the phasor diagram as frequency is varied. At low frequencies, $\omega L \ll R$, the current in the inductor is much larger than the current in the resistor because the inductor reactance approaches a short circuit at dc. This would cause the phase, θ, to approach $-90°$ (current lagging voltage).

As frequency increases, the reactance of the inductor will increase and the resistor gains importance. At $\omega L = R$, the phase is $-45°$, current lagging voltage, and the magnitude of the total current will be $\sqrt{2}$ times the current in the resistor. As higher frequencies are reached, the reactance of the inductor far exceeds that of the resistor, and the impedance of the parallel combination approaches that of the resistor.

In Fig. 4.31 we have converted a parallel circuit to an equivalent series circuit. The series equivalent of $100\,\|\,j100$ is thus $50 + j50$ ohms. The two circuits are equivalent, however, at a single frequency only; they would exhibit very different characteristics if

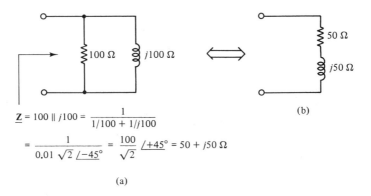

$$\underline{Z} = 100 \parallel j100 = \frac{1}{1/100 + 1/j100}$$

$$= \frac{1}{0.01 \sqrt{2} \; \underline{/-45°}} = \frac{100}{\sqrt{2}} \; \underline{/+45°} = 50 + j50 \; \Omega$$

(a)

Figure 4.31   A parallel circuit and a series circuit can be equivalent. (a) Parallel circuit; (b) series equivalent of parallel circuit.

frequency were varied. But they could be equally valid models for the same device, say, a motor or an induction furnace, at a single frequency.

RC Circuits

Series RC Circuit.   Figure 4.32 shows a series RC circuit. The impedance of the series combination is

$$\underline{Z} = R + \frac{1}{j\omega C} = \sqrt{R^2 + \left(\frac{1}{\omega C}\right)^2} \; \underline{\Big/ -\tan^{-1}\frac{1}{\omega RC}} \tag{4.35}$$

In order to put Eq. (4.35) into the form of Eq. (4.32), we must use the indentity $1/j = j/(j)^2 = -j$:

$$\underline{Z} = R + j\left(-\frac{1}{\omega C}\right) \tag{4.36}$$

Hence we see that the reactance of a capacitor is negative. Figure 4.33 shows the corresponding phasor diagram. We use the current as the phase reference, so the

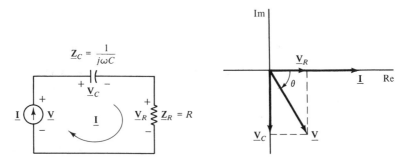

Figure 4.32   Series RC circuit.

Figure 4.33   Current leads voltage in the capacitive circuit.

voltages are shown lagging. However, we would normally say the current leads in a capacitive circuit because voltage would be considered the phase reference for reasons given above.

Frequency Effects in RC Series Circuit.   In this case, we will examine the frequency behavior by starting at high frequencies. At high frequencies, the reactance of the capacitor $(-1/\omega C)$ is very small and the circuit appears resistive. The phase angle is nearly zero, with the current slightly leading the voltage. As frequency is decreased, however, the reactance of the capacitor increases. This increases $\mathbf{V}_C$, which increases the total voltage and the phase angle. At $1/\omega C = R$, for example, the phase is 45°, leading current, and the voltage has increased by $\sqrt{2}$ from its high-frequency value. At low frequencies, the reactance of the capacitor becomes very large, approaching an open circuit at dc. Thus at low frequencies the capacitor dominates the behavior of the series combination, and the current phase approaches 90° leading the voltage. Notice that we always speak of the phase of the current relative to the voltage even though we have drawn the phasor diagram with current as the phase reference.

Parallel RC Circuit.   Figures 4.34 and 4.35 show a parallel *RC* circuit and the corresponding phasor diagram. The currents in the resistor and the capacitor are independent because they are in parallel with an ideal voltage source. Hence the current can be derived directly from consideration of the impedances of each parallel branch.

$$\mathbf{I}_R = \frac{\mathbf{V}}{R} \qquad \mathbf{I}_C = \frac{\mathbf{V}}{1/j\omega C} = j\omega C \mathbf{V} \tag{4.37}$$

$$\mathbf{I} = \mathbf{I}_R + \mathbf{I}_C = \mathbf{V}\left(\frac{1}{R} + j\omega C\right) \qquad \theta = \tan^{-1}(\omega RC) \tag{4.38}$$

We could derive the impedance from Eq. (4.38), as we did for the *RL* parallel circuit in Eq. (4.34), but Eq. (4.38) leads naturally to the definition of "admittance."

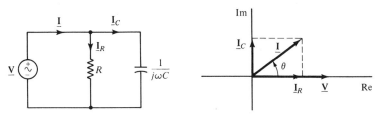

Figure 4.34   Parallel *RC* circuit.   **Figure 4.35**   The capacitor causes the current to lead the voltage.

Admittance.   Recall from Chapter 2 that we found it convenient to introduce the concept of conductance, $G = 1/R$, to simplify discussion of parallel combinations of resistors. There the concept of conductance brought few practical benefits, because modern calculators handle reciprocals without difficulty, but the corresponding concept in ac circuits has considerable theoretical and practical importance. *Admittance*, $\underline{Y}$, is defined as

$$\underline{Y} = \frac{\underline{I}}{\underline{V}} = \frac{1}{\underline{Z}} \tag{4.39}$$

Clearly, the admittance of the parallel RC circuit in Fig. 4.35 is

$$\underline{Y} = \frac{\underline{I}}{\underline{V}} = \frac{1}{R} + j\omega C = G + j\omega C \tag{4.40}$$

where we have introduced the conductance of the resistor, G. Admittance is useful for dealing with parallel circuits and has importance because parallel connections are common in practice. We have a specialized vocabulary associated with admittance. The real part of the admittance is called the *conductive part* or the *conductance*. The imaginary part of the admittance is called the *susceptive part* or the *susceptance*. Thus we say that the conductive part of the admittance of a parallel RC circuit is G and the susceptive part is ωC. The standard symbols for the conductance and susceptance are shown in

$$\underline{Y} = G + jB \tag{4.41}$$

Note that B is a real number. For example the conductive part of the parallel RL circuit in Fig. 4.31 as 0.01 S (siemens) and the susceptive part is $B = -0.01$ S. Note that the susceptance of the inductive circuit is negative, whereas the susceptive part of the admittance of the capacitive circuit is positive—just opposite the signs for the reactance.

RLC Circuits

Series RLC.   Figure 4.36 shows a series RLC circuit. The impedance of the circuit is

$$\underline{Z} = R + j\omega L + \frac{1}{j\omega C} = R + j\left(\omega L - \frac{1}{\omega C}\right) = |\underline{Z}|\underline{/\,\theta} \tag{4.42}$$

where

$$|\underline{Z}| = \sqrt{R^2 + \left(\omega L - \frac{1}{\omega C}\right)^2} \quad \text{and} \quad \theta = \tan^{-1}\frac{\omega L - 1/\omega C}{R}$$

The reactive part of the impedance now combines the effects of the inductor and the capacitor. The frequency response of this reactive term gives this circuit interesting and

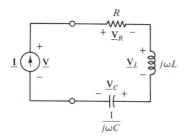

Figure 4.36 Series RLC circuit.

useful characteristics. The phasor diagram is shown in Fig. 4.37. We have drawn the phasors showing the voltage across the inductor greater in magnitude than the voltage across the capacitor. This situation would be appropriate for high frequencies.

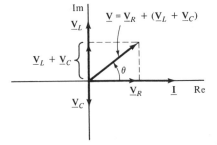

Figure 4.37   Above resonance the inductor dominates, so current lags voltage.

The frequency characteristics of the series RLC circuit combine those of the series RC and RL circuits. At dc the circuit acts as an open circuit because of the capacitor. At low frequencies the reactance of the capacitor dominates and the phase angle would be nearly 90°, with current leading voltage. As frequency increases, however, the inductive reactance becomes significant and at a special frequency grows to the point of canceling the negative reactance of the capacitor. This frequency occurs when $\omega L = 1/\omega C$, or $\omega = 1/\sqrt{LC}$, and is called the frequency of series resonance. At resonance, the inductor and capacitor combination becomes invisible and R is the total impedance of the circuit. As the increasing frequency passes through resonance, the phase changes from leading to lagging (current lagging voltage), and at resonance the phase is zero. At frequencies above resonance, the inductor dominates the circuit characteristics and the phasor diagram of Fig. 4.37 shows the relative voltages. At very high frequencies the phase approaches 90° lagging.

This circuit, with its series resonance, has importance in electronics because it can be used to select one group of frequencies from a broader group. For example, this circuit can be used as part of a radio filter which selects one station for reception, rejecting all others. We will discuss this circuit more fully in Chapter 5 as we discuss energy in ac circuits. We will see that resonance occurs when magnetic and electric energy requirements are equal, just as a mechanical system exhibits resonance when kinetic and potential energy requirements are balanced.

An Example of Series Resonance.   As an example of a circuit exhibiting series resonance we will calculate some characteristic frequencies for the circuit shown in Fig. 4.38. The impedance is

$$\mathbf{Z}(\omega) = 10 + j\left(2 \times 10^{-6}\omega - \frac{10^8}{\omega}\right)\Omega$$

The resistive part of the impedance is constant but the reactive part varies with frequency, being dominated by the capacitive reactance at low frequencies and by the inductive reactance at high frequencies. Resonance occurs when the total reactance is zero:

$$2 \times 10^{-6}\omega = \frac{10^8}{\omega} \Rightarrow \omega = 7.071 \times 10^6 \ (f = 1125 \text{ kHz})$$

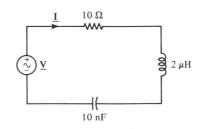

Figure 4.38 Series RLC circuit. We shall find the resonant frequency and the width of the resonance.

At resonance, that is, at a frequency of 1,125 kHz, the impedance of the circuit is 10 Ω resistive. Frequencies will exist below and above resonance where the reactance is equal to the resistance, and the phase will be $\pm45°$. These occur at

$$\left| 2 \times 10^{-6}\omega - \frac{10^8}{\omega} \right| = 10 \tag{4.43}$$

Equation (4.43) leads to two quadratic equations, which yield frequencies of 1592 kHz and 796 kHz. At the lower frequency the circuit would be capacitive, and the angle of the impedance would be negative (leading current). At the higher frequency the circuit would be inductive, and the angle of the impedance would be positive (lagging current). Since these frequencies are in the AM radio band, this circuit might find application as a frequency filter in an AM radio.

What Is Resonance?   We have shown from the mathematics that the series RLC circuit exhibits a resonance at $\omega = 1/\sqrt{LC}$. This occurs when the reactances of the inductor and capacitor cancel each other and the resistor alone is evident to the external source. Resonance implies a balance of stored energy. The inductor and capacitor store energy at different times in the ac cycle and hence can "play catch" with the same energy. In the case of resonance, after the initial energy is impared to the circuit by the source, it no longer has to supply stored energy to the inductor and capacitor. When in the next chapter we explore energy relationships in ac circuits, we will reexamine resonance in greater detail.

The Parallel RLC Circuit.   The parallel RLC circuit (Fig. 4.39) combines the properties on the parallel RL and RC circuits. The admittance of the circuit is

$$\underline{Y} = G + j\omega C + \frac{1}{j\omega L} = G + j\left(\omega C - \frac{1}{\omega L}\right) \tag{4.44}$$

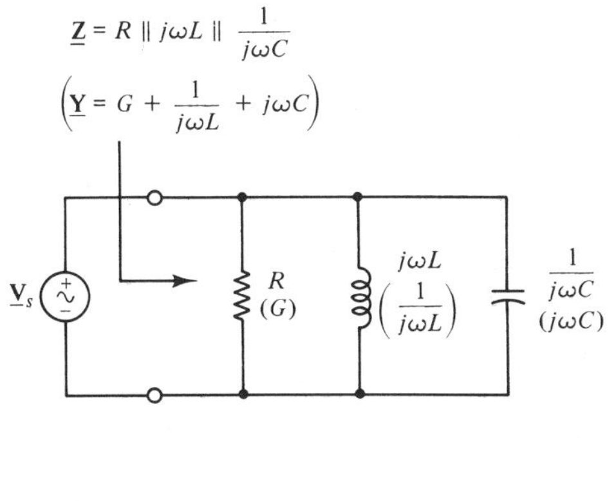

$\underline{Z} = R \parallel j\omega L \parallel \dfrac{1}{j\omega C}$

$\left(\underline{Y} = G + \dfrac{1}{j\omega L} + j\omega C\right)$

R
(G)

$\dfrac{j\omega L}{\left(\dfrac{1}{j\omega L}\right)}$

$\dfrac{1}{j\omega C}$
$(j\omega C)$

$\underline{V}_s$

Figure 4.39 Parallel RLC circuit. Admittance (in parentheses) is natural for parallel circuits.

At low frequencies, the susceptance of the inductor $(-1/\omega L)$ is large and dominates the admittance expression. The admittance is large (infinite at dc) and the phase approaches $-90°$, lagging current. As frequency is increased, the inductive susceptance diminishes and the capacitive susceptance grows until they become equal. This is a parallel resonance and it occurs when $\omega C = 1/\omega L$, or $\omega = 1/\sqrt{LC}$, the same frequency as for series resonance. Thus series and parallel resonance occur at the same frequency for the same ideal inductor and capacitor.

At resonance, the admittance is pure conductance, $\underline{Y} = G = 1/R$. This again can be understood as a balance of stored energy. Once the initial energy needs of the inductor and capacitor are met, they play catch with the same energy, needing it at different times in the ac cycle, and thus become invisible to the source. Because the source must supply only the loss of the resistor, the resistor (or conductance) alone appears in the impedance at resonance.

As the frequency increases above resonance, the capacitive susceptance dominates, and as the frequency approaches very high frequencies, the admittance becomes very large and the phase approaches $+90°$, leading current. Thus the admittance is minimum at resonance and becomes very large at low and high frequencies. Put differently, at low and high frequencies the impedance is very small, approaching a short circuit, but the impedance has a maximum at the frequency of parallel resonance. This contrasts with the series case, where the impedance is minimum at resonance.

4.4 SUMMARY

In this chapter we have presented an efficient method for determining the steady state response of ac circuits. We showed that sinusoidal time functions can be represented by complex numbers called phasors. Circuit elements are represented by complex numbers called impedances, and impedances can be combined in series or parallel like resistors in dc circuits. The DEs describing circuit behavior in the time domain become complex algebraic equations in the frequency domain. Finally, we investigated the effects of frequency variations in series and parallel RL, RC, and RLC circuits.

PROBLEMS

Practice Problems

SECTIONS 4.1.1–4.1.3

P4.1. Find the frequencies in hertz and in radians per second for the following:
   (a) The rotation of the earth on its axis relative to the earth–sun line.
   (b) The rotation of a bike tire (26 in. dia.) at 20 mph.
   (c) The second hand of a watch.
   (d) A dentist's drill going 200,000 rpm.
   (e) An LP phonograph record.

P4.2. A sinusoidal function is shown in Fig. P4.2. Determine the frequency, phase, and amplitude for expressing this sinusoid in the standard form:

$$i(t) = I_p \cos(\omega t + \theta) \qquad \text{amperes}$$

$$\textit{Ans:} \quad i(t) = 9 \cos(5.24 \times 10^6 t - 70.5°) \text{ A}$$

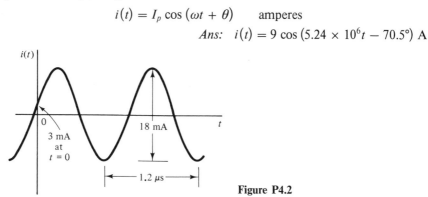

Figure P4.2

P4.3. Sketch the sinusoidal voltage: $v(t) = 60 \cos(100\pi t - 120°)$ V

SECTION 4.2.1

P4.4. Show that Eq. (4.4) is satisfied by a solution of the form

$$i(t) = I_p \cos(\omega t + \theta_I)$$

for specific values of I_p and θ_I. To simplify the algebra of solving for I_p and θ_I, set the phase of the voltage source to zero. Determine I_p and θ by the following procedure: Substitute this form into the DE; expand the left-hand side with the equations $\cos(A + B) = \cos A \cos B - \sin A \sin B$ and $\sin(A + B) = \sin A \cos B + \cos A \sin B$; and equate like terms.

SECTION 4.2.2

P4.5. Given three complex numbers: $z = 2 - j6$, $w = 5\underline{/-135°}$, and $u = 8e^{j2}$:
   (a) Find $|z|$.
   (b) Find u^* in rectangular form.
   (c) Evaluate $z/(w - u)$ and place the result in exponential form.
   (d) Solve for s (a complex number) if $z(s - w) = u$ and express s in polar form.
   $\textit{Ans:}$ (a) 6.32; (b) $-3.32 - j7.27$; (c) $0.585e^{j0.341}$; (d) $6.04\underline{/-142.6°}$

P4.6. For the complex numbers, $z_1 = 2 + j3$ and $z_2 = -1 + j2$:
   (a) Show that $|z_1 \times z_2| = |z_1| \times |z_2|$.
   (b) Show that $|z_1/z_2| = |z_1|/|z_1|$.
   (c) Demonstrate that part (a) and part (b) are generally true for $z_1 = x_1 + jy_1$ and $z_2 = x_2 + jy_2$.

P4.7. Given that $z = x + jy$ is a general complex number:
   (a) Show that Re $\{z\} = (z + z^*)/2$.
   (b) Show that

$$\frac{1}{z} = \frac{x}{x^2 + y^2} - \frac{jy}{x^2 + y^2}$$

   (c) Solve for the times when Re $\{ze^{j\omega t}\} = 0$.

SECTION 4.2.3

P4.8. (a) What is the phasor for $v(t) = 5.2 \cos (100t - 90°)$?
    (b) Sketch one cycle of the time function, $v(t)$, represented by the phasor $\mathbf{V} = 60e^{j\pi/3}$, $f = 60$ Hz.
    (c) Use the trigonometric identity $\cos (A \pm B) = \cos A \cos B \mp \sin A \sin B$ to show that if one has the time-domain function expressed in the sine form (not the cosine form), an additional phase shift of $-90°$ is introduced in the phase of the phasor. That is, prove that $V_p \sin \omega t \Rightarrow \mathbf{V} = -jV_p = V_p\underline{/-90°}$.

Ans: (a) $5.2\underline{/-90°}$

P4.9. Use phasor techniques in the following.
    (a) Find $2 \cos (100t - 45°) - 3 \cos (100t + 60°)$.
    (b) Find $50 \sin (100t) + d \cos (100t - 30°)/dt$.
    (c) Use phasor techniques to evaluate the derivative of $i(t) = 20 \sin 500t$ at $t = 2$ ms. *Hint:* Change to cosine form, then express as a phasor, when use $j\omega$ for the derivative, and finally take the real part of the resulting phasor times $e^{j\omega t}$ term.

Ans: (a) $4.01 \cos (100t - 91.2°)$; (c) 5400 A/s

P4.10. On page 125 it was argued that if

$$\text{Re }\{\mathbf{Z}e^{j\omega t}\} = \text{Re }\{\mathbf{W}e^{j\omega t}\}$$

for all t, then $\mathbf{Z} = \mathbf{W}$. In effect the Re and the $e^{j\omega t}$ can be dropped. Prove this by letting $\mathbf{Z} = Z_r + jZ_i$ and $\mathbf{W} = W_r + jW_i$ and evaluating the equation at the times when $\omega t = 0$ and $\omega t = \pi/2$.

P4.11. Using phasor techniques, prove the trigonometric identities $\cos (A \pm B) = \cos A \cos B \mp \sin A \sin B$ and $\sin (A \pm B) = \sin A \cos B \pm \cos A \sin B$. *Hint:* Start with $\cos (A \pm B) = \text{Re }\{e^{j(A \pm B)}\}$ and work with the right side of the equation.

P4.12. Solve for $v(t)$ in the circuit shown in Fig. P4.12 using the phasor methods described in Section 4.2.3. You may use the "shortened method" if you wish.

Ans: $v(t) = 71.6 \cos (1000t + 46.6°)$ V

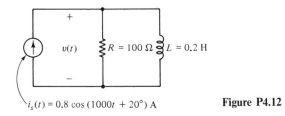

$i_s(t) = 0.8 \cos (1000t + 20°)$ A   **Figure P4.12**

P4.13. Solve for $v_C(t)$ in the circuit shown in Fig. P4.13 using the phasor methods described in Section 4.2.3. You may use the shortened method if you wish.

Ans: $v_C(t) = 6.47 \cos (500\pi t - 105°)$ V

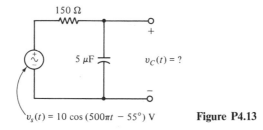

Figure P4.13

P4.14. Rework Problem P4.13 for the total response if the voltage is applied with a switch closure at $t = 0$.

P4.15. The mechanical system shown in Fig. P4.15 is excited by a rotating wheel which gives approximately a sinusoidal displacement for x_1. The differential equation of the displacement x_2 is

$$m\frac{d^2x_2}{dt^2} + D\frac{dx_2}{dt} + Kx_2 = D\frac{dx_1}{dt}$$

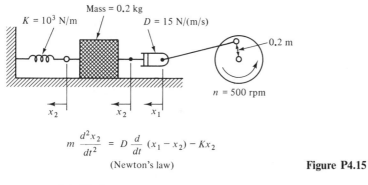

$$m\frac{d^2x_2}{dt^2} = D\frac{d}{dt}(x_1 - x_2) - Kx_2$$

(Newton's law)   **Figure P4.15**

(a) Find ω.
(b) Transform the DE to the frequency domain

$$x_2(t) \Rightarrow \mathbf{X}_2$$

(c) Solve for $\mathbf{X}_2$.
(d) Write $x_2(t)$.

SECTION 4.3.1

P4.16. Make a chart for resistors, capacitors, and inductors with the following columns: name, symbol, time-domain equation, frequency-domain equation, impedance.

P4.17. **(a)** What values of capacitance and inductance have an impedance with a magnitude of 10 Ω at a frequency of 1 kHz?
   (b) What would be the magnitudes of the impedance of this C and this L at 2 kHz?

P4.18. Using the frequency-domain versions of KVL and KCL, show that two impedances in series add like resistors in series, that is, $\mathbf{Z}_{eq} = \mathbf{Z}_1 + \mathbf{Z}_2$. Show also that two impedances in parallel add like resistors in parallel, that is,

$$\mathbf{Z}_{eq} = \mathbf{Z}_1 \| \mathbf{Z}_2 = \frac{1}{\dfrac{1}{\mathbf{Z}_1} + \dfrac{1}{\mathbf{Z}_2}}$$

P4.19. Determine the input impedance of the circuits shown in Fig. P4.19.

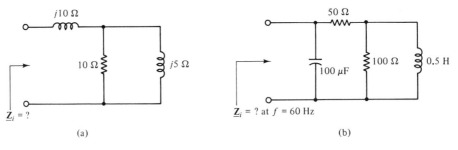

(a)   (b)

Figure P4.19

P4.20. Use the techniques of the frequency domain to solve for $i(t)$ in the circuit shown in Fig. P4.20.

    (a) Find the frequency-domain version of the circuit, using phasors to represent sinusoidal functions, known and unknown, and impedances to represent circuit components.

    (b) Using parallel and series combinations, find $\underline{Z}_{eq}$ as seen by the voltage source.

    (c) Solve for $\underline{I}$.

    (d) Convert back to the time domain.

$$Ans: \quad i(t) = 0.707 \cos\left(1000t - 81.9°\right) \text{ A}$$

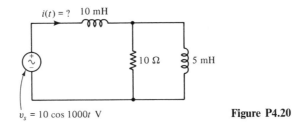

Figure P4.20

P4.21. Consider a 5-μF capacitor and a 10-Ω resistor.

    (a) They are connected in series. At what frequency in hertz is the magnitude of their series impedance 20 Ω?

    (b) If the resistor and capacitor are now placed in parallel, find the frequency at which their combined impedance is 7 Ω.

P4.22. For the circuit shown in Fig. P4.22, determine $v(t)$ using phasor techniques. Sketch $v(t)$ in the time domain.

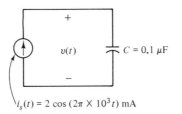

Figure P4.22

P4.23. The circuit shown in Fig. P4.23 is in sinusoidal steady state. Determine the maximum value of the current and the first time after $t = 0$ at which the maximum current occurs.

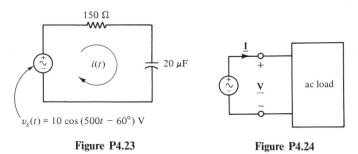

Figure P4.23   Figure P4.24

P4.24. A 60-Hz ac source and load are connected as indicated in Fig. P4.24. The phasor voltage is $\mathbf{V} = 100 + j0$ V and the phasor current is $\mathbf{I} = 8 - j5$ A.
  (a) Find the instantaneous current at $t = 0$.
  (b) What is the impedance of the ac load in rectangular form?
  (c) Assuming that the load is a series resistor and inductor, find the value of the inductance.

P4.25. For the circuit shown in Fig. P4.25:
  (a) Draw a phasor diagram showing $\mathbf{V}_s$, $\mathbf{I}$, $\mathbf{V}_R$, and $\mathbf{V}_C$. The voltages must be drawn to consistent scale and shown to add in accordance with the phasor KVL.
  (b) Find $v_R(t)$ and sketch along with the source voltage.

Ans: $v_R(t) = 14.1 \cos (800\pi t + 38.5°)$

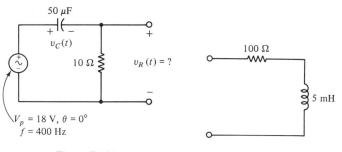

Figure P4.25   Figure P4.26

P4.26. For the circuit shown in Fig. P4.26:
  (a) At what frequency would the magnitude of the input impedance be 160 Ω?
  (b) What is the phase of the impedance at this frequency?
  (c) What value of C should be added in series to make the circuit appear pure resistive at this frequency? In parallel? (The answers to these two questions differ.)

P4.27. Convert the circuit shown in Fig. P4.27 to the equivalent parallel circuit at $f = 1$ kHz.

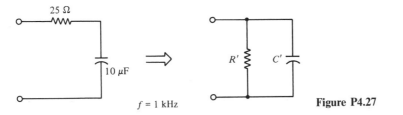

Figure P4.27

P4.28. The circuit in Fig. P4.28 is to be represented by a Norton equivalent circuit. Determine $\underline{I}_N$ and $\underline{Z}_{eq}$.

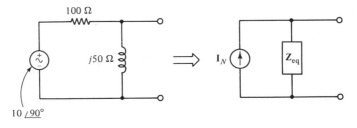

Figure P4.28

POWER IN AC CIRCUITS

5.1 AC POWER AND ENERGY STORAGE: THE TIME-DOMAIN PICTURE

5.1.1 The Importance of Power and Energy

Power and energy are important in ac problems for several reasons. Energy concerns you as a citizen user of electricity because the local electric utility charges you for the energy you use to light, refrigerate, play the radio, and so on. Formerly, this energy cost little, but no longer can we ignore this cost.

You have no doubt been impressed with the role energy plays in describing the behavior of mechanical systems. The roller coaster problem of freshman physics—where you compute the speed at some point on the track from a difference in height—exemplifies how energy considerations often sweep away many details of a problem and lead directly to a useful result. Indeed, the more experience you gain in analyzing physical systems, the more you should be impressed with the importance of energy in revealing the true workings of a system. Some feel, as does your author, that any analysis is incomplete until energy relationships are explored and understood.

In this chapter we investigate energy and power relationships in electrical circuits. We begin by looking at averages because average power is frequently our focus. Then we will examine energy and power relationships in ac circuits from the viewpoint of the time domain. Next comes the frequency-domain viewpoint, and we succeed in expressing energy and power relationships with phasors. Finally, we introduce three topics of eminent practical value: transformers, three-phase circuits, and electrical safety. These will round out our understanding of circuits and complete the circuits section of this book.

5.1.2 The Average Value of an Electrical Signal

What Is an Average?   Everybody knows how to calculate a numerical average. We compute the average of 12, 9, and 15 by adding the numbers (36), dividing by the number of values we are averaging (3), and getting the average (12). In general, the average of n numbers, $x_i, i = 1, 2, \ldots, n$, is

$$X_{\text{average}} = \frac{\sum_{i=1}^{n} x_i}{n} \qquad \text{or} \qquad n(X_{\text{average}}) = \sum_{i=1}^{n} x_i \qquad (5.1)$$

In the second form of Eq. (5.1), we see that the average value multiplied by the number of samples is equal to the sum of the numbers we are averaging.

That is how to compute an arithmetic average, but what is an average in concept? An average is a number that characterizes in some particular aspect a body of information. The arithmetic average, for example, gives us some idea of the size of the numbers, such as the average price of gasoline in Kansas or the average income in Italy. There are many kinds of averages. The grade-point average provides an example near to the heart of most college students. This average characterizes the academic performance of a student even though many important factors, such as course load and difficulty, are ignored. Nevertheless, if we insisted on one indication of excellence, the GPA would be used. Thus an average is a number that characterizes a body of information in one particular way, leaving out all the rest of the information.

Computing Time Averages.   The time average of a periodic function can be defined as a generalization of the arithmetic average. The periodic function in Fig. 5.1 provides an example. Because the function is periodic, the average over all time will be the same as the average over one period; hence we can limit our attention to the time from $t = 0$ to $t = T$, as shown in Fig. 5.2. We can define the time average by analogy with the second form of Eq. (5.1): n becomes T, the period; X_{average} becomes V_{avg}; and the summation of the numbers becomes the summation of all the heights of the function, which is the integral over the time period from 0 to T. Thus we have the definition

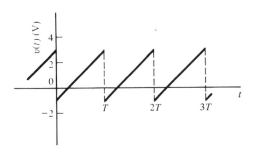

Figure 5.1   Find the time-average voltage.

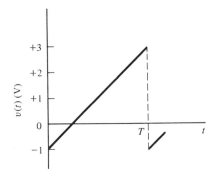

Figure 5.2   One period of the waveform.

of time-average voltage in

$$TV_{\text{avg}} = \int_0^T v(t)\, dt \Rightarrow V_{\text{avg}} = \frac{1}{T}\int_0^T v(t)\, dt \tag{5.2}$$

The first form of Eq. (5.2) can be interpreted in terms of area, as shown in Fig. 5.3. Because V_{avg} is a constant, the product on the left-hand side represents the area on the $v(t)$ graph of a rectangle having base T and height V_{avg}. The right-hand side is the area under the $v(t)$ curve, counting area above the axis as positive and area below the axis as negative. This geometric interpretation is shown in Fig. 5.3. Equation (5.2) requires the two areas to be equal. The second form of Eq. (5.2) serves for computing averages. The present example can be handled as the areas of triangles, but we will use calculus to illustrate the more general method. First we derive the equation of $v(t)$ in the slope-intercept form: the intercept is -1; the slope is $3 - (-1)$ divided by 10 ms, or 400 V/s. Thus the equation is

$$v(t) = -1 + 400t \qquad 0 < t < 10 \text{ ms} \tag{5.3}$$

Because this formula is valid only during the first period of the waveform, we have stated that restriction explicitly. We now substitute into Eq. (5.2), with the result

$$V_{\text{avg}} = \frac{1}{10 \text{ ms}} \int_0^{10\text{ms}} (-1 + 400t)\, dt = 10^2 \left(-t + \frac{400t^2}{2} \right)\Big|_0^{10^{-2}} = +1 \tag{5.4}$$

This is the value indicated on Fig. 5.3 and its validity is apparent.

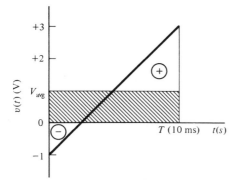

Figure 5.3   The average multiplied by the period has the same area as the original waveform, counting the area below the axis as negative.

Some Special Averages.   Several results follow from the definition of time average. The time average of a dc (constant) voltage or current is the dc value. For this reason the time average value of a signal, such as that shown in Fig. 5.3, is often called the dc value or dc component of the signal. Another obvious result is that the average value of a sinusoidal waveform is zero. The sinusoidal function has equal areas above and below the time axis and thus has zero average (or dc) value.

If we have the sum of two signals, say, two voltage sources connected in series, the average of the sum is the sum of the averages of the component signals. This follows from Eq. (5.2) because the integral distributes to the two functions.

$$(v_1 + v_2)_{\text{avg}} = \frac{1}{T} \int_0^T [v_1(t) + v_2(t)]\, dt$$

$$= \frac{1}{T} \left[\int_0^T v_1(t)\, dt + \int_0^T v_2(t)\, dt \right] = V_{1\,\text{avg}} + V_{2\,\text{avg}}$$

One example would be the sum of a dc and a sinusoidal signal; the average would be the dc value because the sinusoidal part would average to zero.

5.1.3 Effective or Root-Mean-Square Value

Time-Average Power. We will now consider the time-average power in a dc circuit, Fig. 5.4. We have indicated that the resistor is *hot* because the electrical energy into the resistor appears as heat in the thermodynamic realm. From Chapter 2 we know that the power into the resistor is $(V_{\text{dc}})^2/R$, but we will derive this result here from more general considerations, which we shall then apply to the heating of a resistor with an ac source. Voltage is energy/charge and current is charge flow/time; it follows that the instantaneous power (energy/time) into a circuit element is

$$p(t) = v(t)i(t)$$

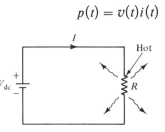

Figure 5.4 The heat is a measure of the average power.

For our dc circuit in Fig. 5.4 both voltage and current are constant, so the instantaneous power into the resistor is $v \times i = (V_{\text{dc}})^2/R$, a constant. The time-average power (P) into R is thus

$$P = \frac{1}{T} \int_0^T p(t)\, dt = \frac{1}{T} \int_0^T \frac{(V_{\text{dc}})^2}{R}\, dt = \frac{(V_{\text{dc}})^2}{R} \tag{5.5}$$

The time-average power determines how hot the resistor will become. The movement of charge through the resistor imparts thermal energy to the material. The input electrical power appears as a heat source internal to the resistor; the actual temperature of the resistor depends on this input and on its thermal coupling to the environment. The more power into the resistor, the hotter it will become. Physical resistors are available for $\frac{1}{4}$ watt, $\frac{1}{2}$ watt, 1 watt, 2 watt, and so on. The power rating indicates how much power the resistor can accept without burning out or changing its resistance value significantly.

Power in AC Circuits.   Figure 5.5 shows the same circuit with an ac source. The instantaneous power is

$$p(t) = v(t)i(t) = V_p \cos \omega t \times \left(\frac{V_p}{R} \cos \omega t \right) = \frac{V_p^2}{R} \cos^2 \omega t \qquad \omega = \frac{2\pi}{T} \qquad (5.6)$$

Figure 5.6 shows a plot of the instantaneous power [Eq. (5.6)]. Although the charges move back and forth in the resistor, the power is always nonnegative; this is shown mathematically from the squaring of the voltage or current in the power relationship. The nonnegative power results physically because the moving electrons heat up the material regardless of their direction of flow. The energy does not flow smoothly into the resistor but flows in spurts, twice each cycle of the ac waveform. Thus the resistor is not heated steadily but cyclically. In most resistors this time variation is unimportant because thermal inertia smooths out the heating variations and hence the temperature remains essentially constant. However, these variations can be a problem in incandescent lighting, where the thermal inertia of the filament is small. The power frequencies of 50 or 60 Hz were established high enough to make the flickering of the light (at 100 or 120 Hz) barely noticeable. Had a lower frequency been chosen, lights might flicker as noticeably as do the pictures of some TV sets.

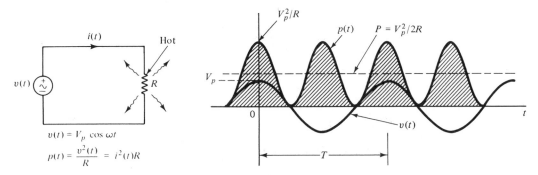

Figure 5.5 The heat is a measure of the time-average power.

Figure 5.6 The energy flows into the resistor in spurts. The power is positive at all times.

The Root-Mean-Square or Effective Value.   The peak value for the power function in Fig. 5.6 is V_p^2/R. From the symmetry of the $p(t)$ function, it is apparent that the average value is half that peak, or $V_p^2/2R$, as indicated on Fig. 5.6. The analytic computation of this result appears in Eq. (5.7). The integral uses the trigonometric identity $\cos^2 \omega t = (1 + \cos 2\omega t)/2$:

$$P = \frac{1}{T} \int_0^T \frac{V_p^2}{R} \cos^2 \omega t \, dt = \frac{V_p^2}{RT} \int_0^T \left(\frac{1}{2} + \frac{1}{2} \cos 2\omega t \right) dt$$

$$= \frac{V_p^2}{RT} \left(\frac{t}{2} + \frac{\sin 2\omega t}{4\omega} \right)_0^T = \frac{V_p^2}{2R} \qquad (5.7)$$

Equation (5.7) leads to the effective value of the ac voltage. When we speak of effective, the "effect" to which we refer is the heating effect, or more generally the time-average energy conversion from electrical to nonelectrical form. In this problem the effect is thus the average power. The effective value of an ac waveform is the equivalent dc value which would heat the resistor as hot as the ac waveform heats it.

$$P = \frac{V_e^2}{R} = \frac{V_p^2}{2R} \Rightarrow V_e = \frac{V_p}{\sqrt{2}} \tag{5.8}$$

Equation (5.8) equates the time-average power from an equivalent dc source with magnitude V_e, the effective value, to the time average of power due to the ac source. This defines the effective value as applied to the specific case of an ac source. The result is that the effective value of the sinusoidal voltage (or current, if we were dealing with current) is $1/\sqrt{2}$ or 0.707 times the peak value. This is an important result because it characterizes the power-producing capability of an ac source, which often is important, after all, for its power-producing ability. For example, an ac voltage with a peak value of 20 V would be equivalent in heating effect to a $20/\sqrt{2} = 14.1$-V battery. But we warn you that Eq. (5.8) applies only for the sinusoidal waveform, as we will illustrate shortly.

The Effective Value in General.   The effective value is often referred to as the root-mean-square value (rms). The general definition of effective value is shown in

$$P = \frac{V_e^2}{R} = \frac{1}{T} \int_0^T \frac{v^2(t)}{R} \, dt \Rightarrow V_e = \sqrt{\frac{1}{T} \int_0^T v^2(t) \, dt} \tag{5.9}$$

Here we have illustrated the definition of rms for voltage, $v(t)$, but a similar expression would apply for the effective value of a current. The effective value is the square root of the mean (i.e., the average, as in "mean sea level") of the square of the function, rms.

The rms value of a voltage or current carries importance because power and energy dominate our concern in many instances. The practical importance of rms values is suggested by the way voltmeters and ammeters are calibrated. If you measured the voltage at the wall outlet in your room with an ac voltmeter, you would expect something around 120 V. Would that be peak or rms or what? If you "looked" at the voltage with an oscilloscope, you would observe the peak-to-peak voltage of the sinusoid to be about 340 V. Consequently, the peak value is 170 V and the rms is about 120 V. Generally, ac meters are calibrated to indicate rms for a sinusoid.

An Example.   To illustrate the computation of the rms value of a nonsinusoidal waveform, we will compute the rms value of the waveform in Fig. 5.1, repeated in Fig. 5.7a. We derived the equation of the voltage during the first period to be

$$v(t) = -1 + 400t \qquad 0 < t < 10 \text{ ms}$$

Substituting into Eq. (5.9) and integrating the function, we obtain

$$V = \sqrt{\frac{1}{10 \text{ ms}} \int_0^{10\text{ms}} (-1 + 400t)^2 \, dt} = 1.53 \text{ V}$$

The rms value is shown in Fig. 5.7a. Notice that the waveform exceeds the effective value for only a small fraction of the cycle. Since the rms computation averages the square of the waveform, the rms value always equals or exceeds the average of the function, even the average of the absolute value of the function. Accordingly, the rms value favors the peak of the sinusoidal waveform, as Fig. 5.7b shows.

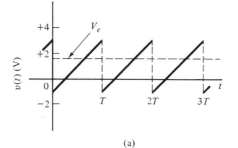

(a)

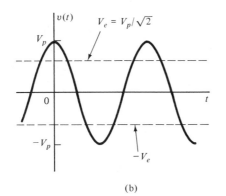

(b)

Figure 5.7 The effective values fall near the peaks. (a) sawtooth waveform; (b) sinusoidal waveform.

Measuring AC Voltage.   Most ac voltmeters would be unlikely to indicate the true rms (1.52 V) of the waveform in Fig. 5.7a. A meter designed to square and average the instantaneous waveform could be complicated and expensive, so usually some other property of the waveform is measured and the rms is inferred from that, assuming a sinusoidal shape. For example, a common type of meter actually responds to the peak-to-peak value of the waveform, but the meter scale is marked to indicate the peak-to-peak value divided by $2\sqrt{2}$, which would be the rms for a sinusoid. Such a meter would indicate $4/2\sqrt{2} = 1.41$ V for the waveform in Fig. 5.7a, clearly not the true rms value for this (nonsinusoidal) waveform. Thus meter readings require careful interpretation when measuring nonsinusoidals.

Summary.   The power-producing capability of a waveform is usually important. We can represent that power-producing capability with an average value called the effective (or rms) value of the waveform. This is defined as the value which, when squared, is equal to the time average of the square of the waveform. For a dc waveform, the effective value is equal to the dc value. For a sinusoidal waveform, the effective value is $1/\sqrt{2}$

times the peak value. For other waveforms, the effective value may be calculated by squaring and averaging. Electrical meters are designed to indicate the effective value of a sinusoidal waveform but are unlikely to indicate the effective value for other wave shapes.

5.1.4 Power and Energy Relations for R, L, and C

Resistance.   We have discussed the power relationship for resistance in deriving the effective value of a sinusoid. As shown in Fig. 5.6, the energy flows unilaterally into the resistor, not smoothly but in spurts. The time-average power into the resistor is

$$P_R = \frac{V_p^2}{2R} = \frac{V_e^2}{R} = I_e^2 R \tag{5.10}$$

where I_e is the effective current. Consider a 120-V, 100-W light bulb. According to Eq. (5.10), this indicates a resistance value of $(120)^2/100 = 144 \ \Omega$ (recall that the 120 V is the rms value). The energy consumed by the bulb, operated for 24 hours, would be 24 hours × 0.100 kW or 2.4 kWh (kilowatthours). This represents the total energy consumed, and at 6 cents/kWh the bulb would cost you slightly less than 15 cents per day to operate.

Inductance.   We will calculate the instantaneous and time-average magnetic energy stored by an inductor. The current and voltage for an inductor are

$$i_L(t) = I_p \cos \omega t$$

$$v_L(t) = L \frac{di}{dt} = -\omega L I_p \sin \omega t$$

These are shown in Fig. 5.8. The instantaneous power into the inductor is the product of voltage and current:

$$p_L(t) = v_L(t) i_L(t) = (-\omega L I_p \sin \omega t)(I_p \cos \omega t)$$

$$= -\frac{\omega L I_p^2}{2} \sin 2\omega t \tag{5.11}$$

where we have used the trigonometric identity $(\sin \omega t)(\cos \omega t) = (\sin 2\omega t)/2$. The

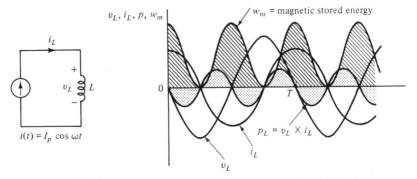

Figure 5.8   Voltage, current, stored energy, and power into the inductor.

power into the inductor is shown in Fig. 5.8 as the lightly shaded area. We notice that the average power is zero, for the inductor loses no energy. Thus an inductor gives back on the average as much energy as it receives. (This is strictly true only for an ideal inductor; a real inductor made of wire would have resistive losses.) Being the product of v_L and i_L, the power is zero four times each cycle, when either the voltage or the current is zero. For a 60-Hz source, the energy would pulsate in and out of the inductor 120 times per second because it makes two round trips from source to inductor each cycle. For this reason, heavy electrical equipment, such as transformers and motors, often hums audibly at 120 Hz.

This pulsation at twice the ac frequency applies to the stored energy also. The magnetic energy stored in an inductor is given in Eq. (3.3) as

$$w_m(t) = \tfrac{1}{2} L i_L^2(t)$$

For the sinusoidal source, we find the instantaneous stored energy to be

$$w_m(t) = \tfrac{1}{2} L (I_p \cos \omega t)^2 = \frac{L I_p^2}{4}(1 + \cos 2\omega t) \tag{5.12}$$

This stored energy is shown as the heavily shaded area in Fig. 5.8. The stored energy is nonnegative and pulsates at twice the ac source frequency. This is shown mathematically by the squaring of the negative current peaks. While the stored energy is increasing, the power into the inductor is positive. During this period of time the ac source supplies energy and the inductor acts as a load. While the stored energy is decreasing, the power into the inductor is negative, indicating that the inductor now acts as a source, returning energy to the ac source.

The time-average stored energy is one-half of $\tfrac{1}{2} L I_p^2$, the peak stored energy shown by Eq. (5.12). This can be written

$$W_m = \tfrac{1}{2} L I_e^2 \tag{5.13}$$

where W_m is the average stored magnetic energy and I_e is the effective value of the current. The time-average energy is important for two reasons. Although the time-average power into the inductor is zero, the time-average stored energy must be supplied to the inductor when the ac source is originally connected to the inductor. The analysis of this startup transient involves the transient and steady-state solution; and we will not examine the details, except to inform you that the time-average energy must be supplied at the beginning. Additionally, the time-average stored energy has importance because it is a measure of the magnitude of the energy pulsations in the inductor. Specifically, we know that the ac source must lend twice this amount of energy to the inductor twice each cycle. If the "ac source" is your local electric utility, they might charge you extra for lending you this energy, but more about that later.

Capacitance.   We will calculate the instantaneous and time average electric energy stored by a capacitor. The voltage and current for a capacitor are

$$v_C(t) = V_p \cos \omega t$$

$$i_C(t) = C \frac{dv}{dt} = -\omega C V_p \sin \omega t$$

These are plotted in Fig. 5.9. The instantaneous power into the capacitor is the product of voltage and current:

$$p_C(t) = v_C(t)i_C(t) = (V_p \cos \omega t)(-\omega C V_p \sin \omega t)$$

$$= -\frac{\omega C V_p^2}{2} \sin 2\omega t \tag{5.14}$$

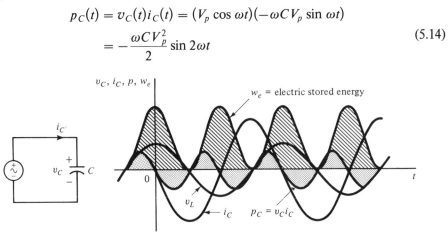

Figure 5.9   Voltage, current, stored energy, and power into the capacitor.

where we have again used the trigonometric identity $(\sin \omega t)(\cos \omega t) = (\sin 2\omega t)/2$. The power into the capacitor is shown in Fig. 5.9 as the lightly shaded area. We notice that the average power is zero, which indicates that, like the inductor, the capacitor is lossless. (This is true only for an ideal capacitor; a real capacitor would have some loss, though not as much as a real inductor.) Being the product of v_C and i_C, the power is zero four times each cycle, when either the voltage or the current is zero.

The pulsation of the power at twice the ac frequency corresponds to the shuttling of electric energy between source and capacitor. The stored electric energy in a capacitor is given in Eq. (3.8) as

$$w_e(t) = \tfrac{1}{2} C v_C^2(t) \tag{5.15}$$

For the sinusoidal source, we find the instantaneous stored energy to be

$$w_e(t) = \tfrac{1}{2} C (V_p \cos \omega t)^2 = \frac{C V_p^2}{4}(1 + \cos 2\omega t) \tag{5.16}$$

This stored electric energy is shown as the heavily shaded area in Fig. 5.9. The stored energy remains nonnegative and pulsates at twice the ac source frequency. This is shown mathematically by the squaring of the negative half-cycle of the voltage. While the stored energy is increasing, the power into the capacitor is positive. During this period the ac source supplies energy and the capacitor acts as a load. While the stored energy is decreasing, the power into the capacitor goes negative, indicating that the capacitor is acting as a source, returning energy to the ac voltage source.

The time-average stored energy is one-half of $\tfrac{1}{2} C V_p^2$, the peak stored energy. This can be written

$$W_e = \tfrac{1}{2} C V_e^2 \tag{5.17}$$

where W_e is the average stored electric energy and V_e is the effective value of the voltage.

5.1.5 The General Case for Power in an AC Circuit

We have dealt with resistance, inductance, and capacitance separately and shown the role of each in power and energy relationships. We will now consider the general case of circuits containing combinations of resistors, inductors, and capacitors. We will think in terms of the circuit shown in Fig. 5.10, a parallel RLC circuit, although our results and interpretations will apply generally to all ac circuits.

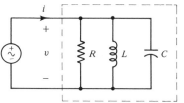

Figure 5.10   General RLC load.

This circuit is analyzed with frequency-domain techniques on page 141. The impedance, $\mathbf{Z} = |\mathbf{Z}| \underline{/\theta}$, is the reciprocal of the admittance given in Eq. (4.44). We are dealing in this section with power and energy in the time domain, so we will express the relationship between voltage and current generally

$$i(t) = I_p \cos \omega t \quad \text{and} \quad v(t) = V_p \cos (\omega t + \theta) \tag{5.18}$$

where $V_p = |\mathbf{Z}| I_p$. Although impedance is a frequency-domain concept and is computed in the frequency domain, the magnitude of the impedance is a real number and hence the previous is a legitimate time-domain equation. We have for convenience used the current as the phase reference, but this in no way limits the generality of our results.

Of immediate interest is the time-average power into the RLC load. The instantaneous power is

$$p(t) = v(t)i(t) = V_p \cos (\omega t + \theta)I_p \cos (\omega t)$$
$$= \frac{V_p I_p}{2}[\cos \theta + \cos (2\omega t + \theta)] \tag{5.19}$$

The second form of Eq. (5.19) follows the application of the trigonometric identity,

$$(\cos A)(\cos B) = \frac{\cos (A - B) + \cos (A + B)}{2}$$

We have plotted Eqs. (5.18) and (5.19) for $\theta = 56°$ in Fig. 5.11. The power function still must go through zero when either v or i is zero, but here the picture lies intermediate between those for pure resistance and for a lossless energy storage element. As Eq. (5.19) indicates in the second form, the power can be considered a combination of a constant term plus a fluctuating term at twice the frequency of the ac source. Because the average of the fluctuating term is zero, the time-average power is the constant term

$$P = \frac{V_p I_p}{2} \cos \theta = V_e I_e \cos \theta \tag{5.20}$$

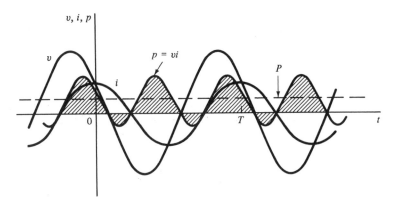

Figure 5.11 Voltage, current, and power into a general load.

The second form of Eq. (5.20) is preferred because meters indicate effective values, and because we usually speak in terms of effective values when discussing power relationships. Equation (5.20) reminds us of the power in the dc case, with effective values for voltage and current, except that we now have the $\cos \theta$ factor. Figure 5.12 shows the effect of the $\cos \theta$ factor, which is called the *power factor*. When voltage and current have the same phase, that is, when θ is zero, the power factor is unity and the power into the load is maximum, $V_e I_e$. When the voltage has a different phase from the current, the power factor is less than unity and the time-average power is decreased proportionally. This occurs symmetrically, whether the circuit is inductive (positive θ) or capacitive (negative θ). When the phase is $\pm 90°$, no time-average power is transferred to the load.

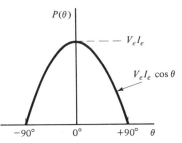

Figure 5.12 Average power versus θ.

    The fluctuating term in Eq. (5.19) shares aspects of the resistive load shown in Fig. 5.6 (where we interpreted the fluctuations in terms of pulsating energy flow) and the lossless load (L or C), where we interpreted the fluctuations as energy being shuttled back and forth between source and load. That is, if the power factor were near unity (small θ), the fluctuations would indicate pulsating energy flow, but if the power factor were low (θ near $\pm 90°$), the fluctuations would indicate shuttled energy. For all power factors less than unity, there exists a part of each cycle during which the instantaneous power is negative and the load acts as a source for a brief period of time.

    The analysis of the RLC load pictured in Fig. 5.10 will be resumed in the next section. Here for simplicity we will work out power and energy calculations for the RL

circuit pictured in Fig. 5.13. We are working in the time domain for our power calculations, but frequency-domain techniques are appropriate for finding the voltage and current which we need in the power calculations. We will use the voltage source as our phase reference, so it would be represented by the phasor $120\sqrt{2}\,\underline{/\,0°}$. In order to calculate the current, we need the impedance,

$$\underline{Z} = 10 + j120\pi \times 0.04 = 10 + j15.1 = 18.1\underline{/56.4°}\ \Omega$$

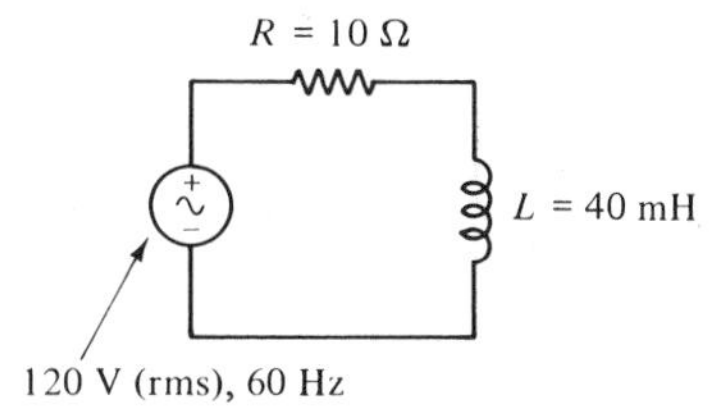

Figure 5.13   Series RL circuit.

The magnitude of the current is thus $120\sqrt{2}/18.1 = 6.63\sqrt{2}$ and the phase angle of the current is $-56.4°$. Thus the time-domain voltage and current are

$$v(t) = 120\sqrt{2}\,\cos{(120\pi t)}$$

$$i(t) = 6.63\sqrt{2}\,\cos{(120\pi t - 56.4°)}$$

The numbers in this example correspond to the plots in Fig. 5.11, except that here we have used the voltage as the phase reference. This phase change corresponds to moving the time origin in Fig. 5.11 back to the peak of the voltage. From Eq. (5.20) we compute the time-average power delivered to the RL load by the ac voltage source to be

$$P = 120 \times 6.63 \cos{(+56.4°)} = 440\ \text{W}$$

We know from physical considerations that this power represents electrical energy converted to thermal energy in the resistor. This interpretation is confirmed by direct calculation of the power into the resistor.

$$P_R = I_e^2 R = (6.63)^2(10) = 440\ \text{W}$$

The magnetic energy storage represented by the inductance has an average value of

$$W_m = \tfrac{1}{2}LI_e^2 = \tfrac{1}{2}(0.04)(6.63)^2 = 0.880\ \text{J} \tag{5.21}$$

The instantaneous magnetic energy storage fluctuates between zero and twice the average energy calculated in Eq. (5.21). This energy must be loaned twice each cycle to the load by the source. This energy may seem small (1 J of energy would lift 1 kG of mass about 10 cm), but the presence of shuttled energy is an important consideration in power systems.

You may have noticed that we used rms values in every power and energy calculation in the foregoing example, and we started out with the rms value of the source voltage because that is what a meter would indicate. However, we were careful to use peak values for phasors and for time functions. This required inserting and taking out a lot of $\sqrt{2}$'s which never entered in the calculation. In order to be consistent with our definition of phasors, we must carry these $\sqrt{2}$'s in the notation. Indeed, many texts and most practitioners drop them out and use rms values for phasors, but we favor the more explicit approach and will maintain it in this book.

We could push our investigation of power and energy in the time domain a little further but we would rather move on to the frequency domain. The important question we ask is: Can such power calculations be made in the frequency domain without explicitly considering the time functions? The next section shows the answer to be yes. As before, the frequency-domain viewpoint serves not only for efficiency in calculation but suggests new insights.

5.2 POWER AND ENERGY IN THE FREQUENCY DOMAIN

5.2.1 Time-Average or Real Power

The time-average power in an ac circuit is given in Eq. (5.20). This involves the peak (or rms) values of the voltage and current and the power factor, which is the cosine of the phase angle between the voltage and current. All of these quantities can be expressed through frequency-domain concepts; indeed, the frequency-domain concept of impedance provides the most sensible way to determine the magnitude of the current (given the voltage) and the phase angle. Thus in considering time-average power there remains no necessity of considering the time-domain voltage and current.

Let us reconsider the RL circuit in Fig. 5.13 wholly in the frequency domain (Fig. 5.14). The time-average power into the load is

$$P = \frac{V_p I_p}{2} \cos \theta = \tfrac{1}{2} |\underline{V}| \underbrace{|\underline{I}| \cos \theta}_{\text{``in-phase'' current}} \qquad (5.22)$$

Figure 5.14 suggests an interpretation of the power factor. If we associate the cos θ with the magnitude of $\underline{I}$, the term $|\underline{I}| \cos \theta$ can be interpreted as the projection of the current phasor on the voltage phasor. This part of the current is "in phase" with the voltage, and we can thus say that the time-average power is given by the product of the voltage and the current in phase with the voltage.

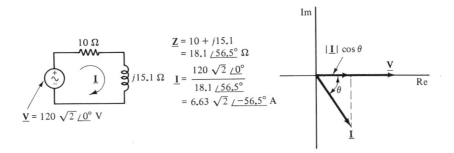

Figure 5.14 Solution in the frequency domain. The phasor diagram shows the "in-phase" component of the current.

The power calculated by Eq. (5.22) is called the *real power* for two reasons. It is real because of the geometric interpretation of the phasor diagram, because it is the product of the phasor voltage and the real component of the current (with the voltage considered the phase reference, of course). But it is also called real because this is the time-average power registered by a wattmeter—the power that makes your "electric meter" revolve and affects in your electric bill. We give this power a special name to distinguish it from other kinds of ac power which we define below.

5.2.2 Reactive Power

The Definition of Reactive Power.   We have an interpretation of the product of the voltage with the real component of the current in Fig. 5.14. Can we place a meaningful interpretation on the product of the voltage with the "out-of-phase" component of the current?

$$Q = \tfrac{1}{2}|\underline{V}| \times \text{"out-of-phase current"}$$

$$= \tfrac{1}{2}|\underline{V}|\,|\underline{I}|\sin\theta = ?$$

where Q is called the reactive power, to be explained below. In order to interpret the product of the voltage and the out-of-phase current, we will return to the parallel RLC circuit, repeated in Fig. 5.15. We can calculate the current by multiplying the voltage by the admittance of the circuit, given in Eq. (4.44).

$$\underline{I} = \frac{\underline{V}}{R} + j\underline{V}\left(\omega C - \frac{1}{\omega L}\right) = |\underline{I}|\cos\theta + j|\underline{I}|\sin\theta \qquad (5.23)$$

$$\underbrace{\phantom{\frac{\underline{V}}{R}}}_{\text{in phase}}\quad \underbrace{\phantom{j\underline{V}\left(\omega C - \frac{1}{\omega L}\right)}}_{\text{out of phase}}$$

$$\underline{V} = V_p\,\underline{/0°} = V_p + j0$$

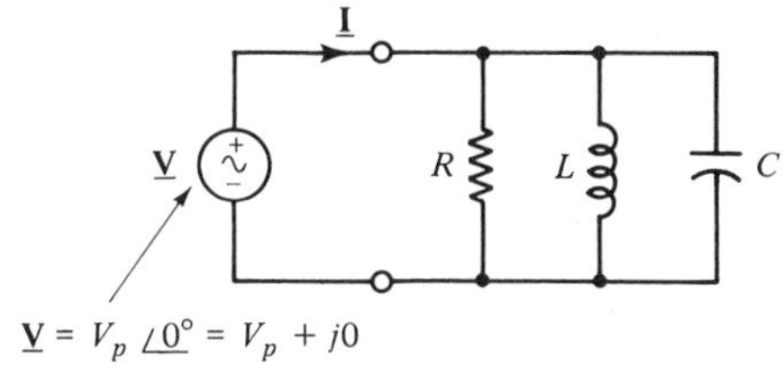

Figure 5.15   General RLC load.

Because $\underline{V}$ is a real quantity, we can express both the real power and the out-of-phase power mathematically by multiplying Eq. (5.23) by $\tfrac{1}{2}\underline{V}$. We will seek an interpretation of the out-of-phase term by juggling the resulting mathematical expressions until we have everything expressed in terms of stored energy. Hence we mutliply both sides of Eq. (5.23) by $\tfrac{1}{2}\underline{V}$ and work on the result.

$$\tfrac{1}{2}\underline{V}\underline{I} = \frac{|\underline{V}|^2}{2R} + j\frac{|\underline{V}|^2}{2}\left(\omega C - \frac{1}{\omega L}\right)$$

$$= \underbrace{\frac{|\underline{V}|^2}{2R}}_{\text{I}} + j\left(\underbrace{\frac{|\underline{V}|^2\omega C}{2}}_{\text{II}} - \underbrace{\frac{|\underline{V}|^2}{2\omega L}}_{\text{III}}\right) \qquad (5.24)$$

Although $\underline{\mathbf{V}}$ is a real quantity, we have used $|\underline{\mathbf{V}}|^2 = \underline{\mathbf{V}}^2$ for generality. The first term (I) we have already interpreted; this is the real power, P, the power which makes the electric meter revolve. The second term (II), from the capacitor, can be interpreted in terms of the peak stored energy in the capacitor. Equation (5.16) can be introduced to yield

$$\frac{|\underline{\mathbf{V}}|^2 \omega C}{2} = \omega \times W_{ep}$$

where W_{ep} is the peak electric energy in the capacitor. The third term (III), from the inductor, requires some juggling. In order to express the magnetic stored energy, we need the current in the inductor. Recall that the peak magnetic energy storage in the inductor is derived from Eq. (5.12) as

$$W_{mp} = \tfrac{1}{2} L |\underline{\mathbf{I}}|^2 \tag{5.25}$$

To introduce the current we require the impedance of the inductor

$$|\underline{\mathbf{V}}| = |j\omega L \underline{\mathbf{I}}| = \omega L |\underline{\mathbf{I}}| \tag{5.26}$$

Note that j goes away because $|j| = |1\underline{/90°}| = 1$. We substitute Eq. (5.26) into term (III) to obtain

$$\frac{|\underline{\mathbf{V}}|^2}{2\omega L} = \omega \left(\frac{|\underline{\mathbf{I}}|^2 L}{2} \right) = \omega \times W_{mp}$$

where W_{mp} is the peak magnetic energy. Putting the two energy terms back into Eq. (5.24), we have

$$\tfrac{1}{2}\underline{\mathbf{V}}\underline{\mathbf{I}} = P + j\omega(W_{ep} - W_{mp}) \tag{5.27}$$

Equation (5.27) is what we sought because it expresses the out-of-phase power in terms of energy relationships, but we will make one change. We arbitrarily change the sign of the imaginary term, for reasons to be explained shortly. The results are

$$\tfrac{1}{2}\underline{\mathbf{V}}\underline{\mathbf{I}} = P + jQ \tag{5.28}$$

where

$$P = \frac{|\underline{\mathbf{V}}|^2}{2R} \qquad \text{the real power} \tag{5.29}$$

and

$$Q = \omega(W_{mp} - W_{ep}) \qquad \text{the reactive power} \tag{5.30}$$

The out-of-phase term, the reactive power Q, is seen to indicate the imbalance between peak magnetic energy storage and the peak electric energy storage in the circuit. We may understand the significance of this term by considering the stored energy requirements of the inductor and capacitor. When magnetic and electric energy is balanced internal to the load, the source does not have to supply any stored energy externally—the load takes care of its own requirements for stored energy. But when these energies are not balanced internally, the source must lend energy cyclically to the load. This is the meaning of the out-of-phase power, the reactive power: reactive power represents energy shuttled between source and load.

The Sign of Reactive Power.   Why did we arbitrarily change the sign of the reactive power term? This change is traditional for historical and practical reasons. When the load stores more magnetic energy than electric energy, the reactive power is positive (after the sign is changed); and when electric energy storage dominates, the reactive power is negative. In a practical power system, magnetic energy requirements usually dominate. This occurs because almost all power equipment is magnetic in its operations—motors, transformers, induction furnaces, even light bulbs. Thus the power company usually lends magnetic energy; that is, the current normally lags the voltage in a power system. Since we prefer positive numbers, it has become traditional to attach a positive number to the normal situation. Hence we call magnetic reactive power positive, and we change the sign in the mathematical expression to conform with this preference.

The reactive power, Q, is so-named because it is associated with the reactance, the imaginary part of the impendance in the circuit. To help distinguish between the real power, P, and the reactive power, Q, we use different units for these quantities. The unit of real power is watts. The unit of reactive power is *volt-amperes reactive*, or VAR (rhymes with "jar"). The kVAR (kiloVAR) is used for large amounts of reactive power, as kW is used for large quantities of real power.

5.2.3 Complex Power

The Definition of Complex Power.   The complex sum of the real and reactive power in Eq. (5.28) is called the *complex power* and can be written

$$\underline{S} = \tfrac{1}{2}\underline{V}\underline{I}^* \tag{5.31}$$

where $\underline{S}$ is the complex power. The complex conjugate of the current changes the sign of the imaginary part and thus introduces mathematically the customary change of sign. The expression for the complex power in Eq. (5.31) also permits us to relax the requirement that the phasor voltage, $\underline{V}$, be a real quantity. Note that if

$$\underline{V} = |\underline{V}|\underline{/\theta_V} \quad \text{and} \quad \underline{Z} = |\underline{Z}|\underline{/\theta}$$

then

$$\underline{I} = \frac{\underline{V}}{\underline{Z}} = |\underline{I}|\underline{/\theta_V - \theta}$$

where θ_V is the phase of the voltage, no longer assumed zero. The complex power, as defined in Eq. (5.31), would be

$$\underline{S} = \tfrac{1}{2}\underline{V}\underline{I}^* = \tfrac{1}{2}|\underline{V}|\underline{/\theta_V} \times |\underline{I}|\underline{/-\theta_V + \theta}$$
$$= \tfrac{1}{2}|\underline{V}||\underline{I}|\underline{/\theta} = \tfrac{1}{2}|\underline{V}||\underline{I}|(\cos\theta + j\sin\theta)$$

because the complex conjugate changes the sign of the current phase angle. Thus the phase of the voltage, or the phase reference generally, drops out with the expression of complex power in Eq. (5.31) and only the phase difference between voltage and current remains.

An Example of Complex Power.   To illustrate these ideas, we shall continue the example of Fig. 5.14. The voltage was $\underline{\mathbf{V}} = 120\sqrt{2}\ \underline{/0°}$ and the current we found to be $\underline{\mathbf{I}} = 6.63\sqrt{2}\ \underline{/-56.4°}$ in polar form or $\underline{\mathbf{I}} = 3.67\sqrt{2} - j5.53\sqrt{2}$ in rectangular form. For this example the complex power is

$$\underline{\mathbf{S}} = \tfrac{1}{2}\underline{\mathbf{V}}\underline{\mathbf{I}}^* = \tfrac{1}{2}(120\sqrt{2} + j0)(3.67\sqrt{2} + j5.53\sqrt{2})$$
$$= 120 \times 3.67 + j120 \times 5.53 = 440 + j663$$

Thus the real power is 440 W and the reactive power is +663 VAR. The positive sign indicates predominant magnetic energy storage, which we could have anticipated in this case by examining the circuit. Note that the $\sqrt{2}$'s all dropped out, as they usually do in power calculations.

Apparent Power.   The complex power is a complex number yielding information about both the flow of time-average power and the shuttling of loaned energy between source and load in an ac circuit. The magnitude of the complex power, $|\underline{\mathbf{S}}|$, also has significance. This magnitude, which is a real number, is called the *apparent power*. The apparent power results when you measure the voltage and current with meters and multiply the measured values without regard for phase.

$$|\underline{\mathbf{S}}| = \tfrac{1}{2}|\underline{\mathbf{V}}|\,|\underline{\mathbf{I}}| = V_e I_e \qquad \text{volt-amperes} \tag{5.32}$$

The units of apparent power are *volt-amperes* (VA). Apparent power is important as a measure of the operating limits in large electrical equipment such as transformers, motors, and generators. Losses in the wires are proportional to the square of the current in the machine, whereas losses in the magnetic materials are roughly proportional to the square of the operating voltage. Because electrical machinery is operated with the voltage more-or-less constant, apparent power limits imply current limits. In the present example the apparent power is $120 \times 6.63 = 796$ VA.

Summary.   The complex power gives concise information about power and energy flow in an ac circuit. The real part of the complex power is the time-average power in watts. The real power heats resistors, turns motors, and makes the electric meter revolve. The imaginary part of the complex power is the reactive power in VARs, which is proportional to the electrical energy lent to the load by the ac source twice each cycle. The reactive power is considered positive when the load is inductive and negative when the load is capacitive, although the latter case is rare in power systems. The magnitude of the complex power is the apparent power in volt-amperes and is a general indicator of the operating level of a power system.

The three units for ac power—watts, volt-amperes reactive, and volt-amperes—all have the proper scientific units for power (J/s). We use different names to clarify communication when speaking of the various kinds of "power" in ac circuits. That such specialized terminology is needed suggests that this aspect of ac circuits can be confusing, as you will perhaps agree. Nevertheless, these distinctions between watts, VARs, and VAs will aid your thought and communication about these related concepts.

Power Triangles.   The apparent power, real power, and reactive power form a right triangle. Figure 5.16 shows this *power triangle* for the present example. The angle at the origin is θ, the angle of impedance ($+56.4°$ in this case). Notice, however, that when the current lags the voltage, the power triangle is drawn above the axis. This reversal is deliberate; we created it by changing the sign on the reactive power term, as discussed above.

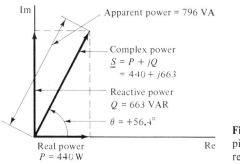

Figure 5.16   The power triangle pictures complex, apparent, real, and reactive power.

Because the various kinds of powers form a triangle, and because the power factor is the cosine of an angle of that triangle, there follow a host of formulas relating power factor (p.f.) with the various types of powers. Below we list several of these, which you can verify from the definitions and common trigonometric identities.

$$P = |\underline{S}|\,\text{p.f.} \tag{5.33}$$

$$|\underline{S}| = \sqrt{P^2 + Q^2} \tag{5.34}$$

$$Q = \pm|\underline{S}|\sqrt{1 - (\text{p.f.})^2}\ (+\ \text{for lagging p.f.}) \tag{5.35}$$

Based on these formulas, and others that can easily be derived, a variety of problems in ac power systems can be formulated, of which the following is typical.

Motor Example.   A 220-V motor has an output of 3 horsepower (hp). The motor efficiency is 89% and the power factor is 0.92 lagging. Find the input current, the apparent power, and the reactive power. Draw a phasor diagram.

The first step in the solution is to convert horsepower to watts. Of course we could look up the conversion factor in a table, but think about Christopher Columbus (1492), divide by 2, and you have it (746 W/hp). Thus the mechanical output is $3 \times 746 = 2.24$ kW. The efficiency would be the mechanical power output divided by the electrical power input; hence the input power is $2.24/0.89 = 2.51$ kW. This would be the real power, so we have stated the units as watts. Next we calculate the current. We know the real power, the voltage, and the power factor, so we can compute the current from Eq. (5.20).

$$P = V_e I_e \times \text{p.f.} \Rightarrow I_e = \frac{2.51\ \text{kW}}{220 \times (0.92)} = 12.4\ \text{A (rms)}$$

From this result follows the apparent power, $|\underline{S}| = 220 \times 12.4 = 2.73$ kVA (kilovolt-amperes). The reactive power can be computed from Eq. (5.34).

$$Q = \sqrt{|\underline{S}|^2 - P^2} = \sqrt{(2.73)^2 - (2.51)^2} = 1.07 \text{ kVAR}$$

(You may not get this exact answer unless you start at the beginning of the problem and retain full accuracy in your calculator. In this book we write only three significant digits, but we calculate results with full precision and then round.) Because the current is lagging, the reactive power would be positive, +1.07 kVAR.

For the phasor diagram in Fig. 5.17 we have used the voltage for the phase reference, which is traditional, and shown the current lagging. The phase angle of the current follows from the power factor, $\theta = \cos^{-1}(0.92) = 23.1°$. The power triangle showing the real, reactive, and apparent powers would have the same angle but would be drawn above the real axis, similar to Fig. 5.16.

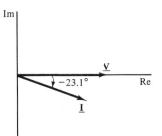

Figure 5.17   Motor voltage and current.

5.2.4 More About Reactive Power

Because the concept of the reactive power is hard to grasp, we will continue to probe its meaning. For this purpose we continue to investigate energy relationships in the circuit in Fig. 5.18, relating these to the physical meaning of reactive power. We have already derived [Eq. (4.44)] the admittance of this circuit to be

$$\underline{Y}(j\omega) = G + j\left(\omega C - \frac{1}{\omega L}\right) = 10^{-3} + j\left(10^{-8}\omega - \frac{10}{\omega}\right) \quad \text{siemens} \quad (5.36)$$

and discussed its frequency response. Here we will investigate the energy aspects of the circuit both at the resonant frequency and off-resonance.

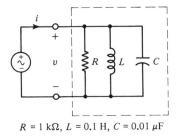

$R = 1 \text{ k}\Omega, L = 0.1 \text{ H}, C = 0.01 \text{ }\mu\text{F}$

Figure 5.18   *RLC* circuit for the example.

Resonance.   Consider first the circuit in Fig. 5.18 at the resonant frequency, $\omega = 1/\sqrt{LC} = 31.6$ krad/s (5.03 kHz). At this frequency, the admittance is real, $\mathbf{Y} = 10^{-3}$ S, and if we assume a voltage of 60 V rms, the current is

$$\mathbf{I} = \mathbf{V}\mathbf{Y} = 60\sqrt{2}\ \underline{/0°} \times 10^{-3}\underline{/0°} = 60\sqrt{2}\ \underline{/0°} \text{ mA}$$

The admittance is real because the currents in the inductor and capacitor are equal in magnitude and opposite in phase. The complex power is

$$\mathbf{S} = \tfrac{1}{2}\mathbf{V}\mathbf{I}^* = \tfrac{1}{2}60\sqrt{2}\ \underline{/0°} \times 60 \times 10^{-3}\sqrt{2}\ \underline{/0°} = 3.60 + j0 \text{ VA}$$

The real power represents the loss in the resistor and the reactive power is zero. This does not mean, however, that energy is absent from the inductor and capacitor. Let us examine the stored magnetic and electric energy in the time domain. The stored electric energy in the capacitor is

$$w_e(t) = \tfrac{1}{2}[v(t)]^2 C = \tfrac{1}{2}(60\sqrt{2}\ \cos \omega t)^2 \times 10^{-8}$$
$$= 36.0 \cos^2 \omega t \qquad \mu J$$

To calculate the magnetic energy stored in the inductor, we need the inductive current, which lags the voltage by 90° and thus is represented by a sine function

$$i_L(t) = \frac{V_p}{\omega L} \sin \omega t = \frac{60\sqrt{2}}{31.6 \times 10^3 (0.1)} \sin \omega t = 19.0\sqrt{2}\ \sin \omega t \text{ mA}$$

Hence the stored magnetic energy would be

$$w_m(t) = \tfrac{1}{2}[i_L(t)]^2 L = \tfrac{1}{2}(19.0\sqrt{2}\ \sin \omega t)^2 (0.1)$$
$$= 36.0 \sin^2 \omega t \qquad \mu J$$

The electric energy storage is shown as the heavily shaded area and the magnetic energy storage is shown as the lightly shaded area in Fig. 5.19. We note that the peaks for the two types of energy storage are equal. This condition corresponds to resonance. Notice that the electric energy storage is maximum when the magnetic energy storage is zero, and vice versa. The instantaneous sum of the two energies is constant at 36 μJ, as is easily verified mathematically. This amount of energy is given to the circuit when the source was first attached, but thereafter the same energy alternates between the inductor and capacitor. As the same energy of a freely swinging pendulum alternates between kinetic and potential form, so the inductor and capacitor of this circuit "play catch" with the same energy. This is why the reactive power is zero; that is, the source no longer exchanges stored energy after it gives the energy to the load when first connected.

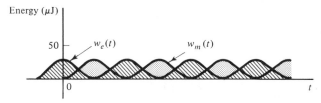

Figure 5.19 Magnetic and electric energy at resonance.

Off-Resonance. We will now lower the source frequency below resonance. We will investigate energy and reactive power at a frequency of 3 kHz. Since the calculations are similar to those detailed above, we will merely give the results. The load admittance at 3 kHz is

$$\mathbf{Y} = 10^{-3} - j3.42 \times 10^{-4} = 1.06 \times 10^{-3} \underline{/-18.9°} \text{ S}$$

and thus the current into the load is

$$\mathbf{I} = \mathbf{V}\mathbf{Y} = 60\sqrt{2} \underline{/0°} \times 1.06 \times 10^{-3} \underline{/-18.9°} = 63.4\sqrt{2} \underline{/-18.9°} \text{ mA}$$

The complex power is

$$\mathbf{S} = \tfrac{1}{2}\mathbf{V}\mathbf{I^*} = \tfrac{1}{2}(60\sqrt{2} \underline{/0°})(63.4\sqrt{2} \times 10^{-3} \underline{/+18.9°})$$
$$= 3.80\underline{/+18.9°} = 3.60 + j1.23 \text{ VA}$$

The real part again represents the loss of the resistor. The reactive power, +1.23 VARs, represents magnetic energy loaned periodically to the entire load by the source. To see why the loan of magnetic energy is required, we will again examine the stored energies in the time domain. The instantaneous stored energy in the capacitor,

$$w_e(t) = \tfrac{1}{2}[v_C(t)]^2 C = 36.0 \cos^2 \omega t \qquad \text{joules}$$

is unchanged because the voltage is unchanged. The instantaneous stored energy in the inductor has increased to

$$w_m(t) = \tfrac{1}{2}[i_L(t)]^2 L = 101.3 \sin^2 \omega t \qquad \text{joules}$$

The resulting energy imbalance is indicated by Fig. 5.20. The stored energies still peak up alternately, but now the inductor requires more energy than the capacitor. The total

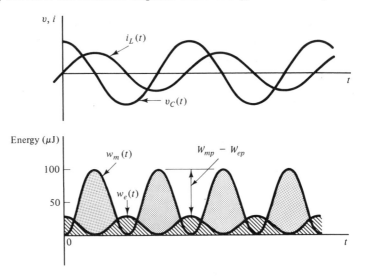

Figure 5.20 Below resonance, magnetic energy requirements exceed electric energy requirements.

instantaneous stored energy in the circuit is no longer constant but is

$$w_e(t) + w_m(t) = 36.0 + (101.3 - 36.0)\sin^2 \omega t \qquad \mu J \qquad (5.37)$$

The energy that resides in the load remains at 36.0 μJ (the smaller of the two) and the remainder of the energy required for the inductor, 65.3 μJ, must be loaned to the load by the source: this loaning of magnetic energy is represented by the reactive power.

The Meaning of Reactive Power.   Thus we can describe the energy flow as comprised of two components: energy that flows in one direction, yielding time-average power, and energy that flows alternately in both directions, as required by the energy storage needs of the circuit. The power associated with two-way flow of stored energy, $p_Q(t)$, would be the time derivative of a generalized version of Eq. (5.37):

$$p_Q(t) = \frac{d}{dt}\left[w_e(t) + w_m(t)\right] = \underbrace{\omega(W_{mp} - W_{ep})}_{Q} \sin 2\omega t \qquad (5.38)$$

where we have utilized Eq. (5.30) to show that the peak value of this two-way energy flow is the reactive power.

For more complicated circuits containing many inductors and capacitors, the appropriate formula would be

$$Q = \omega(\sum_{\substack{\text{all} \\ L\text{'s}}} W_{mp} - \sum_{\substack{\text{all} \\ C\text{'s}}} W_{ep})$$

Because in the more complicated case the peaks of the stored energies do not alternate nicely as in Fig. 5.20, the interpretation of the physical meaning of reactive power is more complicated than for the circuit in Fig. 5.18. Nevertheless, our interpretation remains valid: reactive power is a measure of energy loaned by the source to the load twice each cycle.

The Importance of Reactive Power.   Power companies have to be careful about the reactive power load on their systems. As stated above, the limits of larger power equipment such as generators and transformers are described by the apparent power, the Pythagorean sum of the real and reactive powers. Thus if the reactive power becomes large, a piece of equipment may become overloaded even though the load of real power is moderate. Also, the reactive energy must be transported from generator to user, often over great distances. Reactive power increases line current, and hence increases line losses. The losses in the transmission system are not charged directly to the consumer, since the watthour meter is placed at the load. For this reason, the power company might penalize customers whose requirements for reactive power are great. Often the industrial consumer can save money by placing a bank of capacitors in parallel with an inductive load to store energy locally. In effect, they receive the stored energy from the power company only once and then keep it "in house" with the capacitors. The industrial customer thus "corrects" his power factor by creating a resonance between the electric energy stored by the added capacitance and the magnetic energy used by motors or other heavy equipment.

5.2.5 Reactive Power in Electronics

Reactive power flow also plays an important role in electronics. Here we normally do not deal with large amounts of power; indeed, the goal often is to make full use of the small amounts of power which are available. To see the role of reactive power, let us examine the conditions for maximum power transfer in an ac circuit.

Thévenin and Norton Equivalent Circuits for AC.   In Chapter 2 we justified the Thévenin equivalent circuit by using the concepts of linearity and superposition. In Chapter 4 we reduced ac problems to equivalent dc problems through the use of phasors, through which sources and circuits are represented by complex numbers. All the techniques we developed for dc circuits remain valid for solving ac circuits, including Thévenin and Norton equivalent circuits. In Fig. 5.21 we show a Thévenin equivalent circuit with a phasor voltage source and an output impedance, $\underline{Z}_{eq}$, which we have expressed in terms of a resistive and reactive part for benefit of the derivation that follows. Recall that the circuit replaced by the Thévenin equivalent circuit can be arbitrarily complicated.

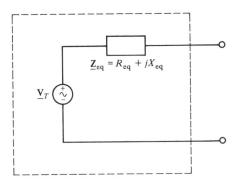

Figure 5.21   For an ac circuit the Thévenin voltage is a phasor and the output resistance becomes an impedance.

Maximum Power Transfer for AC Sources.   Let us consider a typical situation which might arise in electronics, that of getting maximum power out of a radio antenna. Specifically, consider the telescoping AM radio antenna on an automobile, as suggested in Fig. 5.22. Radio waves are radiated by a commercial station, perhaps at considerable distance, and these waves interact with the antenna to give a small voltage, typically 10 mV rms, between the fender and the end of the antenna. The derivation of the output impedance of such an antenna lies beyond the scope of this text, but we can say a little based on our understanding of electric and magnetic energy. As a circuit element, the antenna represents a wire that leads nowhere, that is, an open circuit. The radio waves tend to make current flow on the wire, but a small current produces a buildup of charge on the wire and the current stops. Thus we anticipate that an antenna of this type would build up charge but carry little current. This suggests that electric energy storage would dominate magnetic energy storage and that the output impedance would be capacitive. We have shown typical values in Fig. 5.23. The 10-Ω resistive part of the output impedance represents the resistance of the antenna and the energy reradiated by the antenna currents. Our task is to specify the input impedance of the radio, $\underline{Z}_L = R + jX$,

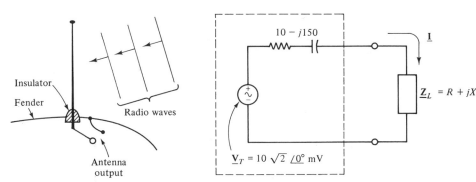

Figure 5.22   The antenna plus the radio station can be represented by a Thévenin equivalent circuit.

Figure 5.23   Find R and X to maximize power in R.

to receive maximum power out of the antenna. The average power into the load would be $|\underline{I}|^2 R/2$, where $\underline{I}$ is the current in the load. This current can be calculated as the voltage divided by the total impedance in the circuit, the sum of $\underline{Z}_{eq}$ and $\underline{Z}_L$.

$$\underline{I} = \frac{\underline{V}_T}{\underline{Z}_{eq} + \underline{Z}_L} = \frac{10\sqrt{2}\,\underline{/0^\circ}}{(10 + R) + j(X - 150)} \quad \text{mA}$$

Because the phase does not matter in this power calculation, we will consider only the magnitude

$$|\underline{I}| = \frac{10\sqrt{2}}{\sqrt{(10 + R)^2 + (X - 150)^2}} \quad \text{mA}$$

Consequently, the average power delivered to the radio by the antenna would be

$$P(R, X) = \frac{|\underline{I}|^2 R}{2} = \frac{2 \times (10^{-2})^2 R/2}{(10 + R)^2 + (X - 150)^2}$$

This is the function to be maximized, so we might make an assault using the methods of differential calculus. Before taking derivatives, however, we should note the effect of the load reactance, X. Being in the denominator and being squared, the total reactance can only decrease the power. Clearly, the best we can do is set X to $+150\ \Omega$. Once we do that, the power is a function of R only:

$$P(R, +150) = \frac{10^{-4} R}{(10 + R)^2} \Rightarrow R = 10\ \Omega \text{ for maximum power}$$

We have written by inspection the value of R which gives maximum power because we recognize that the problem has been reduced to that solved back in Chapter 2. That is, once the load reactance (X) is adjusted to balance the reactance in the source, the maximum power follows from setting the resistance of the load equal to the resistive part of the output impedance of the source.

In general, the maximum power transfer will occur when the load has the same resistance as the source but the opposite reactance. We therefore create a resonance from the viewpoint of the load, in effect balancing the equivalent stored energies. In the case solved above, you can confirm that the power delivered to the radio by the antenna is 2.5×10^{-6} W. Not much power, but we know that it must be adequate since AM radios work. In Section 6.5.4 we will show how transistors are used to amplify these small signals to a level where loudspeakers produce audible sound.

5.3 TRANSFORMERS

5.3.1 Introduction

What Is a Transformer?   A transformer is a highly efficient device for changing ac voltage from one value to another, for example, from 120 V to 6 V. Transformers come in all sizes, from the enormous transformers used in power substations to the small transformers we use for doorbells. (Sometimes the word "transformer" is used for a device which employs a transformer (in the sense above) but includes other controls or devices, such as the transformers used to charge calculators or to control model railroad trains.)

The transformer gives to ac a feature lacking in dc power systems. Using a transformer, we can efficiently change ac voltage from small amplitudes to large amplitudes, or vice versa. Such changes are not simply accomplished with dc voltage.

Figure 5.24 shows a simple transformer. It consists of two coils and an iron core, which enhances the magnetic coupling between the coils. Let us say that we construct such a device, connect one coil to an ac voltage source, and connect the other coil to a resistive load. In this case, we find that the resistive load gets hot, suggesting the flow of electrical energy from the ac source through the transformer and into the load. Furthermore, we find that the transformer does not get very hot, suggesting that the transformer is an efficient device for coupling load and source.

The coil connected to the ac source is called the *primary*, and that connected to the load is called the *secondary*. There is nothing special about the two sides, for the transformer can convey power both ways and in some applications does. But in most applications the power flows in only one direction and hence these names are useful.

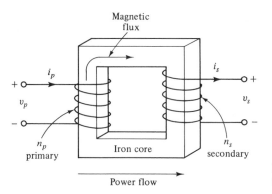

Figure 5.24   Simple electrical transformer.

Approximate Theory of Transformers.   Let us consider what happens when we connect the transformer to the ac source. The primary voltage, v_p, is equal to the source voltage, which causes an ac current, i_p, to flow in the primary. The primary current causes a sinusoidal magnetic flux, Φ_p, to exist in the primary coil and the iron core directs most of this magnetic flux to pass also through the secondary coil, Φ_s. This sinusoidally varying magnetic flux in the secondary coil induces an ac voltage, v_s, in the secondary coil, and this causes an ac current to flow in the secondary coil and the resistive load. In this way, the transformer couples electrical energy from the ac source to the load.

Faraday's law of induction states that a changing magnetic flux produces a voltage in a coil:

$$v = n \frac{d\Phi(t)}{dt} \tag{5.39}$$

where Φ is the magnetic flux, n the number of turns in the coil, and v the induced voltage. On the primary side, the induced voltage will be equal to the input ac voltage, which is the same as v_p.

$$v_p = n_p \frac{d\Phi_p(t)}{dt} \tag{5.40}$$

where n_p is the number of turns in the primary coil, as shown in Fig. 5.24, and Φ_p is the magnetic flux in the primary coil. Similarly, the secondary voltage is generated by the secondary magnetic flux

$$v_s = n_s \frac{d\Phi_s(t)}{dt} \tag{5.41}$$

where n_s is the number of turns in the secondary coil and Φ_s is the magnetic flux in the secondary coil. Because the iron promotes efficient magnetic coupling between primary and secondary coils, the magnetic fluxes are nearly equal:

$$\Phi_p(t) \approx \Phi_s(t)$$

Hence from Eqs. (5.40) and (5.41) it follows that the voltages in primary and secondary are proportional to the turns in the coils:

$$\frac{v_p}{n_p} \approx \frac{v_s}{n_s} \tag{5.42}$$

Thus if the secondary has 10 times the turns of the primary, the secondary voltage will be about 10 times as great. Often in practice only the ratio of secondary to primary turns is known, and Eq. (5.42) is written in the form

$$v_s \approx \frac{n_s}{n_p} v_p \tag{5.43}$$

where n_s/n_p can be determined experimentally if necessary.

The primary and secondary current are also related to the turns ratio. Note first in Fig. 5.24 that the primary voltage and current reference directions are a load set and the secondary reference directions a source set. Thus the input power to the transformer primary is $p_{\text{in}} = +v_p i_p$, and the output power from the secondary is $p_{\text{out}} = +v_s i_s$. As stated above, a well-designed transformer is highly efficient; indeed, 95% is a typical efficiency. Thus input and output power are nearly equal:

$$p_{in} \approx p_{out} \tag{5.44}$$

From Eqs. (5.42) and (5.44) we conclude that primary and secondary current must be related as in

$$n_p i_p \approx n_s i_s \tag{5.45}$$

Equations (5.42) and (5.45) describe the ratios of primary and secondary voltage and current in a good transformer.

A Mechanical Analogy.   The lever shown in Fig. 5.25 is a mechanical analog to an electrical transformer. Balance of torque requires

$$f_1 \ell_1 = f_2 \ell_2 \tag{5.46}$$

and the velocities are related by

$$\frac{u_1}{\ell_1} = \frac{u_2}{\ell_2} \tag{5.47}$$

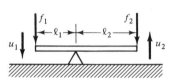

Figure 5.25   The lever is a mechanical transformer and is an analog of the electrical transformer.

If we compare Eqs. (5.46) and (5.47) with Eqs. (5.42) and (5.45), and associate force with voltage and velocity with current, we see that the lever and the transformer are analogs provided we associate lengths with the reciprocal of turns:

$$f \Rightarrow v \qquad u \Rightarrow i \qquad \ell \Rightarrow \frac{1}{n}$$

One weakness of this analogy is that the mechanical system supports a constant force, whereas the electrical system *must* have a time-varying voltage to operate. This follows from Faraday's law of induction [Eq. (5.39)], which requires a time-varying magnetic field in order to generate a voltage. This weakness is inherent in the physical operation of the two systems and cannot be eliminated.

Another weakness occurs when the transformer secondary is open-circuited. This corresponds to locking the position of the lever output. In the mechanical system, this would also lock the input position, but in the electrical system we would expect some current to flow in the input coil. This ac current would be required to furnish the magnetic stored energy associated with the sinusoidal magnetic field. In other words, we would expect the input coil to act as an inductor even with the output open-circuited. This second weakness of the analogy results from the idealization of the mechanical system, which we discuss below.

The Ideal Transformer.   You might object to our analogy because the electrical equations are approximate equalities, whereas the mechanical equations are exact. This has occurred because we idealized the mechanical system but dealt with a realistic electrical system. If we consider the mass of the lever, Eq. (5.46) becomes approximate; and if we consider the compliance of the beam and its losses due to flexure, the velocity equation becomes approximate. With a realistic beam, it is possible to have motion at the input with the output locked. Thus the analogy improves when we unidealize the mechanical system.

Another way to improve the analogy is to idealize the transformer. The equations of the ideal transformer are defined to be

$$\frac{v_p}{n_p} = \frac{v_s}{n_s} \quad \text{and} \quad n_p i_p = n_s i_s \tag{5.48}$$

These equations imply perfect magnetic coupling and lossless operation. Real transformers can be modeled as ideal transformers in many applications.

The Impedance Transformer.   Equations (5.48) lead to a very useful property of transformers—resistance (or impedance) transformation. Figure 5.26 shows the common symbol for the ideal transformer: two coils with parallel lines between them to suggest the iron core. We have connected a load resistor, R_L, to the secondary and defined an equivalent resistance, R_{eq}, into the primary. This section shows that such an equivalent resistance is meaningful and that its value depends on the load resistance and the turns ratio of the transformer. The equations of the circuit are those of the ideal transformer in Eq. (5.48) plus Ohm's law for R_L. If we divide the voltage equation of the ideal transformer by its current equation, we obtain

$$\frac{1}{n_p^2}\frac{v_p}{i_p} = \frac{1}{n_s^2}\frac{v_s}{i_s} \tag{5.49}$$

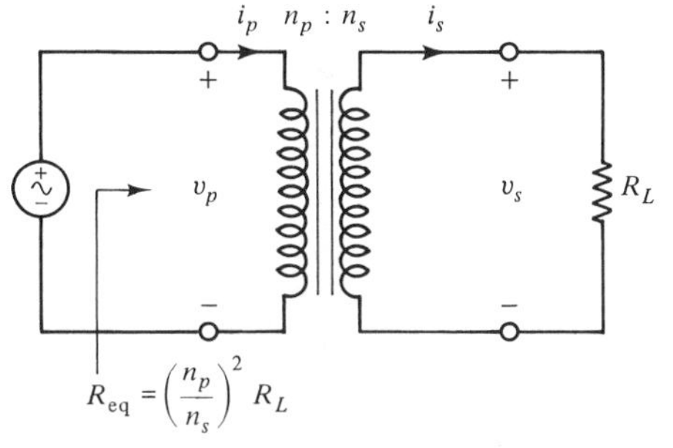

$$R_{eq} = \left(\frac{n_p}{n_s}\right)^2 R_L$$

Figure 5.26   The transformer will change the effective value of R_L to R_{eq}.

But v_s/i_s is the load resistor (R_L) and v_p/i_p defines the equivalent resistance (R_{eq}) into the primary, so Eq. (5.49) leads to the value of the equivalent resistance

$$R_{eq} = \left(\frac{n_p}{n_s}\right)^2 R_L \tag{5.50}$$

Thus the load resistance is transformed by the square of the turns ratio. By using a transformer, we can make a large resistor appear small, or we can make a small resistor appear large. For example, if we wish to derive maximum power out of a source with an output resistance of 100 Ω, we must use a load resistance equal to the output resistance of the source (page 173). If we are required to furnish this power to a load resistance of 500 Ω, as shown in Fig. 5.27, we can accomplish the task with a transformer having $\sqrt{100/500}$ for a turns ratio. Because we wish to make the 500-Ω resistor look smaller, we must connect it to the high-voltage side of the transformer, the side with the more turns, and look into the low-voltage side to see the smaller resistance.

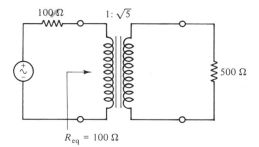
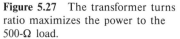

Figure 5.27   The transformer turns ratio maximizes the power to the 500-Ω load.

Their ability to transform impedance provides a valuable tool for analyzing circuits containing transformers. Figure 5.28a shows an ac circuit: we are to solve for $\mathbf{I}_L$. One approach would be to write the circuit equations, plus those of the ideal transformer, and proceed to eliminate variables. Without defining transformer primary and secondary variables, however, we may transform the load impedance into the primary and solve directly for the primary current. Figure 5.28b shows the result of impedance transformation, and the primary current is easily determined as shown in the figure. The secondary current, which is the load current, is greater by the turns ratio, twice as great in this case. Note that we have twice as many primary turns as secondary turns; hence the secondary voltage is half the primary voltage and the secondary current is twice the primary current, the product (the power) remaining constant. Thus the secondary current is $2.14\sqrt{2}\,\underline{/0°}$ A.

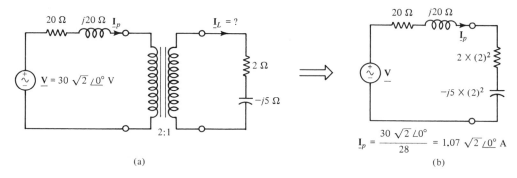

Figure 5.28   (a) Solve for I_L; (b) equivalent circuit.

5.3.2 Transformer Applications

Voltage and Current Transformation.   We have shown that transformers change voltage levels, current levels, and impedance levels. Each of these properties leads to important applications. Voltage changes are frequently required to convert the standard power distribution voltages, usually 120 V, into higher or lower ac voltages as required by specific devices. A TV set normally contains one or more transformers to furnish ac voltage to several internal "power supplies," a subject we shall study in Chapter 6. The "coil" in an automotive ignition system is a transformer whose function is to provide high voltage to the spark plugs, where we want sparks, and to keep the voltage low at the points, where we do not want sparks. Transformers are also used to reduce 120 V to lower values for safety. Domestic doorbell and thermostat circuits provide examples.

A few transformer applications arise out of need for high current levels. Arc welders and ac electromagnets are applications where transformers are used to product large currents. One is also tempted to list the hand-held electric soldering gun in this category, but it probably belongs in the third category.

Impedance Transformation.   The third, and probably the most important, class of transformer applications involves impedance transformation. We have already illustrated the use of transformers in effecting maximum power transfer, a common application in electronics. The most important transformer applications of this class, however, lie in power distribution. Figure 5.29 suggests the generation and delivery of electrical power to a distant user over a transmission line. Although we have shown identical transformers at each end, the two transformers would not necessarily be identical, for the power would not be generated at the low voltage required by the user. We have made the transformers identical to simplify the analysis.

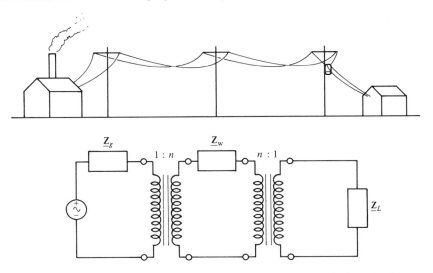

Figure 5.29   Simple power distribution system and its equivalent circuit.

A power distribution system should offer constant voltage and high efficiency. The user wants constant voltage, independent of load, because his equipment is designed to operate at standard voltage. He requires, for example, that the voltage remain reasonably near 120 V whether he uses 1 kW or 10 kW of power. This requires, in turn, that the equivalent impedance of the source, as seen from the user's point of view, be as low as possible. Of course, an ideal voltage source has zero output impedance, but the generator has an inherent output impedance, $\underline{Z}_g$, and the wire has an impedance, $\underline{Z}_w$, as shown in Fig. 5.29. The generator output impedance can be made quite small by good design, but the resistance and reactance of the transmission line can only be reduced within limits due to the large distances.

The transmission system is required to be efficient to reduce costs and energy waste. We shall use the impedance-transforming properties of the transformers to show how the system characteristics are improved by their presence. First we shall transform $\underline{Z}_L$ with the transformer at the load. This raises the load impedance by a factor of n^2 and places it in series with $\underline{Z}_w$. We now can transform this series combination with the transformer at the generator, which lowers the impedances by n^2. The resulting equivalent circuit is shown in Fig. 5.30. Note that the load impedance appears with its true value in the final equivalent circuit, as does the generator impedance, but the impedance of the transmission line is reduced by the factor n^2. This will reduce its losses by this factor, as compared with the losses without the transformers.

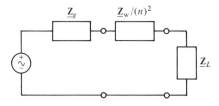

Figure 5.30   Simplified equivalent circuit for the power distribution system.

This reduction in transmission line losses can be understood by considering the current in the transmission line. The transformer at the generator increases the voltage and decreases the current by its turns ratio, the total power being unchanged. The transmission-line losses will be $I^2 R_w$, where I is the rms current in the line and R_w is the wire resistance. Reduction of the line current by n therefore reduces the losses by n^2, and this reduction is reflected in the equivalent circuit in Fig. 5.30. For this reason, electrical power is distributed at extremely high voltages, up to 750 kV.

Figure 5.30 also suggests that the voltage regulation at the load (constant voltage, independent of load) is improved by the use of a high-voltage transmission line, which is achieved through the use of transformers. That the load impedance appears with its true value rightly suggests that the same equivalent circuit would have resulted had we transformed all impedances to the load end of the circuit. Thus the output impedance of the generator–transmission line system is $\underline{Z}_g + \underline{Z}_w/n^2$. The effect of the transmission-line impedance is therefore reduced by the square of the turns ratio, to the improvement of load voltage regulation.

A more realistic circuit would have the power generated at higher voltage, and hence identical transformers would not be used on each end of the transmission line. Furthermore, the distribution system would likely use an intermediate voltage for local distribution, between the high-voltage long-distance transmission line and the pole transformer in the alley. Finally, a realistic power generation and distribution system would use a three-phase system. Nevertheless, the principles we have demonstrated are still valid, and high-voltage distribution systems are universally used.

5.4 THREE-PHASE POWER SYSTEMS

5.4.1 Introduction

The Importance of Three-Phase Systems.   If you looked out of your window at this moment, you would probably see some power lines. Count the wires and you will likely find there to be four. Go examine a pole closely and you will see that at each pole one of the four wires is connected to a conductor which comes down the pole and enters the ground.

When you are driving cross country and see a large electrical transmission line, you will again see four wires. One of them, usually running along the top of the towers, will be noticeably smaller than the other three. If you looked closely, you would again see that the small wire is grounded at every tower. Four wires—what can it mean?

These observations suggest that there is more to ac power circuits that we have described hitherto. We have been discussing what are called single-phase circuits, one ac generator connected to a load with two wires. The three ungrounded wires in the transmission systems we have just described are driven by three ac generators. The grounded wire exists to increase safety and protection from lightning. Power is conveyed by the remaining three wires in the form of three-phase electric power. The overwhelming majority of the world's electric power is generated and distributed as three-phase power. If you were to examine a catalogue of industrial-grade motors, you would discover that all the larger motors, bigger than a few horsepower, would be three-phase motors.

What Is Three Phase Power?   Physically, there are three wires which carry the power, and often a fourth wire, called the *neutral*, which is grounded. In enclosed cables, the active wires are colored red, black, and green; and the neutral, if present, will be white. The phases are traditionally designated A, B, and C, and the voltages between them are as shown in Fig. 5.31. The Greek lowercase letter ϕ (phi) is a common abbreviation for "phase"; hence 3ϕ means three phase and 1ϕ means single phase. The voltages are expressed mathematically as

$$v_{AB}(t) = V_p \cos \omega t$$
$$v_{BC}(t) = V_p \cos (\omega t - 120°) \qquad (5.51)$$
$$v_{CA}(t) = V_p \cos (\omega t - 240°)$$

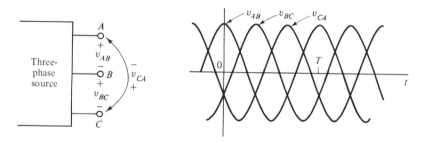

Figure 5.31   Voltages of a three-phase system in the time domain.

The frequency-domain picture for a three-phase system is shown in Fig. 5.32. We have used $\underline{V}_{AB}$ as our phase reference, and shown $\underline{V}_{BC}$ following by 120°, then $\underline{V}_{CA}$. This is known as an *ABC* phase sequence and corresponds to the time-domain representation in Fig. 5.31 and Eqs. (5.51).

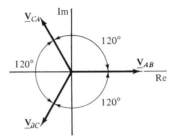

Figure 5.32   Voltages of a three-phase system in the frequency domain.

The Advantages of Three-Phase Power.   We have shown in Fig. 5.6 that single-phase power produces a pulsating flow of energy. A smooth flow of energy from source to load is achieved by a balanced three-phase system. If we have identical resistive loads connected between the three phases, the instantaneous flow of power would be given by Eq. (5.52). We use the trigonometric identity $\cos^2 \alpha = (1 + \cos 2\alpha)/2$ to derive the second and third forms.

$$
\begin{aligned}
p(t) &= \frac{v_{AB}^2(t)}{R} + \frac{v_{BC}^2(t)}{R} + \frac{v_{CA}^2(t)}{R} \\
&= \frac{V_p^2}{2R}[1 + \cos 2(\omega t) + 1 + \cos 2(\omega t - 120°) + 1 + \cos 2(\omega t - 240°)] \qquad (5.52) \\
&= \frac{3V_p^2}{2R} + \frac{V_p^2}{2R}[\cos (2\omega t) + \cos (2\omega t - 240°) + \cos (2\omega t - 480°)]
\end{aligned}
$$

We see a constant term and a term that appears to be time varying at twice the source frequency. Actually, the second term adds to zero at all times. This is easily shown by a phasor diagram; indeed, the phasors representing these terms give the same phasor diagram as Fig. 5.32, since a phase of $-480°$ is the same as $-120°$. Clearly, the phasor sum of the three symmetrical phasors is zero, and hence the fluctuating power term is also zero at all times. Thus Eq. (5.52) reduces to

$$
p(t) = \frac{3V_p^2}{2R} \qquad \text{(a constant)}
$$

This constant flow of energy effects general smoothness of operation in three-phase electrical equipment. A rough analogy is suggested by comparing an engine having one cylinder with an engine having many cylinders—clearly, the multicylinder engine will run smoother.

Compared with a single-phase system, distribution losses are proportionally less for a three-phase system. Additionally, three-phase motors offer advantages over single-phase motors in both startup and run characteristics. In short, three-phase systems are supremely importa.ıt for the generation, distribution, and use of electrical power, particularly in industrial settings.

5.4.2 How Is Three-Phase Power Generated?

Single-Phase Generation.   How, for that matter, is single-phase power generated? We will consider three-phase after showing how single-phase AC voltage is generated. Figure 5.33a shows the essentials of a single-phase generator: a magnetic field supported by an iron pole structure, a rotating loop of wire, and some slip rings to connect the generated voltage to stationary terminals. Figure 5.33b shows this arrangement from the side and introduces the angle α, which increases with time as the coil rotates.

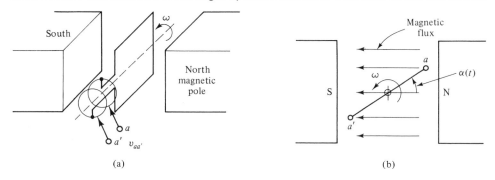

(a)   (b)

Figure 5.33   (a) Simple one-turn generator; (b) side view of the generator.

The operation of the generator can be understood as an interaction between the moving wire and the magnetic flux existing between the poles. The electrons in the wire are moving through the magnetic field and experience a force. This force displaces them slightly (toward a' at the instant pictured) and results in a voltage between a and a'. If the circuit were completed through a resistor, electrons would move back and forth through the circuit and energy would be conveyed to the resistor from the resulting ac current. Whatever was turning the coil would, of course, be the source of this energy.

A mathematical analysis follows from Faraday's law of induction,

$$v(t) = \frac{d\Phi(t)}{dt}$$

where $\Phi(t)$ (an uppercase phi) represents the total magnetic flux passing through the coil, and $v(t)$ is the voltage appearing at the coil terminals. Faraday's law is often written with a minus sign, but we will ignore signs because our only goal in this discussion is to indicate how an ac voltage is generated. In the rotating coil shown in Fig. 5.33, voltage is produced by the rotation of the coil in the magnetic field. The amount of flux through the coil will be zero when $\alpha = 0$ and will be maximum when $\alpha = 90°$. The flux will be

$$\Phi(t) = \Phi_m \sin \alpha(t) = \Phi_m \sin \omega t \qquad \text{webers}$$

where Φ_m is the maximum amount of magnetic flux through the coil. Applying Faraday's law, we find the voltage at the coil terminals to be

$$v_{aa'}(t) = \frac{d\Phi(t)}{dt} = \omega\Phi_m \cos\omega t = V_p \cos\omega t$$

Thus an ac voltage is generated with peak value $V_p = \omega\Phi_m$.

Figure 5.34 indicates some of the practical details. The magnetic flux is produced by a dc electromagnet. The flux is supported by an iron structure, called the *stator*. The coils are wound on an iron rotor, which also supports the magnetic flux. Notice that we show several coils. These are wound on the rotor in such a way that the voltages of the coils add; that is, the coils are connected in series. Because these coils are spread out over a considerable angular extent of the rotor, the voltages induces in them have different phases, as shown in Fig. 5.35. However, since they are close together, they add constructively and together generate the voltage appearing at the slip rings for a and a'.

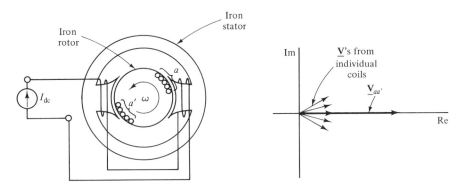

Figure 5.34 A generator has coils distributed around the rotor.

Figure 5.35 The voltage in each coil has a different phase but voltages add constructively when connected in series.

In a single-phase ac generator, the coils would cover the entire rotor, and thus the phasors for the individual coils shown in Fig. 5.34 would cover the range of angles from $-90°$ to $+90°$. We have not covered the entire rotor because we are leaving room for the coils of the other two phases of a three-phase generator.

Three-Phase Generators. Figure 5.36 shows the rotor for a generator which has been wound for three-phase power generation. We have marked the coil terminals with a, a', b, b', c, and c' for the coils of the separate phases so as to retain symmetry, with a, b, and c located 120° apart in space. The purpose of this discussion is to show how to connect these coils together achieve a proper three-phase source. We assume that our generator has six slip rings connected to a, a', b, b', c, and c'.

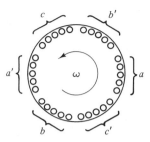

Figure 5.36 The three-phase rotor has three sets of coils.

From the physical angles on the rotor, we can see that the coil voltages would be

$$v_{aa'}(t) = V_p \cos \omega t$$
$$v_{bb'}(t) = V_p \cos(\omega t - 120°) \qquad (5.53)$$
$$v_{cc'}(t) = V_p \cos(\omega t - 240°)$$

These give the phasor diagram shown in Fig. 5.37. There are six slip rings and thus six output terminals to our generator.* In order to have a three-phase generator, however, we need three output wires. This suggests the following problem: Given six terminals as pictured in Fig. 5.38, having voltages described by the phasors in Fig. 5.37, show the connections between the terminals to produce only three external terminals, whose voltages are symmetrically arranged as in Fig. 5.32. This problem has two solutions, which we develop in the following sections.

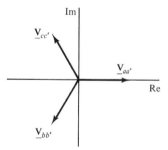

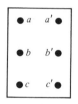

Figure 5.37 Phasor voltages in the three coils. These must be connected externally to make a true three-phase system.

Figure 5.38 External terminals for the three separate phases. These must be connected to produce a three-phase system.

*Standard ac generators are constructed differently from what we have described in the above, although the principles are the same. The dc field is normally placed on the rotor and thus only two slip rings are required. More important, the slip rings carry relatively small currents, and the ac windings in the stator, which carry large currents, have no slip rings. In the above, we put the dc field on the stator because that configuration is easier to visualize.

The Delta (Δ) and Wye (Y) Connections. The symmetry of the desired phasor voltages suggests that we require some sort of symmetrical connection for our three generators represented by the three coils on the rotor in Fig. 5.36. There are only two symmetrical configurations involving three elements connected end-to-end, and these are shown in Fig. 5.39. These configurations bear various names, depending on the context. The closed ring is usually called a *delta* (for the fourth capital letter of the Greek alphabet, Δ), even when it is drawn upside down or on its side. The configuration with the common point is sometimes called a star configuration, but in three-phase terminology it is usually called a *wye configuration* (a phonetic spelling of the letter Y), regardless of orientation.

Figure 5.39 The only two symmetrical configurations of three elements: (a) delta; (b) wye.

These symmetric configurations suggest two solutions of our question posed above. We require three terminals for a source of three-phase power. The delta has three terminals, and hence its application as a possible configuration for the three-phase generators is clear. The wye, on the other hand, has four terminals, counting the common connection in the center. This turns out to give us the fourth wire mentioned earlier, the neutral wire which is grounded. We have to connect the terminals in such a way that they have the geometric symmetry of the delta or wye configurations in Fig. 5.39, but we have also retained the electrical symmetry indicated by the phasor diagram in Fig. 5.32.

The Delta Connection. The delta configuration places the three generators in a closed ring. Figure 5.40b shows one possible connection for the delta. We must be careful, however, if we are to connect c' to a, for we then will have the generators connected in a closed ring. It would be like jump-starting a car having a weak battery, as shown in Fig. 5.40c. The circuit can be closed if the polarities are correct; for there will be at most a small voltage across the gap and only small currents will flow through the batteries with

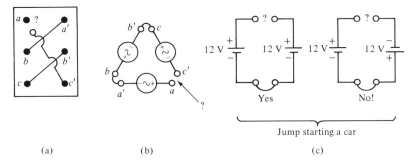

Figure 5.40 Potential delta connection. The voltage across a and c' must be zero if the ring is to be closed.

the connection marked "Yes." But if the polarities are wrong, there will be approximately 24 V across the gap and a huge current will flow if the connection is made.

Similarly, if we are to close the ring of generators in Fig. 5.40b, we require the voltage across the gap to be small. This requires that

$$\underline{\mathbf{V}}_{ac'} = \underline{\mathbf{V}}_{aa'} + \underline{\mathbf{V}}_{bb'} + \underline{\mathbf{V}}_{cc'} = \text{small} \tag{5.54}$$

Notice that Eq. (5.54) follows from the rule for adding subscripted voltages given on page 13 because a' is connected to b and b' is connected to c. That is, the general rule is

$$V_{ab} = V_{ax} + V_{xb}$$

We can extend this to

$$V_{ab} = V_{ax} + V_{yb} \qquad \text{if } x \text{ and } y \text{ are connected}$$

The phasor diagram for the sum in Eq. (5.54) is easily derived from Fig. 5.37; clearly, the sum is small, ideally zero. Thus it is safe to close the ring of generators and bring out the connected terminals as a three-phase source. Figure 5.41b shows the phasor diagram of the final connection. Another possible delta connection would result with a connected to b', b connected to c', and c connected to a'. We leave the investigation of this possibility for a homework problem.

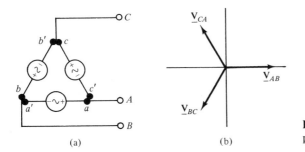

(a)   (b)

Figure 5.41 Delta connection with phasor diagram.

The Wye Connection.   A wye connection results from connecting a', b', and c', as shown in Fig. 5.42a. With this connection the magnitudes of $\underline{\mathbf{V}}_{ac}$, $\underline{\mathbf{V}}_{ba}$, and $\underline{\mathbf{V}}_{cb}$ are $\sqrt{3}$ greater than those of the component voltages, $\underline{\mathbf{V}}_{aa'}$ and $\underline{\mathbf{V}}_{bb'}$, and $\underline{\mathbf{V}}_{cc'}$ and the phase of $\underline{\mathbf{V}}_{ac}$ lies at $-30°$ relative to $\underline{\mathbf{V}}_{aa'}$. These combinations are illustrated in Fig. 5.42b by the form-

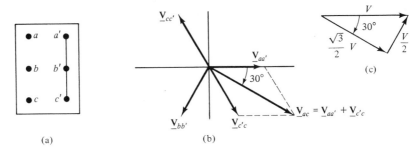

(a)   (b)

Figure 5.42 Possible wye connection.

ing of $\underline{\mathbf{V}}_{ac}$ through addition of $\underline{\mathbf{V}}_{aa'}$ and $\underline{\mathbf{V}}_{c'c}$, which is the negative of $\underline{\mathbf{V}}_{cc'}$. The $\sqrt{3}$ comes from the 30°–60°–90° triangle, as we have shown on the Fig. 5.42c. Unlike the delta connection, the wye connection has all three generators connected to a common point. In Fig. 5.43, we label the three lines, A, B, and C, and the neutral we label N. This is a wye connection even through our Y came out lying down. The wye connection of the generators leads to the four-wire system which we described at the beginning of this section. The wye connection is important in the distribution of electric power.

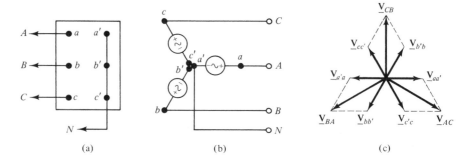

Figure 5.43   Wye connection. The line voltages are $\sqrt{3}$ larger than the phase voltages.

Line Voltage and Phase Voltage.   With the wye, we must distinguish between the line-to-line voltages, usually called *line voltages*, and the line-to-neutral voltages, usually called *phase voltages* for the wye connection. As we have shown, the line voltages are $\sqrt{3}$ times the phase voltages. The line voltages are displaced by 30° from the phase voltages, so we would have to make $\underline{\mathbf{V}}_{AC}$ our phase reference in order to have the line voltages in Fig. 5.43 conform to the picture in Fig. 5.32.

Phase Rotation.   The phase rotation came out ABC in both systems we developed. This is a mere accident of labeling; we could easily relabel the connections to have ACB. That is, we can erase B and C (actually, any two) and switch them to achieve an ACB phase sequence. The notation is arbitrary, but the physical phase rotation is very important. The rotational direction of a three-phase motor, for example, depends on the phase rotation of the input power. Because phase rotation is very important, there exist several techniques for determining the phase rotation of a three-phase power system, the easiest utilizing a device which, when connected to three labeled wires, flashes ABC or ACB with an indicator light.

Other Possible Connections.   We have now shown the two ways for connecting the coils on the three-phase generator of Fig. 5.36 to give a three-wire, symmetrical power system. Actually, we have shown one version of each way, for there are minor variations on the procedures. For example, we can make a, b, and c the neutral for a wye. But we have shown the two basic connections. Each is important; the delta fits some applications of three-phase power, the wye excels for others. We will postpone our discussion of typical applications until we have explored the ways for connecting three-phase loads.

5.4.3 Three-Phase Loads

Delta-Connected Loads.   Like three-phase generators, three-phase loads can be connected in delta or wye. In Fig. 5.44 we show a balanced three-phase resistive load connected as a delta. The source of the three-phase power is not shown; we shall assume ABC phase rotation and a line voltage $\sqrt{2}\ V_L = |\underline{\mathbf{V}}_{AB}| = |\underline{\mathbf{V}}_{BC}| = |\underline{\mathbf{V}}_{CA}|$. We have shown no neutral because the load offers no place for connecting a neutral. In practice, however, the three-phase load would be housed in a physical structure of some kind, and this structure would normally be grounded either directly to earth ground or through the neutral. For example, the housing of a three-phase motor would be grounded through the neutral.

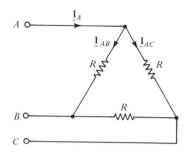

Figure 5.44   Delta-connected load.

With a load connected as a delta, we must distinguish between the line currents and the phase currents flowing in the resistors. Figure 5.45 shows the phase currents, $\underline{\mathbf{I}}_{AB}$ and $\underline{\mathbf{I}}_{BC}$ and $\underline{\mathbf{I}}_{CA}$. These currents are in phase with the line voltages. The line current, say, $\underline{\mathbf{I}}_A$, can be determined by phasor addition of the phase currents. Kirchhoff's current law at the top node is

$$\underline{\mathbf{I}}_A = \underline{\mathbf{I}}_{AB} + \underline{\mathbf{I}}_{AC}$$

but

$$\underline{\mathbf{I}}_{AC} = -\underline{\mathbf{I}}_{CA}$$

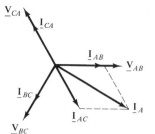

Figure 5.45   Phase current addition to yield line current.

and hence the currents add as in Fig. 5.45. The other line currents could be determined similarly; indeed, the picture develops like that of the wye generator connection in Fig. 5.43. We see that the line currents are $\sqrt{3}$ greater than the phase currents. For the resistive load, the phase of the line current in A is in phase with the average phase between $\underline{\mathbf{V}}_{AB}$ and $\underline{\mathbf{V}}_{AC}$. The phase relationships are shown in Fig. 5.46.

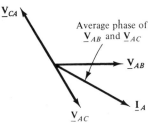

Figure 5.46 The line current is in phase with the average phase of the voltages to the other two lines.

With balanced loads that are not resistive, the phasor diagram shown in Fig. 5.45 changes only slightly. The magnitude of the phase current is computed by dividing the line voltages (which are also the phase voltages) by the magnitude of the phase impedance. The phase currents will lead or lag the line voltages according to the angle of the impedance. Hence the line currents will also be shifted in phase by the angle of the impedance. We give an example below.

Power in Three-Phase Delta Connections. We are, as always, interested in the power in the system. Consider first a resistive load. The total power to the load would be the sum of the powers delivered to the three resistors; and this would be

$$P = 3P_R = 3(V_\phi I_\phi) \tag{5.55}$$

because the power factor is unity for a resistor. In Eq. (5.55) V_ϕ and I_ϕ represent the rms values of the phase voltage and current. We desire, however, to express the total power in terms of line voltage and current. The phase currents are often inaccessible for measurements, but the line voltage and current can always be measured. Consequently, we introduce the line variables

$$P = 3V_L \frac{I_L}{\sqrt{3}} = \sqrt{3}\, V_L I_L \tag{5.56}$$

where V_L and I_L are the rms line voltage and current. Hence with three identical resistors connected as a delta, you measure the line voltage and line current, take their product and then, because it is three phase, you multiple by $\sqrt{3}$.

If the load were not resistive, you would multiply also by the power factor. The more general formula is thus

$$P = \sqrt{3}\, V_L I_L \times \text{p.f.} \tag{5.57}$$

In applying Eq. (5.57), the power factor is the cosine of the angle of the phase impedance and is not the phase angle between line current and line voltage. The power factor angle is, however, the angle between the phase of the line current and the average phase of the voltages to the two other lines, as defined in Fig. 5.46.

An Example. A 220-V three-phase power system supplies 2000 W to a delta-connected, balanced load with a power factor of 0.9, lagging. Determine the line currents, the phase currents in each phase of the load, and the phase impedance. Draw a phasor diagram.

First we shall calculate the magnitude of the line currents from Eq. (5.57).

$$P = \sqrt{3}\ V_L I_L \times \text{p.f.} \Rightarrow I_L = \frac{2000}{\sqrt{3}\ (220)(0.9)} = 5.83 \text{ A (rms)}$$

The phase currents are smaller by $\sqrt{3}$, so

$$I_\phi = \frac{I_L}{\sqrt{3}} = \frac{5.83}{\sqrt{3}} = 3.37 \text{ A}$$

This allows us to calculate the impedance in each phase of the delta. The angle of the impedance is implied by the power factor: $\theta = \cos^{-1}(0.9) = +25.8°$, $+$ because the current is lagging (inductive).

$$\underline{\mathbf{Z}}_\phi = \frac{V_\phi}{I_\phi} \underline{/\cos^{-1}(\text{p.f.})} = \frac{220}{3.37} \underline{/\cos^{-1}(0.9)} = 65.3 \underline{/+25.8°}$$

We can now draw the phasor diagram (Fig. 5.47). We will use $\underline{\mathbf{V}}_{AB}$ for the phase reference, with the other line voltages placed symmetrically in ABC sequence. The phase currents lag by 25.8°, as shown. The line currents can be computed by phasor addition of the phase currents, as we did in Fig. 5.45, but another approach is to use our earlier results to place the line currents behind the phase currents by 30° and greater by $\sqrt{3}$. Whichever way is chosen, only one line current need be determined, $\underline{\mathbf{I}}_A$ for example, and the other two can be constructed by symmetry.

Wye-Connected Loads.   A load is said to be connected in a wye when it is connected into a neutral, as shown in Fig. 5.48. In Fig. 5.48 we have labeled the input connections for the three phase power a, b, and c, and the neutral, n. We have shown no

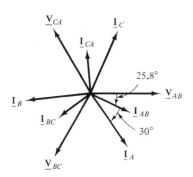

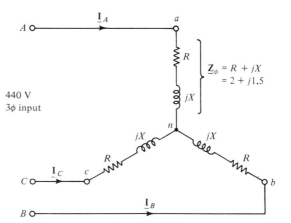

Figure 5.47 Line voltages, phase currents, and line currents. The line currents are $\sqrt{3}$ greater than the phase currents.

Figure 5.48 Wye-connected load. To find the line currents we must determine the line--to-neutral voltages.

connection to the neutral, but there would often be a connection between the load neutral and the source neutral, if such existed. For a perfectly balanced load no current would flow in the neutral because the three line currents add to zero. Otherwise, the type of connection of the generators producing the three-phase power (A, B, and C) is unimportant. We shall calculate the line-to-neutral voltages, $\underline{V}_{an}$, $\underline{V}_{bn}$, and $\underline{V}_{cn}$ and the line currents, $\underline{I}_A$, $\underline{I}_B$, and $\underline{I}_C$.

First we shall solve for the line-to-neutral voltages. The phase impedances are given, $\underline{Z}_\phi$; hence it is clear that the line currents can be determined once the line-to-neutral voltages are known; for example,

$$\underline{I}_A = \frac{\underline{V}_{an}}{\underline{Z}_\phi} \qquad (5.58)$$

The line-to-neutral voltages can be determined from consideration of the symmetry of the circuit. It is convenient to reason as if we know the line-to-neutral voltages (which we do not) and wish to determine from them the line-to-line voltages. The relationship between these two sets of three-phase voltages then becomes known and we henceforth can deduce either set of voltages from the other. We assume the ABC phase sequence; hence the line-to-neutral voltages must appear as in Fig. 5.49, assuming that we make $\underline{V}_{an}$ the phase reference. First we shall determine $\underline{V}_{AB}$. We can express the $\underline{V}_{AB}$ in terms of $\underline{V}_{an}$ and $\underline{V}_{nb}$:

$$\underline{V}_{AB} = \underline{V}_{an} + \underline{V}_{nb} \qquad (5.59)$$

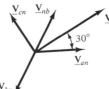

Figure 5.49 Determining line-to-line voltages from line-to-neutral voltages.

We can use A and a, and so on, interchangeably, because these indicate the same points in the circuit. Equation (5.59) becomes more useful when we reverse the subscripts on $\underline{V}_{nb}$, which we may do by changing the sign of that term:

$$\underline{V}_{bn} = -\underline{V}_{nb} \Rightarrow \underline{V}_{AB} = \underline{V}_{an} - \underline{V}_{bn} \qquad (5.60)$$

Equation (5.60) is represented in Fig. 5.49, with the negative of $\underline{V}_{bn}$ drawn and added to $\underline{V}_{an}$. We note that $\underline{V}_{AB}$ leads $\underline{V}_{an}$ by 30°, and is somewhat greater in magnitude. The phasor addition is identical to that as shown in Fig. 5.43 and the magnitudes of phase and line voltages have the ratio $\sqrt{3}$, just as the currents in the delta-connected load. The remaining line-to-line voltages, $\underline{V}_{BC}$ and $\underline{V}_{CA}$ may be determined by similar reasoning, or more directly by arranging them in ABC sequence, each 120° from $\underline{V}_{AB}$. Hence we arrive at the picture shown in Fig. 5.50. The line-to-neutral voltages lag the corresponding line-to-line voltages by 30°, when one considers the two voltages with, say, A (and a) written first, like $\underline{V}_{AB}$ and $\underline{V}_{an}$. But a better way to think about the phase is to realize that the phase of the corresponding line-to-neutral voltage lies between the phases of the two line-to-line voltages which connect to the same point. That is, $\underline{V}_{an}$ will lie halfway be-

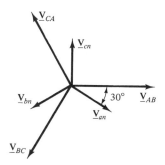

Figure 5.50 The line-to-neutral voltages are smaller by $\sqrt{3}\cdot$ and lag line-to-line voltages by 30°.

tween $\underline{\mathbf{V}}_{AB}$ and $\underline{\mathbf{V}}_{AC}$. This phase relation, together with the magnitude ratio of $1/\sqrt{3}$, allows us to determine easily the line-to-neutral voltages from the set of line-to-line voltages, or vice versa.

We now can solve the problem pictured in Fig. 5.48. We shall assume that $\underline{\mathbf{V}}_{AB}$ is the phase reference, that is, $\underline{\mathbf{V}}_{AB} = 440\sqrt{2}\,\underline{/0°}$. From Fig. 5.50 we see that $\underline{\mathbf{V}}_{an}$ will lie at $-30°$ and be smaller by $\sqrt{3}$, or $\underline{\mathbf{V}}_{an} = 254\sqrt{2}\,\underline{/-30°}$. Hence the current in line A will be

$$\underline{\mathbf{I}}_A = \frac{\underline{\mathbf{V}}_{an}}{\underline{\mathbf{Z}}_\phi} = \frac{254\sqrt{2}\,\underline{/-30°}}{2 + j1.5} = \frac{254\sqrt{2}\,\underline{/-30°}}{2.50\,\underline{/36.9°}} = 101.6\sqrt{2}\,\underline{/-66.9°}\ \text{A}$$

The other line currents can be determined similarly or by symmetry from $\underline{\mathbf{I}}_A$.

Power in Wye-Connected Loads. The total power to the wye-connected load will be three times the power to each phase of the load (P_ϕ). Thus we can compute the total power with

$$P_{3\phi} = 3P_\phi = 3V_\phi I_\phi \times \text{p.f.} \tag{5.61}$$

where V_ϕ is the phase rms voltage, the line-to-neutral voltage in this instance, I_ϕ is the phase rms current, also the line current in this instance, and p.f. is the power factor of the phase impedance. Equation (5.61) is similar to Eq. (5.55), derived for the delta-connected load, but is applied differently. For the delta-connected load, the phase voltage is identical to the line-to-line voltage, but the phase current is smaller than the line current by $1/\sqrt{3}$. For the wye-connected load, the phase current is identical to the line current, but the phase voltage is smaller than the line voltage by $1/\sqrt{3}$. For the present example, the total power to the three-phase load would be

$$P_{3\phi} = 3(254)(101.6)(\cos 36.9°) = 62.0\ \text{kW}$$

The neutral of the wye-connected load might not be accessible for voltage measurement; hence it is desirable to express the total power in terms of the line voltage and current. Using the $\sqrt{3}$ ratio between phase voltage and line voltage, we may convert Eq. (5.61) to

$$P_{3\phi} = 3\left(\frac{V_L}{\sqrt{3}}\right)(I_L) \times \text{p.f.} = \sqrt{3}\ V_L I_L \times \text{p.f.} \tag{5.62}$$

where the L subscript indicates a line rms current or a line-to-line rms voltage. In Eq. (5.62) the power factor is the cosine of the angle between the line current and the average phase angle of the two line voltages involving that line. Thus Fig. 5.46 applies to this case as well as the delta-connected load because the phase voltage, $\underline{\mathbf{V}}_{an}$ in this case, would lie symmetrically between $\underline{\mathbf{V}}_{AB}$ and $\underline{\mathbf{V}}_{AC}$. Of course, the power factor can also be determined from the angle of the phase impedances if they are known.

Equation (5.62) is identical to Eq. (5.57), which was presented for the delta-connected load. Clearly, the two load configurations are indistinguishable to external measurement and can only be identified in practice by examining the internal connections in the three-phase load, be it motor, heater element, or whatever.

Delta–Wye Conversions.   This does not mean, however, that the *same* set of phase impedances are equivalent when connected first in delta and then in wye. Indeed, the appearance of the circuits suggests that the delta gives parallel paths whereas the wye gives series paths. This appearance suggests that the line current for the delta would be larger than for the wye if the same phase impedance were used for each connection. It can be shown that the ratio is $3:1$; that is, three identical resistors will draw three times the current (and three times the power) when connected in delta, as compared to when they are connected in wye.

From the above it follows that the delta and wye are equivalent if the phase impedances differ by a factor of 3, with the delta connection having the higher impedance. This equivalence, shown in Fig. 5.51, is often useful in solving three-phase problems. We leave proof of this equivalence for a problem at the end of this chapter.

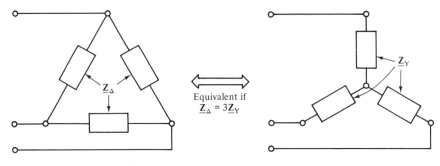

Figure 5.51   Delta and wye equivalence.

To illustrate the usefulness of this delta–wye conversion, we shall solve for the power to the delta-connected load in Fig. 5.52. The 10-Ω resistors represent the load, but we must consider the resistance of the wires leading to the load, which is represented by the 0.5-Ω resistors. The presence of the wire resistance undermines our previous approach for solving delta-connected loads, but if we convert the delta to an equivalent wye load, we can solve the problem. Figure 5.53 shows the circuit after conversion to wye, and it now should be clear how to proceed. The wire resistance can now be combined with the load resistance to yield a phase resistance of 3.83 Ω, and the rms line-to-neutral voltage would be $240/\sqrt{3} = 139$ V. The line current would thus be $139/3.83 = 36.1$ A, and the total power to the wires plus load would be $\sqrt{3}\,(240)(36.1) = 15.0$ kW.

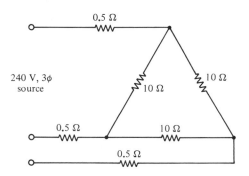

Figure 5.52 Find the power to the delta-connected load.

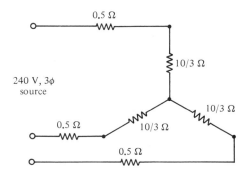

Figure 5.53 After converting the delta to a wye, we can in reality determine the line currents.

The wire losses would be $3(36.1)^2(0.5) = 1960$ W, the rest of the power going to the delta-connected load.

Unsymmetric Loads.   A three-phase load is said to be unbalanced when the phase impedances are not identical. This is considered an undesirable situation and is avoided in practice if possible. When unbalanced loads are connected in delta, calculation of the phase and line currents becomes tedious, though straightforward. All phase and line currents must be calculated individually because symmetry has been lost.

When unbalanced loads are connected in wye, the analysis is straightforward only when the neutral of the load is connected to the neutral of the three-phase source. With the neutral connected, the three loads operate in effect as single-phase loads which share one common wire (the neutral connection). Of course, current will flow in the neutral wire for an unbalanced load.

When there exists no neutral wire in the unbalanced wye connection, complications arise in the calculation of the line-to-neutral voltages and the line currents. Because the neutral of the load is no longer at the same voltage as the neutral of the source, the first step in solving the problem is to calculate the voltage of the neutral of the load. Then one can proceed to solve for the line currents, a tedious though straightforward calculation. This problem lies beyond the scope of this book.

5.4.4 Some Practical Matters

Applications of Delta and Wye Sources.   We have discussed both delta and wye source connections because both find important applications. The wye source connection is used for long-distance transport of electrical power, where the resistive losses of the wires degrade system efficiency. Equation (5.62) suggests the reason: the total power from a source is the product of the line current, the line-to-line voltage, and the power factor. The power required and the power factor are determined by the load on the system, but the levels of the line voltage and current are influenced by the source connections. The wye connection gives a line-to-line voltage which is $\sqrt{3}$ greater than the delta connection and hence, for the same power, the line current will be $\sqrt{3}$ smaller. The losses in the wire vary with the square of the line current and, as a consequence, losses will be one-third as great with a wye source connection, as compared with a delta connection. Thus the high-voltage transformer windings are connected in a wye configuration for power transmission and distribution systems.

The delta source connection finds wide application when three single-phase circuits must be derived from a three-phase source. This three-phase to single-phase conversion is required in residential areas, because single-phase power is required for household appliances and lighting. The power would be brought into the neighborhood at a voltage of at least 12,000 V, line to line, and would be stepped down to 240/120 V with a transformer (more details below). The low-voltage secondaries of the distribution transformer would be connected in a delta, and several households would be served with single-phase power from each side of the delta. This arrangement is advantageous because it minimizes interaction between the three single-phase circuits as compared with the wye connection.

How a House Is Wired.   We do not recommend that you do it, but *if* you sought out the circuit breaker box (or the fuse box in older structures) for your dwelling, and *if* you removed the safety cover, you would discover three wires coming into the box from the transformer on the pole in the alley. One wire would be red, another black, and the third white. Does this mean that you have three-phase power coming into your residence? No, this is not three-phase power; this is 240/120-V single-phase power. Figure 5.54 shows the usual arrangement at the transformer secondary. The white wire is the neutral and is grounded at the transformer by a wire which enters the moist earth and should also be grounded through the household plumbing. The black and red wires are "hot," each carrying 120 V relative to the neutral. As the phasor diagram shows, the two 120-V voltages are opposite in phase and hence add to 240 V. That is, the neutral is at the midpoint between the two lines which carry 240 V and hence 120 V is developed between each hot line and the neutral.

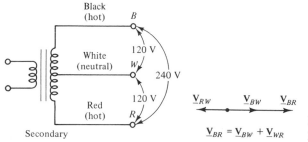

Figure 5.54 A 120/240-V household power system.

Of course, the 240-V power is used for heavy equipment such as air conditioners and certain power tools. Most appliances operate with 120 V, so the lighting and appliance circuits in the house are wired from both hot wires. The red color is not used in the household 120-V wiring: black means hot and white neutral. Modern wiring codes require a third wire for a separate ground. The ground wire does not carry power like the neutral wire but enables equipment to have a separate ground connection independent of the power circuit.

Figure 5.55 shows a modern appliance outlet, which is both *polarized* and grounded with a three-wire system. The outlet and plug are said to be polarized because the hot and neutral connections differ in size. Some loads, such as table lamps, may work equally well, and be equally as safe, plugged in either way. Such a load would have an unpolarized plug, which would fit into the outlet either way. Other loads are equipped with polarized plugs which fit in only one way, thus controlling which wire is neutral. Note that for safety the neutral of the load (the male) cannot make connection with the hot of the source. Incidentally, if you are wiring an appliance outlet, pay close attention to polarity. Usually, the screw to which you should connect the hot (black) wire has a copper color and the screw to which you should connect the neutral (white) wire has a silvery color. If you are installing a lighting fixture in a ceiling, the hot goes to the center of the receptacle and the neutral connects to the screw threads through the wall switch.

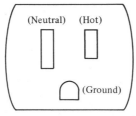

Figure 5.55 Grounded and polarized household outlet.

5.4.5 Electrical Safety

Although use of electrical power underlies much of our modern way of life, the average citizen is often uninformed of the dangers of electrical power. One person might fear to touch the terminals of a 12-V auto battery, while another might think nothing of sticking a finger into a light socket to see if there is any power. (The battery is safe, but please keep your finger out of the light socket.) The purpose of this section is to present some basic information about electrical safety so that the reader will be able to recognize a dangerous situation and hopefully stay out of trouble.

The circuit theory of this subject is simple enough: Ohm's law is the key,

$$I = \frac{V}{R}$$

where R is you. Although we are accustomed to signs warning "DANGER, HIGH VOLTAGE," it is actually the current that affects our bodies. Many thousands of volts will do no more than startle provided that the current is small, as when we feel a small spark of static electricity after sliding across an auto seat. In the following we speak first of the physiological effects of electric current on the human body. We then discuss body resistance and the resistance of typical surroundings. Finally, we will offer some advice about electrical safety.

The Physiological Effects of Current.   Figure 5.56 shows the variety of effects that electrical current may have on the human body and the current levels at which they occur. Injury could be caused indirectly through being startled and losing muscular control, or directly through burns. Death could result from suffocation, through loss of

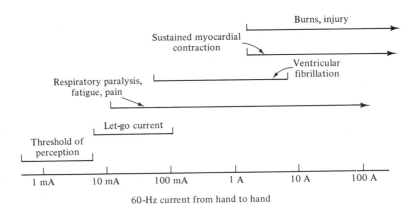

Figure 5.56 Physiological effects of electricity. (Adapted from Webster: Medical Instrumentation, Application and Design, Copyright © 1978 Houghton Mifflin Company. Adapted with permission.)

proper function of the heart, or through severe burns. The large ranges in Fig. 5.56 are represented because of wide variation between individuals in body size, condition, and tolerance to electrical shock. Clearly of importance is the region of the body through which the current passes; current passing through the lower part of the leg, for example, might be painful but would be unlikely to affect heart action.

The surprising aspect of Fig. 5.56 is that small currents can have serious effects. This occurs because the communication system of the human body is electrical in nature and misbehaves under external electrical influence.

When a person is experiencing electrical shock, time becomes an important factor, for the damaging effects are progressive. Hence it is important to remove the source of electrical energy from a shock victim before other action is taken. Particularly serious is the condition of ventricular fibrillation, where the heart loses its synchronized pumping action and enters a useless mode of activity. This condition is very dangerous because the heart may not resume normal action when the source of electrical power is removed; sophisticated medical equipment is required to restore coherent heart action.

The Resistance.   Earlier we stated that the resistance is you. This would be strictly true if you came into simultaneous contact with both wires of an electrical circuit, say, by grabbing a wire with each hand, but most serious electrical shocks occur through another circumstance. The more common dangerous situation is portrayed in Fig. 5.57.

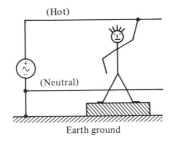

Figure 5.57 Most shocks occur between the hot wire and ground.

Here we have shown the victim in simultaneous contact with the hot wire and with "ground." Ground may be the moist earth, the plumbing of a house, or even a concrete floor which is in contact with the plumbing. In this case the resistance that influences the amount of current through the body includes not only the resistance of the body but also the resistance of the shoes and the resistance between the shoes and earth ground.

To assess the danger of a given situation, we must estimate the resistance of the "circuit" of which our body might become an unhappy part. Table 5.1 shows some basic information which would allow such an estimate of the total resistance to ground. What, for example, would you experience if you were standing on moist ground with leather-soled shoes and you unwittingly grab hold of a 120-V wire? Taking the lowest values in Table 5.1, we estimate the following resistances: 3 kΩ for the grasp, 200 Ω for the body, and 5 kΩ for the feet–shoes. Thus the largest current you might carry would be 120 V/(8.2 kΩ), about 15 mA. From Fig. 5.56 we judge that there is a fair chance that you will be unable to release your grasp and that you might hence be unable to breathe. This is therefore a dangerous situation.

TABLE 5.1   RESISTANCE

(a) For Various Skin-Contact Conditions

Condition (area to situ)	Resistance	
	Dry	Wet
Finger touch	40 kΩ–1 MΩ	4–15 kΩ
Hand holding wire	15–50 kΩ	3–6 kΩ
Finger–thumb grasp	10–30 kΩ	2–5 kΩ
Hand holding pliers	5–10 kΩ	1–3 kΩ
Palm touch	3–8 kΩ	1–2 kΩ
Hand around $1\frac{1}{2}$-in. pipe (or drill handle)	1–3 kΩ	0.1–1.5 kΩ
Two hands around $1\frac{1}{2}$-in. pipe	0.5–1.5 kΩ	250–750 Ω
Hand immersed	—	200–500 Ω
Foot immersed	—	100–300 Ω
Human body, internal, excluding skin = 200–1000 Ω		

(b) For Equal Areas (130 cm^2) of Various Materials

Material	Resistance
Rubber gloves or soles	More than 20 MΩ
Dry concrete above grade	1–5 MΩ
Dry concrete on grade	0.2–1 MΩ
Leather sole, dry, including foot	0.1–0.5 MΩ
Leather sole, damp, including foot	5–20 kΩ
Wet concrete on grade	1–5 kΩ

Source: Adapted from Ralph Lee, Electrical Safety in Industrial Plants, *IEEE* Spectrum, June, 1971.

One factor that Table 5.1 does not contain concerns the breakdown voltage of our skin resistance. At approximately 700 V (ac) the resistance of the skin drops to near zero: in effect, a spark burns a hole in the skin. Thus the resistance of the skin, which might save your life for a lower voltage, becomes ineffective at such high voltages. For this reason, high voltages are seldom used for distribution of power in industrial applications, and 240 V is the highest voltage used in residential wiring.

An interesting, and alarming, calculation the reader might wish to perform is the following: Making the most pessimistic assumptions about body resistance, resistance to ground, and loss of skin resistance (say, cuts or blisters on the hands and feet), calculate the least voltage that might prove fatal. Such a calculation would have you standing in water or on a metal floor without shoes. Although we are describing an unusual situation, we still urge you to make the calculation.

Of more importance are the factors that increase your safety when working around electrical power. Make sure that the power is off before working on any electrical wiring or electrical equipment. Wear gloves and rubber-soled shoes. Avoid standing on a wet surface or on moist ground. Do not work alone around exposed electrical power. Safety is worth the inconvenience it requires.

Grounding. From our discussion of safety, you might notice that danger increases because the electrical power system is grounded. If the circuit were *floating*, that is, not grounded, the only way to get shocked would be to come into contact with both wires simultaneously, an unlikely event. How can we reconcile this viewpoint with the common idea that electrical circuits are grounded for safety? Actually, both ideas are valid, and there is no contradiction; there is more involved in the issue than we have discussed thus far. Consider that the electrical power system in your building is floating. Everything would function correctly and safely until something went wrong with the equipment. But if, say, the wiring of the transformer on the pole became defective and a connection developed between primary and secondary, in this event the 120-V circuits could float at 12,000 V or even 24,000 V. This would endanger virtually every piece of equipment on the line. Also, if the voltage suddenly were floated at 12,000 V, the danger to you becomes much greater. For this reason, the secondary of the transformer is grounded for protection of life and property. For if the secondary is grounded and a fault develops in the transformer, a large current flows immediately, a fuse or circuit breaker opens the circuit, and the source of power becomes disconnected from the offending part of the circuit.

Given that the power system in the building should be grounded, the necessity for grounding the equipment with the three-wire system in Fig. 5.54 becomes apparent. On a piece of equipment, such as a washing machine, if a fault develops between the hot side of the power and the metal chassis, again a fuse or circuit breaker will respond to the large current flowing through the ground connection and will remove power from the circuit containing the faulty connection. Hence the safest way to install the power system involves grounding the power system and the equipment with which you might come into contact. And if you are working with the wiring or equipment, to be safe you must follow the safety rules we outlined earlier.

PROBLEMS

Practice Problems

SECTIONS 5.1.1–5.1.3

P5.1. Compute the time average and the rms value for the waveforms shown in Fig. P5.1.
Ans: (a) 8.0 V average and 8.72 V rms; (b) $2V_p/\pi$ average and $V_p/\sqrt{2}$ rms, just like a pure sinusoid, because the reversal of the bottom half does not matter after the squaring.

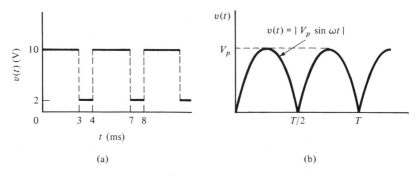

Figure P5.1

P5.2. Show that ac and dc power are additive. That is, if $i(t) = I_{dc} + I_p \cos \omega t$, then the time average of $i^2(t) = I_{dc}^2 + (I_p/\sqrt{2})^2$.

P5.3. For the "sawtooth" waveform in Fig. P5.3:
   (a) Find the average value, V_{dc}.
   (b) Find the effective value, V_e.
   (c) What would be the average power if this voltage appeared across a 10-Ω resistor?

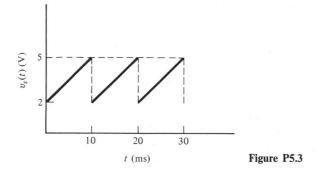

Figure P5.3

SECTION 5.1.4

P5.4. **(a)** A current source having an rms value of 0.75 A is connected to a 10-Ω resistor. What is the power in the resistor?

(b) If the current source in part (a) is found to be sinusoidal with a period of 10 ms, what is the equation of the current as a function of time? (Assume the cosine form and zero phase.)

(c) If the current in part (b) is put through an impedance of $10\underline{/-60°}$ rather than a pure resistance, what now is the time-average power?

SECTION 5.1.5

P5.5. For the circuit shown in Fig. P5.5;

    (a) Solve for $v(t)$. Let the phase of $i(t)$ be zero.

    (b) Compute the time-average power into the circuit using the power factor.

    (c) Show that the time-average power into the entire circuit, P, is equal to the time-average power dissipated in the resistor, P_R.

    (d) Find the time-average electric stored energy.

                     Ans: 319 W average power, 32.0 mJ average energy storage

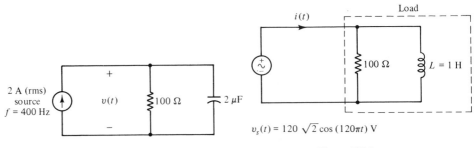

      Figure P5.5   **Figure P5.6**

P5.6. For the circuit in Fig. P5.6.

    (a) Find $i(t)$. Use time- or frequency-domain techniques.

    (b) What is the power factor for the load?

    (c) Compute the time-average power into the load. (You do not have to perform the integrals; use the formula.)

    (d) Show that the time-average power into the load, P, is equal to the power dissipated in the resistor, P_R.

    (e) Calculate the peak and time-average magnetic energy stored in L.

        Ans: (a) $= 1.756 \cos(120\pi t - 14.8°)$; (b) 0.967; (c) 144 W; (e) 0.101 J peak

SECTIONS 5.2.1–5.2.3

P5.7. For the circuit shown in Fig. P5.5, find:

    (a) The complex power.

    (b) The real and reactive power, giving correct units.

    (c) The apparent power, giving correct units.

    (d) Draw a power triangle for this circuit.

    (e) Verify Eq. (5.30) for this circuit.

        Ans: $\underline{S} = 357\underline{/-26.7°} = 320$ W $- j161$ VAR, $W_{ep} = 64.1 \mu$J

P5.8. An electric motor is monitored with an ammeter, voltmeter, and wattmeter, which indicate 5.5 A, 120 V, and 623 W. Assume 60 Hz.

(a) Draw a phasor diagram of the voltage and current, assuming the voltage at zero phase.

(b) What is the reactive power to the motor, including the proper units?

(c) To improve the power factor of the motor, a capacitor is hung directly across the motor terminals. This capacitor will draw leading current and can neutralize the lagging component of the motor current. What value of capacitance will give unity power factor?

Ans: (b) 218 VAR; (c) 40.1 μF

P5.9. A $\frac{1}{3}$-hp motor in a washing machine (120 V) has an efficiency of 78% at full load (rated output power) and a power factor of 0.82 lagging. Find the line current, the reactive, and the apparent power to the motor. Draw a phasor diagram. What capacitor should be placed in parallel with the motor to improve the power factor to 0.95 lagging?

Ans: Current = 3.24 A; reactive power = 223 VAR; C = 21.8 μF

P5.10. For the circuit shown in Fig. P5.10, the wattmeter reads 1200 W, the ammeter reads 10.0 A, and the voltmeter reads 40 V.

(a) What is R?

(b) What would be the peak value of the source voltage, V_s?

Ans: (a) 8.00 Ω; (b) 175.8 V

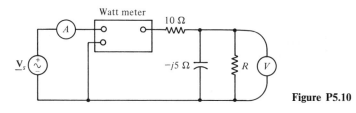

Figure P5.10

SECTION 5.2.4

P5.11. For the circuit shown in Fig. P5.11, the voltage and current are

$$v(t) = 120\sqrt{2}\ \cos\ (120\pi t - 30°)$$

$$i(t) = 5\sqrt{2}\ \cos\ (120\pi t - 60°)$$

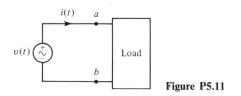

Figure P5.11

(a) Find the complex power into the load, $\frac{1}{2}\mathbf{V}\mathbf{I}^*$.

(b) Draw a power triangle showing the numerical values of the real, apparent, and reactive powers.

(c) Assuming that the load contains no electric energy storage, what is the peak value of the magnetic energy stored in the load?

(d) What value of capacitor connected between a and b will make the power factor to be unity, as seen by the source?

SECTION 5.2.5

P5.12. For the circuit shown in Fig. P5.12, what should be the load impedance to draw maximum power from the circuit in the box? Give component values, not just impedance values. Assume a series circuit for the load. *Hint:* Convert the source to a Thévenin equivalent circuit.

                                   Ans: A 5-Ω resistor in series with a 2-mH inductor

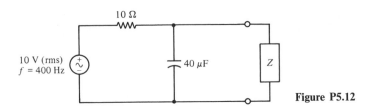

Figure P5.12

P5.13. The output impedance of a typical 120-V appliance wall outlet would be $0.5 + j0.5$. What is the theoretical available power from the outlet? Why would it be a bad idea to actually try to get that much power out?

SECTIONS 5.3.1–5.3.2

P5.14. An ideal transformer has 100 turns on the primary and 300 turns on the secondary. Find the primary and secondary currents for the circuit shown in Fig. P5.14.

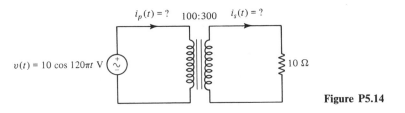

Figure P5.14

P5.15. For the circuit shown in Fig. P5.15, the load impedance is fixed as $100\ \Omega\ \|-j150\ \Omega$, and the source output impedance is fixed at $10\ \Omega\ \|j10\ \Omega$. Maximum power transfer is to be achieved to the load with an ideal transformer with a turns ratio of $1:n$ and a "turning capacitor" on the source side, represented by its reactance $-jX_C$. Find the values for n and X_C that achieve maximum power transfer.

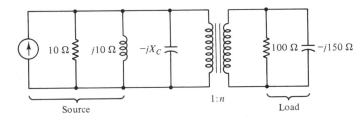

Figure P5.15

P5.16. Power is generated at 480 V and consumed at 120 V. Assume an ideal generator, but the load is at some distance from the generator and hence line losses are appreciable, 1 Ω total for both wires. The equivalent load impedance would draw 10 kW at unity p.f. if 120 V were provided. Calculate the load voltage, the load power, and the efficiency of the transmission system under the following schemes:
  (a) A 4 : 1 transformer at the generator
  (b) A 4 : 1 transformer at the load
  (c) A 1 : 4 transformer at the generator and a 16 : 1 transformer at the load

SECTION 5.4.1

P5.17. Figure P5.17 shows a two-phase power system.
  (a) Prove that such a system produces a smooth flow of power.
  (b) What would be the power in the load if a voltmeter measured 200 V between line A and line B and R were 5 Ω.

$$v_{AN} = V_p \cos(\omega t + 90°)$$
$$v_{BN} = V_p \cos(\omega t)$$

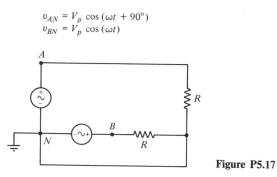

Figure P5.17

P5.18. Pick two values of ωt in Eq. (5.52) and show by direct calculation that the time-varying terms cancel.

P5.19. For the simple generator shown in Fig. 5.33, what should be the rotation speed (rpm) for 60-Hz output power? If the generator had two sets of magnetic poles, arranged at right angles in space, what should be the rotation speed for 60-Hz power?

P5.20. On Fig. 5.40 develop a delta connection by first connecting a to b'. Draw a phasor diagram of the resulting system. Is the phase rotation ABC or ACB?

P5.21. Three 230-V (rms) generators are connected in a three-phase wye configuration to generate three-phase power. The load consists of three balanced impedances, $\underline{Z}_L = 3 + j1.5$ Ω, connected in delta. Find the line current an ammeter would measure, the apparent power, and the real power to the load. What is the phase angle between $\underline{I}_A$ and $\underline{V}_{AB}$, assuming ABC rotation?

P5.22. A balanced, wye-connected three-phase load is shown in Fig. P5.22. The current in line A is $\mathbf{I}_A = 10\sqrt{2}\ \underline{/0°}$. The voltage from b to the neutral point is $\mathbf{V}_{bn} = 100\sqrt{2}\ \underline{/-90°}$.

  (a) Find the line-to-line voltage that a voltmeter would read.

  (b) Find the real and reactive power into the entire three-phase load.

  (c) Determine R and L, assuming 60-Hz operation.

  Ans: (a) 173 V; (b) P is 2.6 kW and Q is $+1.5$ kVAR; (c) R is 8.67 Ω and L is 13.3 mH

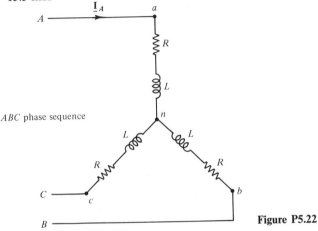

Figure P5.22

P5.23. Demonstrate the equivalence of the delta and wye circuits in Fig. 5.51. Assume an impedance $Z_\Delta\ \underline{/\theta}$ for the delta and an impedance $Z_Y\ \underline{/\theta}$ for the wye and compute the complex power for each. Equate these powers and confirm the 3 : 1 ratio shown in Fig. 5.51.

P5.24. Three identical resistors are placed in a wye configuration and draw a total of 100 W from a three-phase source. What power would the same resistors draw if placed in delta?

P5.25. Using delta–wye transformations, determine the total power given to the delta and wye loads in Fig. P5.25, not counting the losses in the 0.3-Ω resistors, which represent losses in the connecting wires. *Hint:* The neutrals of two balanced wye loads will have the same voltage and hence may be considered as connected.

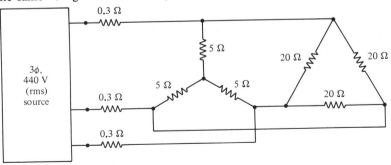

Figure P5.25

Application Problems

A5.1. In Fig. A5.1 we show a transformer with three windings. The equations relating voltages and current can be shown to be

$$\frac{v_1}{n_1} = \frac{v_2}{n_2} = \frac{v_3}{n_3}$$

$$n_1 i_1 = n_2 i_2 + n_3 i_3$$

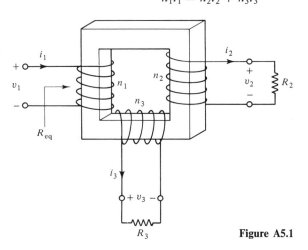

Figure A5.1

(a) Show that power is conserved.
(b) Find the equivalent resistance, R_{eq}, as shown in Fig. A5.1, with load resistors R_2 and R_3 attached.
(c) From the result in part (b), are the two load resistors effectively in series or parallel, as seen from the input?

A5.2. With the circuit shown in Fig. A5.2, there is no value of the turns ratio, n, which will perfectly "match" the load to the source impedance, in the sense of tuning out all reactance and at the same time making the resistors match. However, there is still an optimum value of n which maximizes the power in the load. Find that value of n.

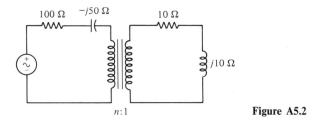

Figure A5.2

A5.3. The circuit shown in Fig. A5.3 is the Thévenin equivalent circuit of a loop antenna operating at a frequency of 570 kHz, at the bottom of the AM radio band. Assuming that the input circuit of the radio consists of a capacitor in parallel with a resistor, what values of R and C will extract maximum power from the antenna. *Hint:* Convert the

Thévenin to a Norton equivalent circuit. Then create a resonance and match the resistors, that is, make them equal for maximum power to the load.

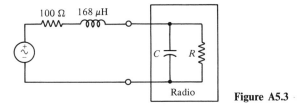

Figure A5.3

6 INTRODUCTION TO ELECTRONICS

6.1 WHAT IS ELECTRONICS?

6.1.1 Electronics and Information

Back in Chapter 1, we distinguished between the two branches of electrical engineering, power and electronics, through their differing uses of electrical energy. In the power industry, electrical energy is used for its own sake: energy is generated from a primary source such as coal or water power, transported long distances with transmission lines, and then converted into heat, illumination, mechanical work, or some other useful form. In electronics, electrical energy is used to symbolize, transport, and process information. Think of a telephone system, the radio and TV industries, the many uses of computers, or a traffic control system—all use electrical energy to convey, process, and use information.

In electronics, voltage and current become electrical signals; that is, they signify something else, as the root word "sign" suggests. The voltage at the appliance outlet in your room delivers no information, except perhaps that you have faithfully paid your electric bill. By contrast, the voltage generated when you speak into a telephone is a signal because it reproduces the acoustic vibrations in the air, these sounds have meaning and information is exchanged in the conversation.

Chapters 6, 7, and 8 address the main ideas that enliven electronics at the present time. As you will see from the historical survey to follow, electronics is a young enterprise. The current rate of progress in the field makes it difficult to anticipate what the future holds. But there are some major themes which, once understood, will give you a good grasp of the nature of electronics. Some of the important factors we address are (1) expanded understanding and use of the frequency domain, (2) application of the electrical properties of semiconductor materials, (3) utilization of nonlinear effects in circuits, (4) feedback, (5) analog and digital representation of information, and (6) analog and digital computers.

6.1.2 The History of Electronics

Before World War II, electronics had commercial importance primarily in radio broadcasting and in the telephone and telegraph industries. Most people who were active in these fields were educated through experience, including some physicists and electrical engineers who were fascinated by radio. In those days, to study electrical engineering in college meant to study about power: motors, generators, transformers, transmission lines.

World War II changed electronics profoundly. In addition to the obvious need to improve radio communication, one of the Allies' top-secret war projects focused on electronics. Everybody knows about the Manhattan Project and the atomic bomb; through the work of the Radiation Laboratory in developing microwave radar, scientists and engineers made an equally important contribution toward ending the conflict. The best technical minds in the country were employed in these two projects and spectacular success crowned both efforts.

The postwar fruit of the Manhattan project was more and bigger bombs, and even today the peaceful application of nuclear technology excites continuing controversy. But widespread and, for the most part, benevolent have been the postwar fruits of the work of the Radiation Laboratory. The cathode ray tubes (CRTs) which were developed as radar displays became TV picture tubes, and soon there were high-fidelity recordings to be enjoyed. Radar techniques developed to detect enemy aircraft allowed commercial aviation to fly in all weather, and you could dial long distance directly and hear the other party without straining.

In the early 1950s the development of the transistor inaugurated the first of several "solid-state" revolutions. The integrated circuit followed and later the microcomputer on a "chip." The limits of the techniques of microelectronics continue to defy the imagination of the most far-seeing sages of the electronics industry.

The idea of electronic computers was conceived before the war, but only modest applications were made, even during the war. An analog computer was developed to control naval gunnery, and some simple "logic" circuits emerged, but practical computers came after the war. The first digital computers were based on vacuum-tube technology and by present standards were huge, slow, and awkward to program. But once the idea of the digital computer merged with that of the transistor and later the integrated circuit, the development of computers became spectacular, and continues so to this day. Your author is typing this book directly into a computer, and although you may not have a personal "word processor" for your term papers and lab reports, certainly coming generations will.

6.1.3 The Nature of Electronics

Electronics is a big bag of tricks. Unlike circuit theory, which submits to an orderly and logical development, electronics employs a diversity of devices, techniques, and processes. It is hard to learn a little bit about electronics, because electronic circuits are complicated. Did you ever, for example, examine the circuit diagram (the schematic) for a TV set? If one fell into your hands at this moment and you set about to apply your newly gained knowledge of Ohm's law, Kirchhoff's laws, and concepts such as impedance and Thévenin equivalent circuits, you would not make much progress in understanding such a circuit diagram.

Our goal in this second part of the book is to investigate the major themes currently active in electronics. We will examine a few of the tricks in the electronics bag at the present time, but only enough to impart a beginning understanding of the nature of the subject. Do not expect to be able to fix your radio or interface a microcomputer after you master this part of the book. But you should understand how these systems work.

6.2 RECTIFIERS AND POWER SUPPLIES

6.2.1 The Ideal Diode

Nonlinear Devices.   The circuit theory we developed in Chapters 1–5 describes the properties of circuits that are linear. The defining equations for resistors, inductors, and capacitors describe linear relationships between voltage and current. Also linear are Kirchhoff's laws describing conservation of energy and charge in electrical circuits. Many of the techniques that we developed are based on the linear properties of the defining equations and Kirchhoff's laws—superposition and Thévenin equivalent circuits being obvious examples.

Electronics uses resistors, inductors, and capacitors also, and certainly Kirchhoff's laws are still important and are still linear. But electronic circuits also employ many devices that are nonlinear in their characteristics. Diodes, transistors, and silicon-controlled rectifiers offer examples of such nonlinear electronic devices. The nonlinearities are not undesirable hindrances to the use of these and other devices; rather, they are useful because of their nonlinear properties.

Because of the importance of nonlinear devices in electronics, we employ more graphical analysis than we used in circuit theory. Many of the circuit solution techniques we develop are based on graphical methods, and much of the information about device characteristics is given in graphical form. In Fig. 6.1 we show the graphical characteristics of a resistor, together with the extremes of an open circuit $(R = \infty)$ and a short circuit $(R = 0)$. Of course, the graphical form of Ohm's law is a straight line.

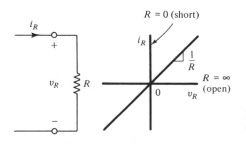

Figure 6.1 Symbol and graphical definition of a resistor.

Ideal Diode Characteristics.   In Fig. 6.2 we show the circuit symbol for a diode, with the associated graphical characteristic, which is strongly nonlinear. The diode characteristic divides into two regions: the vertical region is called the forward bias region and the horizontal region is called the reverse bias region. Comparison with Fig. 6.1 suggests that the diode acts as a short circuit in the forward bias region and an open circuit in the reverse bias region. The behavior of the device in terms of current flow is as follows: As long as the current is positive in the direction of the arrow of the circuit symbol, the diode acts as a short circuit and the current flows without hindrance. The diode is "ON." When the voltage is positive in the direction opposite to the diode arrow, however, the diode acts as an open circuit and no current can flow. The diode is "OFF." Thus the current can flow in the direction of the arrow but cannot flow against the arrow.

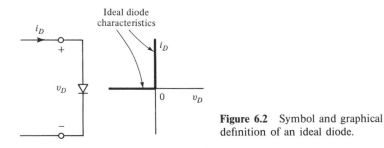

Figure 6.2 Symbol and graphical definition of an ideal diode.

The diode can be considered as the electrical equivalent of the mechanical ratchet, such as is used to tighten the net on a tennis court. The mechanical ratchet allows motion or rotation in one direction only. Similarly, the diode allows charge motion in one direction only. Another analog would be a check valve which allows fluid flow in only one direction.

The diode can also be considered a voltage-actuated switch. As long as the input voltage is positive (with the + at the top of the arrow) the switch is closed, current flows, and the diode is ON. But once the polarity of the voltage reverses and is positive at the other end of the diode, the switch opens, no current flows, and the diode is OFF.

The diode characteristic we have described is that of an ideal diode. Real diodes depart from this ideal characteristic, as will be detailed in Section 6.3.4. But many of the common applications of the diode can be understood in terms of this ideal characteristic;

hence we will move to some of these applications before investigating the physical processes in semiconductor diodes.

6.2.2 Rectifier Circuits

Electronics and DC.   At the beginning of Chapter 4 we mentioned the struggle between dc and ac to see which form of electrical power would dominate the fledgling electrical power industry. We saw that Edison was wrong, Tesla was right, and ac came to be the common mode for the generation, distribution, and consumption of electrical power.

With the ascendency of electronics, however, dc has made a comeback, for almost all electronic circuits require dc power. You might suppose, therefore, that there would be a battery in every electronic device, but this is untrue. Batteries are expensive, heavy, short lived, bulky, and filled with nasty chemicals likely to leak out and destroy flashlights, tape recorders, and the like. Batteries are thus undesirable components and are avoided by designers except where portability is essential.

For this reason, most electronic equipment contains a *power supply* circuit. The function of the power supply is shown in Fig. 6.3. When you plug in and turn on your TV set, the ac power enters the power supply section of the electronic circuit, where it is converted to dc power. From there the dc power flows to other parts of the circuit. Such power supply circuits use the nonlinear properties of diodes.

Figure 6.3   Most electronic circuits require a power supply.

The Half-Wave Rectifier.   Power supplies use diodes to convert ac to dc in "rectifier" circuits; the diodes are said to "rectify" the ac. Figure 6.4 shows a basic rectifier circuit. The ac voltage is represented as a sine function, the rectifier is an ideal diode, and the load is represented as a resistor, although in practice the load would be the electronic circuits which require dc power.

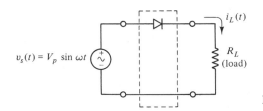

Figure 6.4   Half-wave rectifier circuit.

The rectifying effect of the diode is shown in Fig. 6.5. When the input voltage is positive, the diode turns ON and current flows with a sinusoidal shape, but when the input voltage is negative the diode turns OFF and no current flows. The resulting current flows in spurts, and the voltage across the load is simply that portion of the sinusoidal input which is positive. The diode blocks the negative part of the ac. .

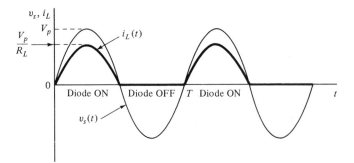

Figure 6.5   Half-wave rectifier waveforms.

Admittedly the output of this simple power supply is a poor approximation of pure dc power. The output, however, does contain a dc component. Indeed, if we define the dc portion of the output as the time average (see Section 5.1.2), we have a dc component of

$$I_{dc} = \frac{1}{T} \int_0^T i_L(t) \, dt = \frac{1}{T} \int_0^{T/2} \frac{V_p}{R_L} \sin \omega t \, dt$$

$$= \frac{V_p}{\pi R_L}$$

The Full-Wave Rectifier.   It is possible to clean up or *filter* the output of the half-wave rectifier in Fig. 6.5 to better approximate a true dc. Before presenting filtering circuits, however, we will discuss another rectifier circuit. The circuit in Fig. 6.6(a) is called a full-wave bridge rectifier. This circuit uses four diodes to accomplish rectification in the following way. When the source voltage is positive, the current tends to flow through the rectifier from a to d. Diode D_1 comes ON but D_2 cannot accommodate current from a to c and will turn OFF. From b the current must flow through the resistor because D_3 will

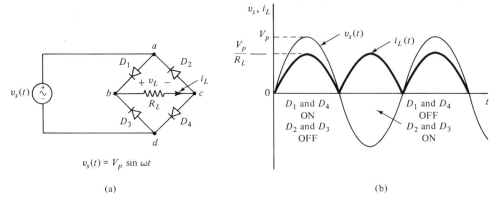

Figure 6.6   (a) Bridge full-wave rectifier circuit; (b) waveforms for the bridge full-wave rectifier.

not permit current flowing directly from b to d. Finally, D_4 will turn ON and the current will flow from c to d. Thus when the source voltage is positive, positive current flows out of the + of the source through D_1, through the load resistor, and finally through D_4. Diodes D_2 and D_3 are OFF during this part of the cycle. When the source voltage is negative, D_2 and D_3 turn ON while D_1 and D_4 turn OFF. Thus during the second half of the cycle, positive current flows out of the − terminal of the source, through D_3 from d to b, through the load resistance, and back to the source through D_2. The two paths for the current are shown in Fig. 6.7. The current flows through the load from b to c during both parts of the cycle, thus the current through the load is as shown in Fig. 6.6(b). The current again flows in spurts, but the full-wave rectifier leaves no idle time between spurts as does the half-wave rectifier.

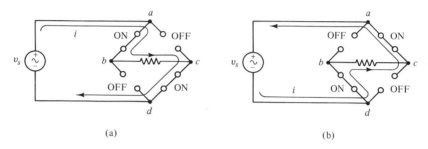

Figure 6.7   Current paths for bridge rectifier: (a) v_s positive; (b) v_s negative.

We can consider the diodes to act as switches which are activated by the voltage. When the source voltage is positive, D_1 and D_4 are turned ON and thus b and c are connected to the + and − terminals of the source, respectively. When the source voltage becomes negative, D_2 and D_3 are turned ON and hence b and c are again connected to the source, this time with b connected to the − and c to the + terminal. Thus b is automatically connected to the physically positive terminal of the source and c is automatically connected to the physically negative terminal. This automatic switching action occurs regardless of the shape of the source voltage—the shape can be sinusoidal, triangular, or an unpredictable communication signal in a radio circuit. The effect of the diode bridge is thus to produce across the load an output voltage which is the absolute value of the input voltage: $v_L(t) = |v_s(t)|$.

As a power supply circuit the full-wave rectifier does better than the half-wave rectifier, but the output is still a poor approximation to a pure dc voltage. By inverting the negative portions on the input voltage, the circuit doubles the dc component in the output; hence the dc component in the output voltage is

$$I_{\text{dc}} = \frac{1}{T} \int_0^T i_L(t)\, dt = \frac{2}{T} \int_0^{T/2} \frac{V_p}{R_L} \sin \omega t\, dt$$

$$= \frac{2V_p}{\pi R_L}$$

A full-wave rectifier circuit using a transformer and two diodes is shown in Fig. 6.8. The transformer secondary is center tapped to supply identical but opposite voltages to the two diodes. Each diode acts as a half-wave rectifier: D_1 supplies the positive part of v_{de} and D_2 supplies the positive part of v_{fe}, as shown in Fig. 6.9. The transformer gives this circuit the versatility to supply any desired dc voltage, depending on the turns ratio.

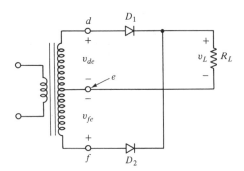

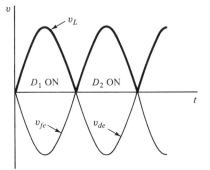

Figure 6.8 Full-wave rectifier using a center-tapped transformer.

Figure 6.9 Waveforms for the full-wave rectifier.

Filtering Out the Ripple. The three rectifier circuits we have described create a dc component in their outputs from an ac input. We may describe their outputs as a desired dc component plus an undesired *ripple*, as shown in Fig. 6.10. We need to eliminate, or at least greatly reduce, the ripple. We need a filter, as in Fig. 6.11, to remove the undesirable ripple component from the output of the rectifier. In a later chapter we will consider filters generally; here we introduce the simplest of filters—we merely connect a capacitor across the load, as shown in Fig. 6.12. The purpose of the capacitor is to stabilize the voltage across the load resistor.

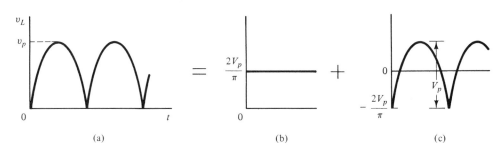

(a) (b) (c)

Figure 6.10 The ripple is the undesirable portion of the output. (a) Rectifier output; (b) dc component; (c) ripple component.

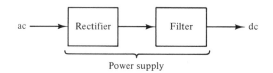

Power supply

Figure 6.11 A filter is used to reduce the ripple.

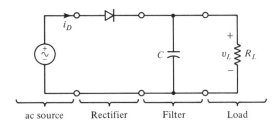

Figure 6.12 Half-wave rectifier with a capacitor filter.

Let us consider that the source is $V_p \sin \omega t$, and thus is increasing through zero volts at $t = 0$, as shown in Fig. 6.13. As the voltage increases, the diode turns ON and current flows through the load as before. Current also flows through the capacitor and charges it to the peak value of the input ac voltage. The first spurt of current is relatively large in Fig. 6.13 due to the initial charging of the capacitor. After its peak, the input voltage drops rapidly. If the voltage of the capacitor were to follow this voltage, a rapid discharge would have to occur. However, the diode prevents discharge through the input source and turns OFF when the input voltage drops below the voltage on the capacitor because at that moment the diode becomes reverse biased. Hence the filter capacitor and the load resistance become disconnected from the source, which continues with the negative part of its waveform. The capacitor thus discharges through the load resistor, as shown in Fig. 6.14.

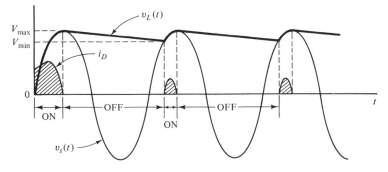

Figure 6.13 Waveforms for the half-wave rectifier with a capacitor filter.

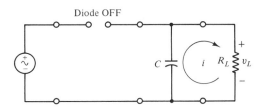

Figure 6.14 When the diode is off, the capacitor discharges through the load.

The discharge of a capacitor through a resistor is a problem we learned to solve in Chapter 3. To write the voltage as a function of time, we need to know the time constant, the initial value, and the final value. The time constant is $R_L C$, the initial value is V_p, and the final value would be zero if the discharge were allowed to go on forever. Thus the voltage of the load (and the capacitor) during the discharge period, when the diode is OFF, would be

$$v_L(t) = 0 + (V_p - 0)e^{-(t/R_L C)} = V_p e^{-(t/R_L C)}$$

where t is measured from the peak.

To function well as a filter, the time constant $(R_L C)$ should be much longer than the period of the input ac voltage (usually $\frac{1}{60}$ s). The load voltage thus decreases only slightly as the input voltage swings negative and again approaches its peak. Figure 6.13 exaggerates the decrease due to the discharge from what it would be in practice. During this time the diode remains OFF until the increasing input voltage becomes equal to the decreasing load voltage. As the input voltage again exceeds the load voltage, the diode turns ON and current again flows through the diode. Most of the current goes to the capacitor, replenishing the charge lost during the discharge part of the cycle. After the initial pulse of current which charges the capacitor, current flows through the diode only during these brief recharging periods, as shown in Fig. 6.13.

If the load voltage decreases only slightly during the period of the ac waveform, as shown in Fig. 6.13, the output voltage of the power supply is approximately equal to the peak value of the input ac waveform. Thus the first benefit of the filter capacitor is to increase the dc output from V_p/π to V_p. Second, the filter capacitor greatly reduces the ripple voltage. For the case we are considering $(R_L C \gg \text{period})$, the exponential decrease of the load voltage is well approximated by a straight line, the leading terms in a series expansion of the exponential.

$$v_L(t) = V_p e^{-(t/R_L C)} = V_p\left(1 - \frac{t}{R_L C} + \frac{(t/R_L C)^2}{2} - \cdots\right) \tag{6.1}$$

where t is measured from the peak. Thus from its maximum value of V_p the voltage decreases to a minimum value of approximately

$$V_{\min} \approx V_p\left(1 - \frac{T}{R_L C}\right) = V_p\left(1 - \frac{1}{f R_L C}\right) \tag{6.2}$$

at the moment when the diode turns ON and permits recharging of the capacitor. In Eq. (6.2), f represents the ac frequency and is the reciprocal of the period. Consequently, the peak-to-peak ripple is

$$V_r = V_{\max} - V_{\min} = \frac{V_p}{f R_L C} \tag{6.3}$$

We note from Fig. 6.13 that a more accurate approximation to the dc component of the filtered output would be the average between the maximum and minimum voltages:

$$V_{dc} \approx \frac{V_{\max} + V_{\min}}{2} = V_p\left(1 - \frac{1}{2f R_L C}\right) \tag{6.4}$$

An Example.   We shall illustrate with the circuit in Fig. 6.15. The input ac has an rms voltage of 12 V at 60 Hz. Thus the peak value, and hence the maximum of the output voltage, is $12\sqrt{2} = 17.0$ V. The time constant is

$$\tau = R_L C = 100 \times 2000 \times 10^{-6} = 0.2 \text{ s}$$

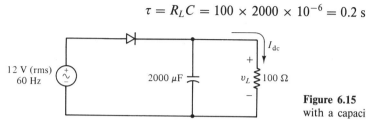

Figure 6.15   Half-wave rectifier circuit with a capacitor filter.

which is long compared to the period of $\frac{1}{60}$ s. The exponential decrease of the output voltage during the period when the diode is OFF can be written

$$v_L(t) = 17.0 e^{-(t/0.2)} \approx 17.0 \left(1 - \frac{t}{0.2}\right)$$

where we have measured time from the peak. At the time of the next peak, the output voltage will have decreased to approximately

$$v_{\min} \approx 17.0 \left[1 - \frac{1}{60(0.2)}\right] = 15.6 \text{ V}$$

Thus the ripple voltage is 1.4 V peak-to-peak and the dc (time-average) voltage at the load is the average between the maximum and minimum, 16.3 V. The dc current in the load is 16.3 V divided by the load resistance, 163 mA.

Better Filters.   We have investigated the benefits of the simplest possible filter. The performance of this filter is adequate for many applications, but high-quality power supplies employ more sophisticated filters. Some filters add inductors and additional capacitors to reduce the ripple; others employ electronic circuits to cancel the ripple.

6.3 THE pn-JUNCTION DIODE

6.3.1 The Reasons for Discussing the Principles of Diode Operation

Diodes Are Important.   The diode finds many uses as an electronic component. We have seen how diodes are used in power supplies to convert ac to dc, but this is only one of the many uses that diodes serve. Diodes are used extensively in analog electronic devices such as radios and audio systems, and are even more important in digital electronic devices such as computers and digital watches.

The diode plays an important role in the study of electronics because it is a simple electronic device and hence presents a good starting place in exploring the bag of electronic tricks. The diode is also important in the teaching of electronics because it offers our first opportunity to investigate the physical basis of semiconductor electronics. This has importance because the development of electronics based upon semiconductor processes has produced essentially all the electronic equipment we require and enjoy in our society. This section describes many of these semiconductor processes.

The Main Question.   What is a physical diode, and how does it work? The answer to that question will require the discussion of a sizable number of physical processes that occur in solids. Your author has a genuine problem in helping you understand how a diode works. It is commonly asserted, and reasonably so, that one must explain the unknown in terms of the known. But your author must explain the unknown (how a diode works) in terms of other unknowns (semiconductor processes such as holes, drift currents, uncovered charges, depletion regions, and the like). Or, put another way, we are required to consider a large number of physical processes before we can understand how a simple *pn*-junction diode operates. Whereas understanding of the diode itself might not merit the effort, our investment is worthwhile because modern electronics is founded on the manipulation of charges in semiconductor materials through such physical processes. In order to understand electrical engineering, you have to understand these details of solid-state phenomena.

6.3.2 Semiconductor Processes and the pn Junction

The Crystalline Nature of Solids.   Matter is considered to exist in four states: solid, liquid, gas, and plasma. These may be understood in terms of the interactions between individual atoms or molecules comprising the matter in question. In a solid, the atoms remain in a fixed position relative to each other. Often the atoms exist in a regular crystalline order, held together by shared electrons in covalent bonds. Figure 6.16 shows a two-dimensional representation of such a lattice. The open circles represent the nuclei, the dots represent the valence electrons, and the curved lines represent the bonding effect of the electrons. The electrons are negatively charged and the nuclei are positively charged. Of course, there are many more electrons associated with each nucleus, but we have not represented the inner shells because these do not enter into the bonding process or the semiconductor processes we are investigating. We have represented the case where there are four electrons in the outer shell, which is typical of a semiconductor such

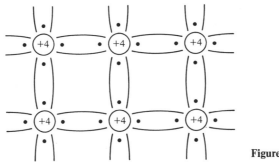

Figure 6.16 Insulator.

as silicon. Because the individual atoms are electrically neutral, all the charge would be neutralized by the inner shells of electrons except for a positive charge of four times the electronic charge, which is thus the effective charge of the nucleus. This is indicated by the +4 in Fig. 6.16.

Insulators, Conductors, and Semiconductors.   At zero absolute temperature, solids are either electrical conductors or insulators. The material represented by Fig. 6.16 would be an insulator because all the electrons are involved in the bonding. If such a material were placed in an electrical circuit, no current would flow because no electrons are free to move. In a conductor, each atom has one excess electron which is not involved in the bonding. Such excess electrons are known as conduction electrons and they move freely in response to electrical forces, such as a battery might exert on an electrical circuit. Metals such as copper and aluminum are thus good electrical conductors.

Thermal energy has a strong effect on the conducting properties of solids. Thermal energy is distributed throughout the electrons and nuclei of the materials and this energy is stored, among other ways, in the physical movement, or vibration, of the electrons and nuclei. The vibrations may cause some of the bonding electrons to break loose from their bonding positions to become conduction electrons.

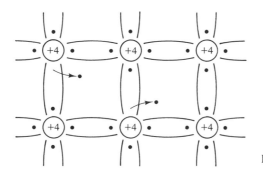

Figure 6.17   Semiconductor.

Materials that have no excess electrons and retain all their electrons in bonds at normal temperatures remain good insulators. Carbon in a diamond crystalline order and many plastics and ceramics furnish examples of good insulators at normal temperatures. In other materials which are good insulators at absolute zero, however, the electrons are not tightly held in the bonds, and in these materials some of the electrons escape their bonds at normal temperatures to become conduction electrons. We have represented this condition in Fig. 6.17 with two electrons out of their bonding positions. Such a material is known as a semiconductor because it becomes a conductor at normal temperatures due to the electrons that have escaped from their bonding positions.

Holes and Hole Movement.   When an electron leaves its bonding position, it leaves behind a vacant position, which is called a *hole*. We may think of the hole as having a positive charge because the nucleus adjacent to it now has charge not neutralized by the bonding electrons. Holes are also free to move under the influence of electrical forces because other bound electrons may move into the associated vacant location. For example, if in Fig. 6.17 there were an electrical force tending to move electrons from left to right, the conduction electrons would move in response to such a force. But the electron next to the vacant position will also tend to move to the right and may leave one bonding position for a vacant position toward the right. Thus we can envision the process represented in Fig. 6.18. First the electron breaks its bond and becomes a conduction electron, thus creating a hole. Because there is an electrical force tending to move electrons toward the right, there will also be a tendency for electrons remaining in bonds to go left to right; hence the transitions we have marked (2) through (5) are likely to follow. The hole clearly moves toward the left; in effect a positive charge moves right to left because there is excess positive charge associated with the hole. If a semiconductor were placed in a circuit where the voltage created a current, the current would be carried by both holes and electrons, which are said to be *carriers*. Note that the currents carried by the movement of holes and electrons are additive: holes moving toward the left carry a positive current toward the left; and electrons moving to the right carry a negative current toward the right, which is a positive current toward the left. In a typical semiconductor, the holes move about one-third as fast as the electrons and hence carry about one-fourth the total current.

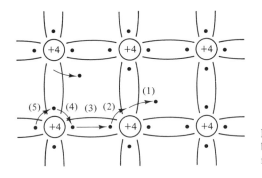

Figure 6.18 The orderly movement of bound electrons produces hole movement.

       The Importance of Thermal Energy.   We have stated that hole–electron pairs are created because of thermal energy, and we have shown how electrons changing bonding positions due to thermal vibrations act like mobile positive charges. But we have described these processes as if they were orderly and sedate, which they are not. To get a feeling for what is happening in a semiconductor, you have to understand that the thermal processes are rapid and violent.

       In order to hint at the violence of these processes, we shall compute a representative velocity of a conduction electron using ideal gas law theory. From statistical thermodynamics we learn that, in a thermal system in equilibrium, the energy will distribute throughout the available energy storage modes of the system, and the energy per mode will be proportional to the temperature. The constant of proportionality is one-half of Boltzmann's constant, $k = 1.38 \times 10^{-23}$ J/K. Consider an "electron gas" consisting of the electrons moving around as conduction electrons in a semiconductor material. They can store kinetic energy in three directions of motion; hence their total kinetic energy will be

$$\tfrac{1}{2}m_e \times \text{time average of } (u_x^2 + u_y^2 + u_z^2) = 3 \times \tfrac{1}{2}kT$$

$$u = \sqrt{\frac{3kT}{m_e}} = 117 \text{ km/s} \qquad (6.5)$$

where u is the Pythagorean sum of the x, y, and z components of velocity and would be the rms speed of the electron in three-dimensional space. When we set T to be 300 K and the mass of the electron, m_e, to its value of 9.11×10^{-31} kg, we calculate the rms velocity to be 117 km/s. Of course, this calculation is not highly accurate because the electrons are not an ideal gas, but the conclusion offered is valid—those electrons are really moving around. When you consider the small physical scale of the lattice structure, you must conclude that the time processes occur quite rapidly. For pure silicon at 300 K, there are approximately 1.5×10^{16} conduction electrons/m^3 and there are approximately 10^{22} hole–electron pairs/m^3 created and eliminated through recombination per second through thermal action.

Hence emerges a picture of violent, random thermal motion of conduction electrons in a semiconductor. Many hole–electron pairs are created and many recombinations occur during short periods of time.

Doping.   We can increase the concentration of carriers by adding small amounts of *impurities* to the pure (called *intrinsic*) semiconductor. Consider that we have intrinsic silicon, which has four bonding electrons per atom. If we add a small amount of an element which has five electrons in its valence shell, such as phosphorus, the impurity nuclei will bond into the lattice with one electron left over. This electron will be a conduction electron, as shown in Fig. 6.19, and there will also be one additional positive charge fixed into the lattice of nuclei because of the additional charge of the nucleus of the impurity atom. This process of adding impurities is called *doping*; Fig. 6.19 shows *n*-type doping, so-called because of the additional negative carriers. This would make an *n*-type semiconductor because it has an increased concentration of electrons. The impurity in this case would be called a *donor* atom because it donates an additional conduction electron to the semiconductor.

Similarly, if we add an impurity with three electrons in its valence shell, we would

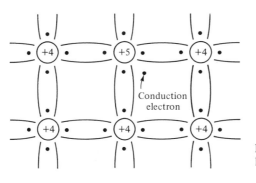

Figure 6.19   An *n*-type semiconductor has extra conduction electrons.

create a hole for each impurity atom, as shown in Fig. 6.20. Here we have shown the hole and we have indicated that the nucleus is deficient one positive charge with a $+3$. If the hole were filled by a conduction electron, that region would in effect have a negative charge built into the lattice structure of the semiconductor. Such an impurity is called an *acceptor* atom because it accepts an electron from the conduction electrons in the semiconductor and an extra hole is created. A semiconductor that is doped with acceptor atoms is called a *p*-type semiconductor because it has an excess of holes, which act like mobile positive charges.

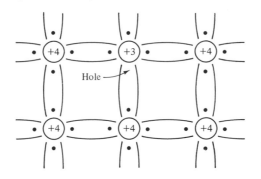

Figure 6.20 A *p*-type semiconductor has extra holes.

The electrons donated by the donor atoms and the holes created by the acceptor atoms do not remain near their associated nuclei but participate in the random thermal processes of all carriers. By doping the semiconductor, we have the ability to increase the carrier concentration (either holes or electrons) and to control what type of carriers will dominate the conduction processes in the semiconductor.

Drift and Diffusion Currents.   We have already described what happens when a semiconductor is placed in an electric circuit. A current would flow through movement of both holes and electrons. If the semiconductor were *p*-type, most of the current would be carried by the holes. If the semiconductor were *n*-type, most of the current would be carried by the electrons. The type of current we have described is called a *drift current* because the carriers drift in a certain direction as dictated by an external voltage. We may think of this type of current as being caused by an orderly process.

However, if we had an excess of carriers in a certain region, they would tend to distribute uniformly throughout the material because of their random thermal movement. Such a flow is known as a *diffusion flow*. A diffusion flow occurs when a bottle of a smelly chemical is opened in a room filled with calm air. People near the open bottle would smell it first, then those farther away. Eventually, everyone in the room would smell the chemical because the molecules would be distributed uniformly throughout the room by their thermal motion. In a similar way, carriers move away from regions of concentration due to their thermal motion. Such carrier movements are known as *diffusion currents*.

Summary.   An intrinsic semiconductor will have many, but equal numbers, of holes and electrons because thermal energy causes electrons to leave their bonding positions, become conduction electrons, and leave behind a hole. We can create *p*-type or *n*-type semiconductors by doping the pure material with either acceptor or donor impurity atoms. The additional holes or electrons participate in the violent, random thermal motion of the carriers in the material. The carriers form currents either by the orderly process of a drift current or by the disorderly process of diffusion. There are also charges bound into the lattice structure associated with the acceptor or donor nuclei. The acceptor atoms act as stationary negative charges because their nuclei are deficient one positive charge relative to the other nuclei in the vicinity. The donor atoms act as stationary positive charges because their nuclei have one additional positive charge relative to the surrounding nuclei. The semiconductors are neutral in total charge because the additional mobile carriers neutralize the bound charges of the impurity atoms. All of these processes occur in a *pn* junction, to which we now turn.

6.3.3 The pn Junction

Structure.   A *pn* junction is formed at the boundary between regions of *p*-type semiconductor and *n*-type semiconductor, as shown by Fig. 6.21. When such a junction is formed, the sequence of events diagrammed in Fig. 6.22 occurs rapidly.

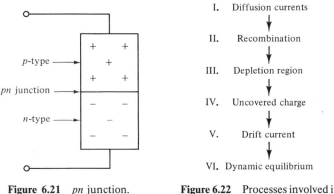

Figure 6.21   *pn* junction.

Figure 6.22   Processes involved in *pn*-junction formation.

I. Electrons in the *n*-type material will diffuse toward the *p*-type material and holes in the *p*-type material will diffuse toward the *n*-type material. These strong diffusion currents will occur because the concentrations are unequal and the carriers are in violent thermal motion.

II. Recombinations will occur immediately as the electrons which have diffused into the *p*-type material find holes to fill and holes which have diffused into the *n*-type material are filled by conduction electrons.

III. Thus there will be a region on both sides of the junction which have deficiencies of carriers due to recombinations. This region is called a *depletion region*. We might anticipate a continual pouring of carriers into this depletion region were it not for another process which comes into play.

IV. On the *p*-type material side of the junction in the depletion region, the acceptor atoms bound into the lattice structure are now *uncovered*; that is, because the electrons from the *n* side have recombined with many of the holes, the deficiency of charges in the nuclei of the acceptor atoms acts as an excess of negative charges fixed in this region. Similarly, the excess positive charges of the donor nuclei will act as positive charges bound into the lattice structure, now uncovered because the electrons that formerly neutralized them have recombined with holes from the *p* side. Thus we have uncovered charges bound into the lattice structure in the depletion region, as shown by Fig. 6.23. This charge distribution will act similar to a capacitor, as shown in Fig. 6.24. Here we have shown a capacitor with a charge placed on it by a battery. The bottom plate of the capacitor will have positive charges and the top will have negative charges. If an electron were in the region between the plates, it would move downward, attracted by the positive charges below and repelled by the negative charges above. Similarly, a positive charge between the plates would move upward. This would be an orderly process and would create a drift current in this region if there were electrons or positive charges between the plates. Similarly, in Fig. 6.23 excess carriers in the depletion region experience forces from the uncovered, bound charges.

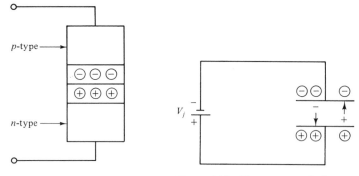

Figure 6.23   Uncovered charges.   **Figure 6.24**   The uncovered charges act like a charged capacitor.

V. These forces would tend to create a drift current upward as electrons are sent back toward the *n*-type material and holes are sent back toward the *p*-type material. Thus we can envision the holes and electrons moving across the junction due to diffusion but moving the other way by the drift current caused by the uncovered charges.

VI. These two processes rapidly reach dynamic equilibrium.

Perhaps a better way to view this dynamic equilibrium is to consider that the uncovered charges constitute an internal battery–capacitor effect which automatically adjusts itself to stop the diffusion currents. Hence we have a dynamic equilibrium between a disorderly, random process (the diffusion process) and an orderly process (the battery–capacitor effect). Due to the uncovered charges, the *pn* junction diode will have a voltage of approximately 0.7 V across its junction, but no current will flow unless the equilibrium is disturbed by an external force.

The Forward Bias Characteristic.   If we add an external battery to the diode, as shown by Fig. 6.25, and if the polarity of the external battery is such as to oppose the internal battery–capacitor effect, the equilibrium is disturbed and current will flow continually through the diode. In this case the battery will cause the holes in the *p*-type material to move downward toward the junction and the electrons in the *n*-type material to move upward toward the junction. Because the internal orderly process tending to prevent the flow of carriers through the depletion region has been partly neutralized by the external battery, current is carried across the junction by the diffusion currents. The current increases dramatically with increases in the externally applied voltage because it is driven by the energetic thermal motion of the carriers. If the external battery has a voltage that exceeds the voltage of the internal battery–capacitor effect, large currents will flow. Because this voltage is only about 0.7 V for a silicon diode, the forward bias region of the diode approximates that of an ideal diode quite well. A germanium diode requires about 0.2 V to cause substantial conduction and hence is more nearly ideal that a silicon diode.

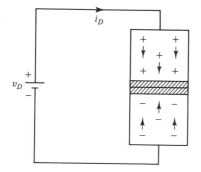

Figure 6.25   Forward-biased *pn* junction.

The Reverse Bias Characteristic.   If the polarity of the external battery reinforces the influence of the internal battery–capacitor effect, the holes in the *p*-type semiconductor will tend to move upward and the electrons in the *n*-type semiconductor will tend to move downward, as shown in Fig. 6.26. The externally applied voltage merely widens the depletion region and strengthens the restraining effect of the internal battery–capacitor effect. For this polarity of the external battery, very little current will flow through the diode and an ideal diode is well approximated. Actually, the reverse-biased *pn* junction

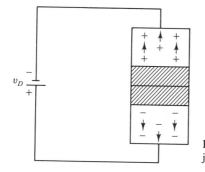

Figure 6.26   Reverse-biased *pn* junction.

behaves as a capacitor due to the charge separation associated with the uncovered charges. In the next section we will give the equation of a *pn* junction diode and see how well ideal diode behavior is approximated by the semiconductor *pn* junction.

6.3.4 The Physical Properties of Real Diodes

The pn-Junction Equation.   The voltage–current characteristic of the *pn*-junction diode, as described in the previous sections, is well described by

$$i_D = I_0 \exp\left(\frac{qv_D}{\eta kT} - 1\right) \tag{6.6}$$

where I_0 = a constant called the *reverse saturation current*; depends on the semiconductor materials, manner of junction formation, and junction size

$q = |e|$, the magnitude of the electronic charge, 1.60×10^{-19} C

k = Boltzmann's constant, 1.38×10^{-23} J/K

T = absolute temperature, K

η = a constant between 1 and 2, called the *ideality factor*, which depends on junction materials and method of formation

Figure 6.27 is a plot of Eq. (6.6) for the parameters $\eta = 1.5$ and $I_0 = 10^{-9}$ A. These would be typical of a silicon diode at room temperature. If you were required to plot a diode characteristic using Eq. (6.6) for a given current range, you would soon discover that it is hard to find a voltage which gives a current in the required range. The exponential function, which dominates the behavior in the forward bias region, rises so rapidly that every voltage you assume leads to a current which is outside the prescribed range, either too big or too small. If you wish to plot the *pn*-junction equation, it works better to invert the equation so that v_D is a function of i_D, and calculate from this equation. The resulting equation is

$$v_D = \eta V_T \ln\left(\frac{i_D}{I_0} + 1\right) \tag{6.7}$$

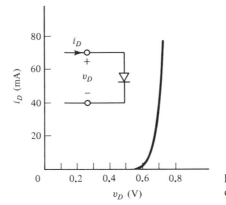

Figure 6.27 Typical current–voltage characteristic for a silicon diode.

where $V_T = kT/q$ is called the *voltage equivalent of temperature* and has the magnitude of about 26 mV at $T = 300$ K.

If you compare the characteristics of a *pn*-junction diode [Fig. 6.27 or Eq. (6.6)] with those of an ideal diode (Fig. 6.2), you might experience some disappointment, for the properties of a real diode appear quite nonideal. That impression is created by our plotting the *pn* junction characteristic on an expanded voltage scale. Figure 6.28 shows the diode characteristics on the same scale as a 1-kΩ resistor; here the *pn*-junction diode shows promise.

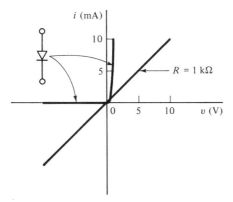

Figure 6.28   Diode characteristic compared with a 1-kΩ resistor.

The properties of the *pn*-junction diode may be summarized in this fashion: Essentially, zero current flows in the reverse bias region. Negligible current flows in the forward bias region until a small threshold voltage is reached, after which the magnitude of the current rises rapidly. Once the current begins to rise, the voltage remains fairly constant. The threshold voltage for Fig. 6.27 is about 0.7 V, which is typical for a silicon diode. A germanium diode has a lower threshold voltage, typically 0.2 V. The *pn*-junction equation [Eq. (6.6)] gives the precise characteristic, but the concept of a threshold voltage is used to simplify analysis for many diode applications. For the rectifier circuits presented ᴄ﹅rlier in this chapter, we may assume that the diodes will have a voltage drop of about 0.7 V when they are ON and thus the rectifier output voltages and currents will be reduced accordingly. When the power supplies produce dc voltages in excess of, say, 10 V, this represents a minor correction. When characterized by a threshold voltage, the voltage drop of the diode is easy to take into account in design and analysis of diode circuits. A threshold voltage also plays an important role in design and analysis of transistor circuits.

Power Limits and Heat Transfer.   Temperature strongly affects the semiconductor processes upon which diode operation depends. In Eq. (6.6) the temperature appears explicitly in the exponential term, but the reverse saturation current, I_0, is also affected by temperature. For this reason, the electronics designer must prevent the diode from getting too hot. Heat is produced by the electrical power given to the diode. The power into a device is

$$p = vi$$

The ideal diode requires no power because in the forward bias region the voltage is zero and in the reverse bias region the current is zero; hence the product of voltage and current is always zero. A real diode inherently receives little power because it also has small voltage when the current is high (the forward bias region) and low current when the voltage is high (the reverse bias region). Even so, the small amount of power that is given to the diode is important because the thermal energy is generated in the junction region, which is physically small. Thus a small power can cause a significant rise in the junction temperature and affect diode performance. For this reason diodes in power supplies are designed to have good heat conduction between the junction and the outer case of the diode, and the diode is mounted in such a way as to enhance heat transfer to the ambiance. Often "heat sinks" and even fans are used to improve cooling of the diode. Occasionally, water cooling is used on large power supplies. Figure 6.29 shows a diode mounted on its heat sink.

Figure 6.29 Power diode mounted on a heat sink.

Breakdown. A physical diode is also limited by the amount of voltage it can withstand in the reverse bias region. Conduction is prevented in the reverse bias region by the reinforcement of the effect of the uncovered charges in the depletion region, as suggested in Fig. 6.26. When too much voltage appears across the depletion region, however, the forces on the bound electrons in that region become so great that they can be torn from their locations. In this condition several processes are involved, and a rapid buildup of current occurs, as shown in Fig. 6.30. The rapid buildup of current at the large reverse breakdown voltage, V_B, dumps a large amount of energy into the junction region, and the diode is quickly destroyed.

Breakdown must be avoided in power supply operation. Designers have available diode types with breakdown voltages (peak inverse voltage, or PIV) ranging in excess of -1000 V. In the unfiltered half-wave rectifier shown in Fig. 6.4, the maximum diode voltage is $-V_p$, which occurs when the diode is OFF and the source at its maximum negative value. For successful operation in such a power supply, the PIV of the diode must exceed this voltage.

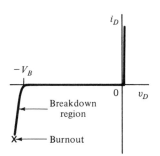

Figure 6.30 When reverse-bias voltage exceeds the breakdown voltage, the current increases rapidly.

In the filtered power supply in Fig. 6.12, the maximum voltage across the diode is approximately $-2V_p$, as shown in Fig. 6.31. This maximum occurs because the capacitor holds the load voltage at approximately $+V_p$ while the voltage source swings negative to $-V_p$. To operate successfully in such a power supply, the diode must be able to withstand $-2V_p$.

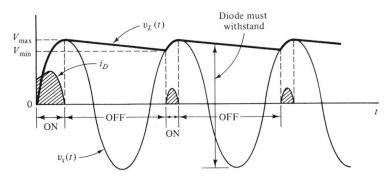

Figure 6.31 To operate successfully in a rectifier, a diode must never have its breakdown voltage exceeded.

The diode fails immediately in the breakdown region because it is not designed to handle the thermal energy generated. On the other hand, the rapid increase of current at a fairly constant voltage is potentially a useful feature, provided that the diode can handle the heat. Diodes that are designed to function in the breakdown region are known as *zener diodes*. The properties and some applications of zener diodes are presented in the next section.

6.4. ZENER DIODES AND SILICON-CONTROLLED RECTIFIERS

6.4.1 Zener Diodes

Zener Diode Characteristics.   Figure 6.32 shows the symbol and characteristic of a zener diode. The symbol is the usual diode symbol with a broken line added, but notice that the voltage and current are defined with reference directions opposite to those for an ordinary diode. This reversal is introduced because the zener diode is designed to operate

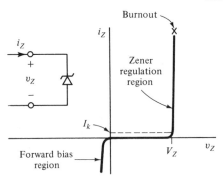

Figure 6.32 Symbol and graphical characteristic of a zener diode.

in the breakdown region as a dc voltage regulator. The zener diode characteristic resembles that of an ordinary diode except that voltage and current reference directions are reversed: The forward bias region appears at small negative voltage, and the breakdown region appears at positive voltage. One can purchase zener diodes with breakdown voltages from a few volts to several hundred volts. The semiconductor designer balances several semiconductor processes to maintain the zener voltage reasonably constant over a wide range of currents and temperatures.

A Zener Voltage Regulator.   In Fig. 6.32 we have indicated a *knee current, I_k,* that current which the zener diode must have to maintain its zener voltage. We will now analyze a zener regulator to show the role of the knee current in the regulating process.

In Fig. 6.33 we show a variable input voltage, as indicated by a battery symbol with an arrow through it; the regulator consists of a 100-Ω resistor and a zener diode. The output voltage is taken in parallel with the zener. We shall vary the input voltage and observe the output voltage. For an input of zero volts, we have no output voltage. If we increase the input voltage to, say, 3 V, negligible current will flow because the zener remains in its reverse bias region below breakdown. Hence there is no voltage across the resistor and KVL requires that the 3 V appear across the zener (point *a*). As the input voltage increases, the operating point on the zener moves out on the voltage axis and a small current begins to flow as breakdown is approached. When we reach the knee current of 10 mA (point *b*) the zener voltage is 6 V. The input must be 7 V to place operation at this point because the voltage across the resistor rises to $100 \times 0.01 = 1$ V. As we increase the input voltage further, the operating point will move up the zener characteristic at

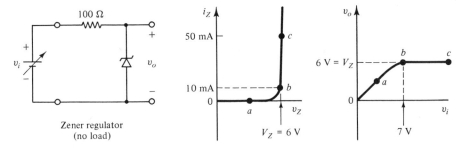

Figure 6.33 The zener knee current must be exceeded for regulation to occur.

constant voltage (Fig. 6.33) and thus the output voltage is controlled by the zener. At point c for example, the input is 11 V and the current has risen to 50 mA because the resistor voltage must now be 5 V.

As we continue to increase the input voltage, the output voltage will remain at 6 V. As the current increases, the resistor will get hot, the zener will get hot, and the input source may get hot; but the output voltage will remain at 6 V until some component fails. The zener will regulate the output voltage as long as it remains in its regulating region and does not burn out.

The Zener Regulator with Load.   Figure 6.34 shows the regulator of Fig. 6.33 with a load. Specifically, we will assume that the input voltage comes from an automotive electrical system which can vary from 12 to 14 V, and we assume that the load is a tape recorder which requires 6 V for proper operation. The current drawn by the tape recorder can vary from zero (when off) to 50 mA when the recorder is rapidly advancing the tape. The object of this example is to show that the regulator functions properly. Also, we shall consider the power requirements that we should specify for the components of the regulator.

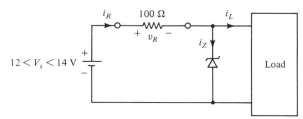

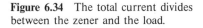

Figure 6.34   The total current divides between the zener and the load.

In order to understand the effect of load current, consider first that the input voltage is at its nominal value of 13 V and that the load draws no current. In this case we would assume the zener to be in its regulating region; thus the current through the resistor can be calculated from KVL and Ohm's law:

$$-13 + v_R + 6 = 0 \Rightarrow i_R = \frac{v_R}{100} = 70 \text{ mA} \tag{6.8}$$

This confirms the zener to be in its regulating region, which requires only 10 mA (the knee current) through the zener.

Consider now that the load begins to draw current. The current through the resistor will remain constant as long as the zener continues regulating and the input voltage is constant; hence as the load current increases, the zener current will decrease accordingly. In effect, the load takes current away from the zener. If the load current were 50 mA, for example, the zener current would be 20 mA. If the load current were increased beyond 60 mA, however, insufficient current would remain to provide the 10 mA knee current to the zener, and the zener would not remain in its regulating region. Generally, the voltage would drop and performance of the regulator would be unsatisfactory if the load required more than 60 mA.

So far we have discussed the nominal performance of the regulator, which is satisfactory. Let us now examine the extremes to verify that performance remains satisfactory under all conditions. One worst case would occur if the input voltage were minimum and the load current were maximum because this condition results in minimum current to the zener. This minimum current must exceed the knee current for proper regulation by the zener. Using the analysis shown in Eq. (6.8), you can confirm that in this case the resistor current is 60 mA, the maximum load current is 50 mA, and the zener current is 10 mA, just enough to keep in the regulating region. Thus the regulator can function properly at this extreme.

The other extreme occurs when the source voltage is maximum. This extreme will establish the power requirements for the resistor and the zener diode. For an input voltage of 14 V, the resistor current rises to 80 mA; hence the resistor must dissipate $(0.08)^2(100) = 0.64$ W. A 1-W resistor would provide sufficient margin of safety. The worst case for the zener occurs for maximum input voltage and no load current, because the zener must take the entire 80 mA in this situation. The maximum power into the zener, therefore, would be $(0.08)(6) = 0.48$ W. To be safe, we should specify a 600-mW zener.

A heftier regulator would employ a high-wattage resistor with smaller resistance and a zener capable of dissipating more power. For example, a regulator employing a 5-W 6-V zener and a 10-Ω high-wattage resistor would deliver over 0.5 A at 6 V to a load, assuming the same source voltage range as above.

6.4.2 Silicon-Controlled Rectifiers

The Circuit.   In this section we shall examine the operation of the circuit shown in Fig. 6.35. The ac source represents a standard 120-V 60-Hz supply. The circuit in the box is a power controller: the resistor with the arrow through it represents a variable resistor, and the load resistor, R_L, represents the load receiving the power. The two unfamiliar symbols represent semiconductor devices called a *thyristor* and a *silicon-controlled rectifier* (SCR).

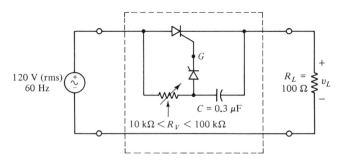

Figure 6.35   Typical SCR circuit.

Common Applications for this Circuit.   Circuits of this type are used in light dimmers. In this application, the variable resistor is adjusted by a rotary mechanism, and an ON-OFF switch is normally built into the same mechanism. Of course, the load would be a lighting fixture or spot light. A variable-speed drill is another application of this or a similar circuit. Here the variable resistor is controlled by the squeeze trigger on the drill handle and the load is a common type of electric motor.

The Thyristor Characteristic.   The thyristor is also called a four-layer diode, for reasons suggested by Fig. 6.36. The device consists of four alternating layers of p-type and n-type semiconductor, forming three pn junctions. Two of the pn junctions face the same direction, and the middle one faces the opposite direction. Thus you might anticipate that the device will operate as three diodes in series, with the middle one turned around. Hence no current ought to flow in either direction, because at least one of the diodes will always be reversed biased when voltage is applied. The characteristics displayed in Fig. 6.36 reveal that no current flows for negative voltages, where two pn junctions are reverse biased. Neither does current flow for positive voltages, where only one of the pn junctions is reverse biased, until a threshold voltage, V_{th}, is reached. After this threshold voltage is exceeded, the thyristor begins to conduct freely (it "fires"): it acts as if the middle pn junction disappears. The threshold phenomenon occurs because the doping levels in the four layers differ greatly. The outside p and n materials are doped heavily and hence have many carrier holes and electrons, respectively, available to diffuse into the middle n and p regions, which are lightly doped. Once the breakdown occurs in the middle junction at V_{th}, the holes from above and the electrons from below flood into the depletion region of the middle junction, where the internal battery–capacitor effect reinforces their movement across the junction. Thus this middle depletion region effectively disappears due to the carriers from the forward bias junctions, and we are left with two conducting junctions, two forward-biased diodes in series. The voltage of 0.5 V for the thyristor in the ON state results from the contributions from each ON diode. Less than 0.7 is required for each pn junction because the excess carriers from each help the other.

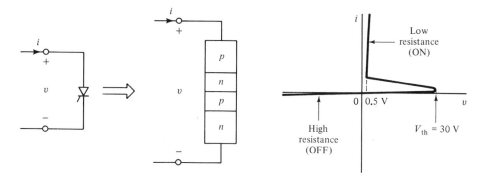

Figure 6.36 Thyristor symbol, structure, and current–voltage characteristic.

Thus the thyristor has two states. It is OFF with negative input voltage, and remains OFF for positive input voltage until the threshold voltage is reached, after which it turns ON. It will remain ON until the current is reduced to a small value, after which it will turn OFF again. The value of threshold voltage, V_{th}, can be controlled over a wide range by the semiconductor designer. One can purchase thyristors with firing voltages from 6 V to 32 V. We have shown 30 V, a typical value. Thus this device, like the diode, is a voltage-controlled switch, except that the thyristor requires much more than 0.7 V to turn ON.

The SCR Characteristics. As Fig. 6.37 reveals, the SCR is constructed like a thyristor, except that an additional external connection has been made to one of the internal layers. The extra input is called the *gate* and, as the name suggests, its function is to turn the device ON. With no gate current, the SCR characteristic is that of the thyristor, as shown. The important difference, however, is that the breakdown threshold occurs at a much higher voltage; indeed, the SCR should never fire because the input voltage exceeds its threshold. On the contrary, the SCR should fire when a pulse of current is delivered to the gate. This is shown by the $i_G > 0$ characteristics in Fig. 6.37; in effect, the threshold is reduced to a very small value when the gate conducts. Physically, this occurs because the gate injects holes into the lightly doped p region and floods the depletion region with carriers, thus initiating breakdown.

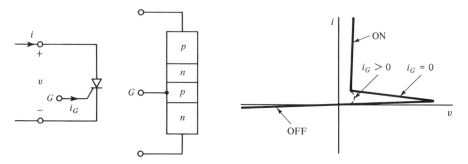

Figure 6.37 SCR symbol, structure, and current–voltage characteristic.

To summarize, the SCR has an OFF and an ON state. The device remains OFF until it has positive voltage and also receives current at the gate. Once turned ON, it will remain ON until the current is reduced below that necessary to sustain the ON state.

Circuit Operation. To explain the operation of the circuit in Fig. 6.35 we shall begin with the simplier circuit shown in Fig. 6.38. We begin with a discharged capacitor. Closing the switch at $t = 0$ will allow current to flow and the capacitor voltage will increase as charge builds up. The thyristor will remain OFF and no current will flow through the small resistor until the voltage across the capacitor reaches the threshold voltage of 30 V. Thus until the thyristor fires, we have a simple RC circuit. The time constant is $R_V C$, the initial value of the voltage across the capacitor is 0 V, and the final value would be 170 V if the transient reached completion. Using the techniques from

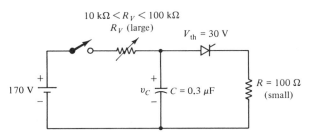

Figure 6.38 The thyristor will fire when $v_C = 30$ V.

Chapter 3, we write the voltage across the capacitor as

$$v_C(t) = 170(1 - e^{-t/R_V C}) \tag{6.9}$$

The capacitor voltage will increase according to Eq. (6.9) until the threshold voltage of the thyristor is reached. At this point the thyristor turns ON and allows current to flow through the small resistance. Current from the dc source is limited by the large value of R_V, but the current from the capacitor discharge can be quite large. The time at which the discharge will occur will be

$$30 = 170(1 - e^{-t_f/R_V C}) \Rightarrow t_f = -R_V C \times \ln\left(1 - \frac{30}{170}\right)$$

$$= +0.194 R_V C$$

The capacitor value is 0.3 μF and the value of the variable resistor lies between 10 and 100 kΩ. Thus the minimum firing time will be about 0.6 ms, and the maximum firing time about 6 ms. The firing of the thyristor results in a current pulse through the small resistor. The peak magnitude of this current pulse is about $(30 - 0.5)/100 = 0.295$ A, and its duration would be approximately the time constant of the RC circuit containing the capacitor and the smaller resistor, about 30 μs. After the discharge, the thyristor will turn OFF, there being insufficient current flowing to keep it ON, and the cycle will begin again. Hence the capacitor voltage and the current through the small resistor will be as shown in Fig. 6.39, where we have shown a 5-ms cycle time.

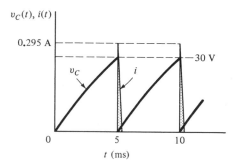

Figure 6.39 Capacitor voltage and load current waveforms.

We will now consider what would happen if the dc source were replaced by an ac source, as shown in Fig. 6.40. We have drawn the input voltage as a sine function and shown the capacitor voltage as v_C, as it would be if the thyristor did not conduct. If the capacitor were initially uncharged, and if the thyristor never fired, there would be a start-up transient, and then the voltage would lag the input voltage with a phase shift between 0 and 90°, depending on the value of $R_V C$. However, the thyristor will fire at the time, labeled t_f, when the capacitor voltage reaches the threshold of 30 V. The calculation of t_f is somewhat more complicated than the calculation from Eq. (6.9), but for the values of R_V and C which we show, the values lie between 30 and 180° expressed in phase relative to the beginning of the cycle. Hence we can control the firing time, or *firing angle*, with the value of R_V much as we did with the dc source. When the thyristor fires, the capaci-

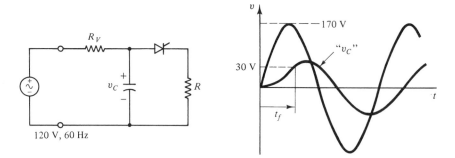

Figure 6.40 Circuit response with a sinusoidal source.

tor will discharge through the small resistor with a short time constant, as before. After the capacitor discharges, the thyristor turns OFF, there being insufficient current passing through R_V to keep it conducting. The capacitor voltage again begins building up and may fire the thyristor again before the input voltage goes negative, but once the input becomes negative the thyristor will not fire again until the input again goes positive.

Summary.   We can create a pulse of current with the thyristor. The timing of this pulse of current is controlled by the R_VC time constant and can be varied over a wide range. We now can understand the operation of the circuit in Fig. 6.35, repeated in Fig. 6.41. We see that the SCR gate is the "load" to which the thyristor delivers its current pulses. The load of the SCR, R_L, is presumably a low resistance and does not affect greatly the charging of the capacitor. Hence, when the input voltage goes positive, the capacitor begins charging through R_V and R_L in series, controlled by R_V. When the capacitor voltage builds up to the threshold voltage, the thyristor fires and the current pulse passes through the SCR gate, turning ON the SCR. The voltage to the load would thus be as shown in Fig. 6.42. By adjusting the firing time with R_V, we can control the power to the load. The current pulse may be too great for the gate, which typically requires only 10 mA or so to fire the SCR. A resistor would be inserted in series with the thyristor to limit the gate current to safe levels. With the SCR gate for a load, repeated firing of the thyristor would not matter. Once the SCR is turned ON, it will remain ON until the input voltage goes negative.

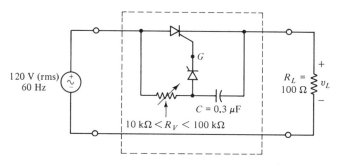

Figure 6.41 Typical SCR circuit (identical with Fig. 6.35).

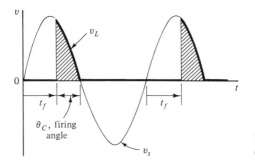

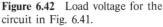

Figure 6.42 Load voltage for the circuit in Fig. 6.41.

Diacs and Triacs.   A similar circuit is shown in Fig. 6.43. The device replacing the thyristor is called a *diac*. It is like two thyristors turned in opposite directions: it will fire in either direction. Similarly, the device replacing the SCR is called a *triac* and it functions like two parallel SCRs facing opposite directions. This circuit operates like the circuit in Fig. 6.41, except that it performs no rectification. Hence the load voltage would be as shown in Fig. 6.44. This is actually a better circuit for light dimmers and variable speed drills, which do not require dc voltage.

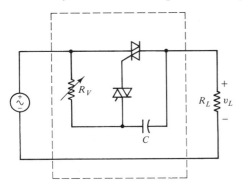

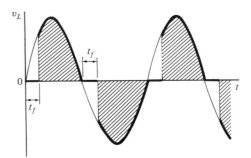

Figure 6.43 Diacs and triacs fire in both directions.

Figure 6.44 Load voltage for the triac circuit.

6.5 TRANSISTOR OPERATION

6.5.1 The Importance of the Transistor

From your earlier experience with circuit theory, you will recall that one can do a fair amount of work solving a two- or three-loop ac (or even dc) circuit, and circuits with more than three loops or nodes cause one to run to the computer for help. That being true, did you ever wonder how electrical engineers can design circuits containing literally thousands of loops and nodes? For example, if you ever examine the circuit diagram for a relatively simple piece of electronics, say, a TV set or an FM radio, you would see hundreds of resistors and capacitors, not to mention diodes and transistors and other strange components. How do electrical engineers produce such complicated circuits?

The design of such complicated circuits is simplified by the unidirectional properties of transistors. The transistor is a three-terminal device connected normally as shown
in Fig. 6.45. The transistor has an input and an output, suggesting that the cause–effect
relationship goes from left to right. That is, the input affects the output but the output has
little effect on the input. This one-directional causality we have indicated by marking an
arrow from left to right while crossing out the arrow from right to left. We might say that
the transistor is a "one-way street" to the electrical signal. This one-directional property
of the transistor isolates the output from the input and allows electrical engineers to build
complicated circuits such as TV and radio receivers and computers. This complexity can
be mastered because circuits can be designed (or analyzed) one part at a time, unlike the
two- or three-loop circuits that we tackled in circuit theory, which must be analyzed all at
once.

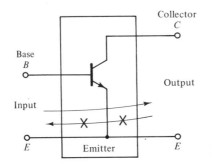

Figure 6.45   *npn* transistor symbol.

In addition to this isolating property, transistors have other important properties.
Transistors can give signal gain, thus allowing small signals (such as a voltage induced in
a car radio antenna) to be amplified by stages until large enough to power a radio speaker.
Transistors can also be used to couple circuits of greatly differing impedance levels, allowing more efficient transfer of signals between them. Transistors are used in digital circuits
as electrically controlled switches. Finally, transistors offer a variety of nonlinear effects
which are used in communication circuits for manipulating signals in the frequency
domain. In this section, however, we limit our attention to the isolating and amplifying
properties of the transistor.

There are currently many types of transistors, all of which accomplish the purposes
stated above. There are *npn* and *pnp* bipolar junction transistors (BJTs) and there are
p-channel and *n*-channel field-effect transistors (FETs). FETs can be either junction
field-effect transistors (JFETs) or they can be metal-oxide semiconductor field-effect
transistors (MOSFETs). In this book we will deal with only one type—the *npn* bipolar
junction transistor, which is symbolized in Fig. 6.45. The principles would be the same
for all types of transistors.

6.5.2 Transistor Characteristics

Input–Output Characteristics.   As shown in Fig. 6.45, the transistor is a three-
terminal device which is connected with one terminal in common between the input and
output circuits. The parts of the transistor are the emitter (E), the base (B), and the col-

lector (C), each with appropriately labeled terminals (Fig. 6.45). Our goal in this section is to describe the input and output characteristics of a typical *npn* transistor. Figure 6.46 shows how input and output circuit variables are normally defined for such a device, the device being represented in this case by a box.

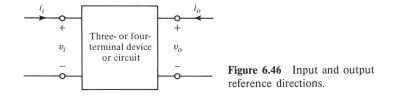

Figure 6.46 Input and output reference directions.

Because this is a new approach to circuits for you, we have drawn in Fig. 6.47 a simple resistive circuit in the box to demonstrate this method of description. First we shall derive the output characteristics, that is, the relationship between voltage and current at the output. Clearly, we cannot consider the relationship between v_0 and i_0 without somehow taking account of the effect of the input voltage and current. We will, therefore, place a dc current source (i_i) at the input and determine the relationship between the output voltage and current, with i_i held constant. This output relationship is most easily determined by applying KCL equation to the Norton equivalent circuit. The equivalent circuit is shown in Fig. 6.48. Of course, the input has disappeared, but this does not matter because we are interested presently in the output characteristics. The KCL for the Norton equivalent circuit is

$$-I_N + \frac{v_0}{R_{\text{eq}}} - i_0 = 0 \Rightarrow i_0 = -I_N + \frac{v_0}{R_{\text{eq}}} \qquad (6.10)$$

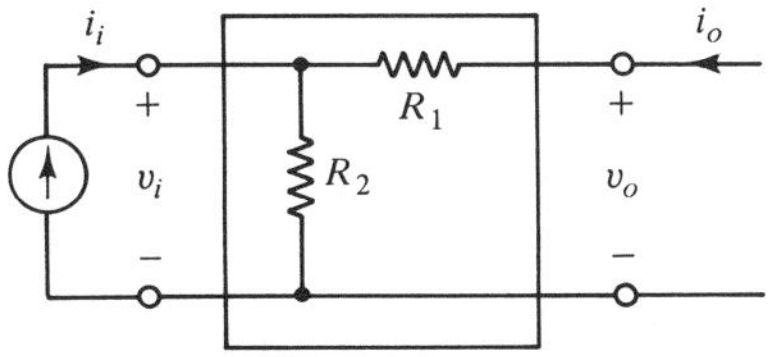

Figure 6.47 Resistive circuit with input and output.

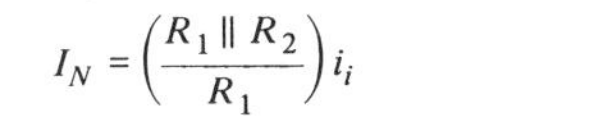

Figure 6.48 Norton equivalent output circuit for resistive circuit in Fig. 6.47.

$$R_{\text{eq}} = R_1 + R_2$$

$$I_N = \left(\frac{R_1 \parallel R_2}{R_1}\right) i_i$$

Equation (6.10) reveals a linear relationship between v_0 and i_0, which is plotted in Fig. 6.49. Clearly, the output characteristics depend on the current at the input, as anticipated. The output characteristics are linear because the equations governing the circuit are linear equations. To summarize, the output characteristic of the resistive circuit in Fig. 6.47 is shown in Fig. 6.49 to be linear and dependent on the input current.

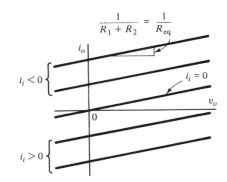

Figure 6.49 Output characteristics of circuit with i_i constant.

We could also derive the input characteristic of this circuit. The input characteristic would depend on the output voltage and current. In this regard, the resistive circuit differs significantly from the transistor, for an important virtue of the transistor is that its input characteristic does not depend on its output. The transistor is a unilateral device, a one-way street for signals, whereas the resistive circuit in Fig. 6.47 is a bilateral circuit, a two-way street. For this reason we have no interest in the input characteristic of the voltage-divider circuit.

Transistor Input Characteristic.   In Fig. 6.50 we show a simple picture of the physical structure of an *npn* transistor in the common-emitter configuration. The structure is that of a sandwich, with the *p* material like a very thin piece of bologna between two thick pieces of *n*-material bread. The external connections are made through wires which are bonded to the three regions of the transistor. Depletion regions will form at the two *pn* junctions and, in the absence of external applied voltages, the orderly and disorderly processes will rapidly come to equilibrium at both junctions. First we shall consider the effect of placing a voltage at the base–emitter (B–E) junction, the input part of the transistor.

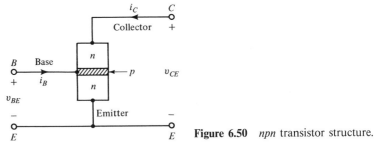

Figure 6.50 *npn* transistor structure.

Ignoring for the present the effect of the collector–base junction, we note that the base–emitter junction forms a *pn* junction diode. We anticipate that the input *i–v* characteristic would be like that of a diode, as shown in Fig. 6.51. The base current will be very small until sufficient voltage exists across the junction to turn it ON, about 0.7 V for a silicon transistor. Once the junction is turned ON, the current increases rapidly, with the voltage remaining constant at about 0.7 V. We can therefore model the base–emitter characteristic as either an open circuit (for $v_{BE} < 0.7$), or else a constant voltage of 0.7 V once the input voltage tries to go above that value. This model is shown in Fig. 6.52. In justifying this model, we have ignored the state of the output circuit (v_{CE} and i_C); but as we stressed above, one virtue of the transistor is that the output has negligible effect on the input.

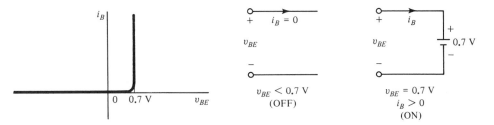

Figure 6.51 Transistor input characteristic.

Figure 6.52 *npn* transistor input circuit model.

Output Characteristics

The Effect of Doping.   Like the input characteristic, the output characteristics depend on whether the collector–base *pn* junction is forward biased or reversed biased. We shall assume that the input current to the base has been fixed at i_B, which requires approximately 0.7 V at the base region, considering the reference node to be the emitter. (Later in this discussion we will present a simple circuit to establish a specified base current.) Now we shall increase the collector–emitter voltage, beginning at zero volts, and observe the collector current.

First we shall consider the holes and electrons in the base region. When we discussed the *pn* junction earlier in describing diode operation, we implied that there is roughly the same density of holes in the *p* region as electrons in the *n* region. In this implied situation, forward biasing the *pn* junction will cause roughly as many holes to diffuse into the *n*-type material as electrons to diffuse into the *p*-type material. But if, say, the *n*-type material were more heavily doped than the *p*-type material, the electron density would greatly exceed the hole density. For this case a forward bias would produce many more electrons diffusing into the *p*-type material than holes diffusing into the *n*-type material. The diode would still work; however, the current would be carried across the junction largely by the electrons and most of the recombinations would occur in the *p*-type material.

When a transistor is made, the emitter is doped more heavily than the base; hence for an *npn* transistor the conditions are those described above—excess electrons diffuse into the base region. For zero volts between collector and emitter, the base–collector junction is also forward biased. Hence electrons also tend to diffuse into the base from the collector region. Although there is a buildup of excess electrons in the base region, the only current which flows is that permitted by recombination of electrons in the base: this current is the i_B we assumed in the base–emitter bias circuit.

The Saturation Region.   If we now increase the collector–emitter voltage, we reinforce the orderly process forcing a drift current of electrons from the base to the collector. The result is a rapid buildup of collector current as excess electrons are permitted to pass into the collector region, where they flow through the collector–emitter circuit. This region of rapid increase in collector current is the "saturation region" marked in Fig. 6.53.

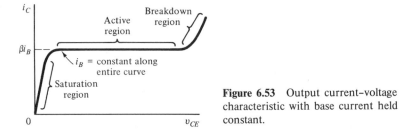

Figure 6.53 Output current–voltage characteristic with base current held constant.

The Active Region.   By the time the collector base voltage is about 0.2 to 0.4 V, all the excess electrons in the base region are being drawn into the collector region and the curve levels off, the active region in Fig. 6.53. In the active region the collector current is controlled by the number of electrons injected into the base region by the emitter. This is controlled by the base–emitter voltage, which is controlled in turn by the amount of base current the base circuit allows to flow. Thus in the active region it is customary to consider the collector current as controlled by the base current.

Let us review the principal processes that occur in the active region. The external bias of about 0.7 V which is applied to the base–emitter junction diminishes the orderly process which might hold back the electrons in the heavily doped emitter region. These electrons diffuse into the base region, where a few, say 1%, combine with holes and create the base current. The remaining 99% diffuse into the collector–base depletion region, where the orderly process of the uncovered bound charges forces them into the collector region. Thus in the active region the collector current is controlled by the base current.

The Current Gain of the Transistor.   If we call α (alpha) the fraction of electrons which diffuse across the narrow base region by a diffusion mechanism, the fraction of electrons which recombine with holes in the base region to create the base current is $1 - \alpha$. The ratios of the base, collector, and emitter currents are thus

$$I_C = \alpha I_E \qquad \text{and} \qquad I_B = (1 - \alpha)I_E \qquad \alpha < 1$$

The value of α is fairly constant throughout the active region for a given transistor and characterizes the current gain of the device. Usually, the current gain is described in terms of the β (beta) of the transistor, defined as

$$\beta = \frac{I_C}{I_B} = \frac{\alpha}{1 - \alpha} \tag{6.11}$$

In terms of β, the ratios of the currents become

$$I_C = \beta I_B \qquad I_E = \frac{\beta}{\beta + 1} I_C = (\beta + 1) I_B \qquad (6.12)$$

We stress that Eqs. (6.12) are valid only in the active region.

   The Breakdown Region.   In Fig. 6.53, we have also marked a breakdown region, where the current increases rapidly with increasing collector–emitter voltage. In this region, the power into the transistor becomes excessive and thermal failure often occurs.

   Normally, the output characteristics are shown for many values of base current, as in Fig. 6.54. Here we have given typical characteristics for a transistor with a β of approximately 100 $(\alpha \approx 0.99)$.

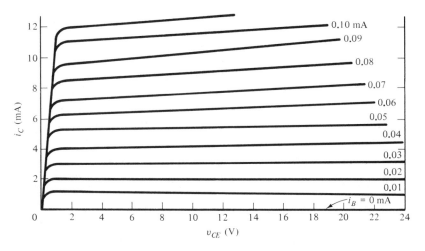

Figure 6.54   Typical *npn* transistor output characteristics.

6.5.3 Transistor Amplifier–Switch Circuit Analysis

   Problem Statement.   The circuit we will analyze in this section is shown in Fig. 6.55. Here is an *npn* transistor in the common-emitter connection, meaning that the emitter of the transistor provides the common terminal between the input and output circuits. The circuit has an input circuit voltage, v_i, and an output circuit with its output voltage, v_o. A dc voltage source (V_{CC}) in the output circuit supplies the energy required by the circuit, and two resistors control the currents and voltages applied to the transistor. The transistor input characteristics are those of a silicon *pn* junction (Fig. 6.51) and the output characteristics are as shown in Fig. 6.54. Our goal is to show how the output voltage depends on the input voltage.

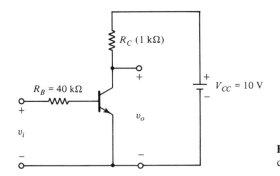

Figure 6.55 Transistor amplifier–switch circuit.

Where to Start?   In a physical world described by cause–effect relationships, *analysis* and *design* activities can be related to available causes and intended effects. In *analysis* we seek to discover the "effects" that result from the "causes" before us. In this case the causes are represented by the circuit to be analyzed. By contrast, we try in *design* to arrange causes to bring about some desired effect or effects (the design specifications). When we *analyze* a circuit, we logically begin at the input, the controlling part of the circuit, and work toward the output, the controlled part. If we are *designing* the circuit, we normally begin at the output, where the "effects" occur, and work backward toward the input, arranging our "causes" as we go. Thus in approaching a problem, an important question is: "Am I designing or analyzing?" The answer to that question usually tells you where to begin on the problem. If analyzing, start at the input; if designing, start at the output.

Input-Circuit Analysis.   In the present instance, we are analyzing the circuit and we begin, therefore, at the input circuit, shown in Fig. 6.56. We wish to determine the base current, i_B, because this current controls the state of the output circuit. Let us consider that we start with a negative value of v_i and increase this voltage to positive values. Comparing the input circuit with the model of the input characteristics in Fig. 6.52, we note that no current will flow until the input voltage becomes at least $+0.7$ V because the *pn* junction will be OFF. This is known as *cutoff*, for no current flows in the base or in the collector. When the input voltage exceeds $+0.7$ V, the base–emitter junction will turn

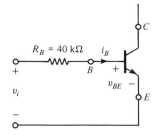

Figure 6.56 Input circuit.

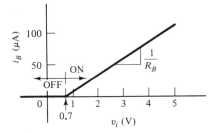

Figure 6.57 Once the base–emitter junction is ON, the base current is limited by the base resistor, R_B.

ON and the second model circuit in Fig. 6.52 applies. Since in this case v_{BE} is constant at 0.7 V, KVL requires the base current to be

$$i_B = \frac{v_i - 0.7}{R_B} \qquad v_i > 0.7 \text{ V} \tag{6.13}$$

This current is graphed in Fig. 6.57. This completes our analysis of the input circuit.

Output-Circuit Load-Line Analysis. The output part of the circuit is controlled by the input circuit. In the active region the base current controls the collector current, which in turn determines v_o. The concept of a *load line* offers an excellent way to understand the interaction of the transistor with the rest of the output circuit.

In Fig. 6.58, we have shown the output circuit broken at the transistor collector-emitter connections. If we stand at that break and look to the left, we see the i–v characteristics of the transistor, which are given in Fig. 6.54. If we look to the right, we see the i–v characteristics of the external circuit, V_{CC} in series with R_C. Using Kirchhoff's voltage law and Ohm's law, we can easily show the external circuit i–v characteristic to be

$$-V_{CC} + iR_C + v = 0 \Rightarrow i = \frac{V_{CC} - v}{R_C} \tag{6.14}$$

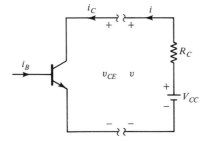

Figure 6.58 The current–voltage characteristics of the left and right halves of the output circuit may be considered separately.

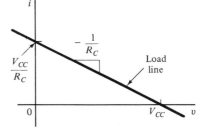

Figure 6.59 The right half of the circuit is characterized by a load line.

If we plot current versus voltage for Eq. (6.14) as in Fig. 6.59, we observe a straight line with a voltage intercept of V_{CC}, a current intercept of V_{CC}/R_C, and a slope of $-1/R_C$. This line, known as the load line, represents the part of the circuit external to the transistor.

Figure 6.60 repeats Fig. 6.54 with the load line drawn for our specific values of V_{CC} (10 V) and R_C (1 kΩ). The mental break in the output circuit was made for the purpose of examining independently the characteristics of the two parts of the circuit. When we mentally reconnect the circuit, the two voltages must be the same and the two currents must be the same. These requirements determine the current and voltage in the output circuit.

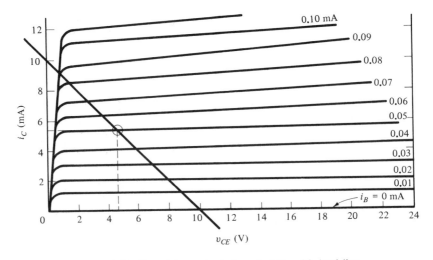

Figure 6.60   Transistor output characteristic with load line.

The Input–Output Characteristic.   Consider now that the base current is 50 μA. According to the transistor output characteristics, the transistor must operate on the line corresponding to that base current. On the other hand, the external circuit must operate on the load line. Both requirements are satisfied at the intersection of the two lines. For 50-μA base current, the intersection occurs at a collector current of 5.3 mA and a collector–emitter voltage of 4.7 V, as indicated in Fig. 6.60. This result demonstrates the method for finding the output voltage from the input voltage. For each input voltage, we can determine the base current [Eq. (6.13) and Fig. 6.57], and for each value of base current we can determine the output voltage by noting the intersection with the load line. We can thus plot the input–output characteristic of the amplifier–switch point by point. Such a plot is shown in Fig. 6.61.

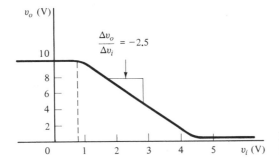

Figure 6.61   Transistor amplifier–switch input–output characteristic.

Several features of Fig. 6.61 merit close attention. When the input voltage is less than +0.7 V, no base current flows and hence no collector current flows. The transistor is cutoff. This cutoff condition fixes the output voltage at V_{CC}, 10 V in this case. Because no current flows in R_C, the full power supply voltage must appear across the transistor.

This is like turning off a valve in a water pipe; the full pressure must be supported by the valve in the absence of flow. As the input voltage increases beyond $+0.7$ V, the base current begins to flow and the transistor moves out of cutoff into the active region. In the active region, the transistor acts as an amplifier to small changes in the input voltage. The incremental gain of the amplifier is the slope of the input–output characteristic in the active region. The amplifier in Fig. 6.55 has an incremental voltage gain of

$$A_v = \frac{-\beta R_C}{R_B} = \frac{-100 \times 1 \text{ k}\Omega}{40 \text{ k}\Omega} = -2.5$$

in this case. The minus sign of the gain signifies the inversion of the incremental changes. That is, a small positive pulse at the input would produce a larger negative pulse at the output.

As the input voltage continues to increase, the base current eventually will saturate the transistor. This is like opening fully a valve in a water pipe; the valve relinquishes control of the flow rate to the capacity of the supply. With the transistor saturated, the output voltage will remain small, 0.2 to 0.4 V, and the collector current will remain at approximately $V_{CC}/R_C = 10$ mA, even though the base current will continue to increase as v_i continues to increase. Thus in the saturation region the output voltage will remain small for increasing values of the input voltage.

6.5.4 Transistor Applications

Operation as a Switch. The circuit shown in Fig. 6.55, whose input–output characteristic is shown in Fig. 6.61, can be used as an electronic switch. If the input voltage is less than 0.7 V, the transistor is cutoff, that is, the electronic switch is OFF. If R_C represented a light bulb, no current would flow through the bulb and it would not glow. If the input voltage exceeded about 4.0 V, the transistor would be saturated and our electronic switch would be ON. If R_C were a light bulb, it would glow. Thus we can use the transistor as a voltage-controlled switch to turn the bulb on and off.

The switching action of the transistor is one of its most valuable properties. The vast majority of all the transistors in the world at this moment are being utilized as switches; relatively few are being used as amplifiers. This is true because all digital circuits—computers, calculators, digital watches, and digital instrumentation—utilize transistors in switching operation. Some of the important applications of transistors in digital circuits are explored in Chapter 7.

Operation as a Large-Signal Amplifier. An amplifier is called a large-signal amplifier when the signal levels require the full transistor characteristics, from near cutoff to near saturation. Such amplifiers can furnish moderate amounts of power to transducers such as loudspeakers or control motors. Our simple amplifier circuit is limited to an output voltage of 10 V, peak to peak, but this limitation can be overcome by increasing the power supply voltage or the circuit complexity. As we shall see in Chapter 8, feedback techniques can be used to reduce distortion and generally improve the characteristics of large-signal amplifiers.

Small-Signal Amplifiers.   Most amplifiers are small-signal amplifiers. Consider, for example, a radio that receives a signal of 10 mV and produces an output voltage of 10 V. Such a radio would require a voltage gain of 1000. This gain would be accomplished in "stages," each transistor amplifier stage taking as its input the output of the previous stage. If each stage had a voltage gain of $\sqrt{10}$, six stages of amplification would provide the necessary gain. Of these, all but the last stage would be a small-signal amplifier.

A small-signal amplifier must have a means (bias circuit) for placing the transistor in its amplifying region, a means for introducing the input signal, and a means for supplying its output signal to the next stage. The circuit shown in Fig. 6.62 is the simplest circuit which accomplishes these functions. This is our basic amplifier with added features to provide bias and coupling at input and output. We have symbolized the power supply with the terminal marked $+V_{CC}$, which appears as an open circuit but actually is connected through a dc power supply to ground. The two resistors R_1 and R_2 comprise a voltage divider to supply dc current to the transistor base in order to bias the transistor to its amplifying region. At the input we have a Thévenin equivalent circuit of the signal source: this might be the previous stage of the amplifier or it might be the ultimate source of the signal such as a microphone or a radio antenna. The load, R_L, represents the input impedance of the next stage of the amplifier. We shall begin our analysis of the small-signal amplifier by examining the role of the input and output coupling capacitors.

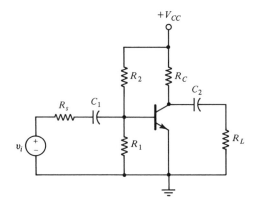

Figure 6.62   Small-signal amplifier.

Coupling Capacitors.   Capacitors block dc current but pass ac current. The infinite impedance of the capacitor at dc will allow the dc state of each amplifier stage to be independent of the adjoining stages. If the impedance of the capacitors is small relative to the resistors in the circuit, the time-varying signals will pass through the capacitors undiminished. Thus the stages are isolated for dc but coupled for ac signals by the coupling capacitors at input and output.

The Bias Voltage Divider.   A dc current is supplied to the transistor base by the voltage divider, R_1 and R_2. Because the coupling capacitors act as blocks to the dc current, the equivalent circuit at dc is as shown in Fig. 6.63a. We have replaced the $+V_{CC}$ symbol with two voltages sources. We draw the circuit this way for two reasons: to emphasize that the power supply acts independently on the bias and collector circuits, and to help you recognize the voltage divider. The circuit in Fig. 6.63a can be reduced to the familiar circuit in Fig. 6.55 by converting the voltage divider to a Thévenin equivalent circuit, as shown in Fig. 6.63b. We have used the symbol V_{BB} for the open-circuit voltage: This would play the role of v_i in Fig. 6.55 and Eq. (6.13).

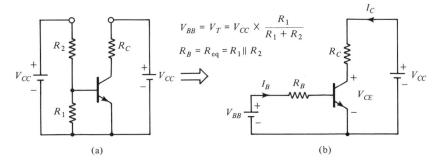

Figure 6.63   (a) Dc bias circuit; (b) Thévenin equivalent circuit of input portion of bias circuit.

We shall continue this development in terms of the specific circuit shown in Fig. 6.64. The base bias voltage is

$$V_{BB} = 10 \times \frac{5.6 \text{ k}\Omega}{50 \text{ k}\Omega + 5.6 \text{ k}\Omega} = 1.0 \text{ V}$$

and the value of R_B is

$$R_B = 5.6 \text{ k}\Omega \,\|\, 50 \text{ k}\Omega = 5 \text{ k}\Omega$$

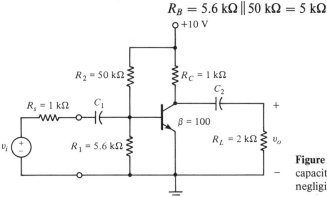

Figure 6.64   Small-signal amplifier. The capacitors are assumed to have negligible impedance to the signal.

Assuming a silicon transistor requiring $V_{BE} = 0.7$ V to turn ON the base–emitter junction, we can use the equivalent circuit in Fig. 6.52. Thus the dc base current is

$$I_B = \frac{V_{BB} - 0.7}{R_B} = \frac{1.0 - 0.7}{5.0 \text{ k}\Omega} = 60 \ \mu\text{A}$$

Assuming the transistor to be in the active region, we calculate the collector dc current

to be

$$I_C = \beta I_B = 100 \times 60 \ \mu\text{A} = 6.0 \ \text{mA}$$

To confirm that the transistor is in the active region (not saturated), we must compute the collector–emitter voltage, V_{CE}. We could use a load line, but KVL around the collector–emitter loop in Fig. 6.63b will do as well.

$$-V_{CC} + I_C R_C + V_{CE} = 0 \Rightarrow V_{CE} = 10 - (6.0 \ \text{mA})(1 \ \text{k}\Omega) = 4.0 \ \text{V}$$

Thus the transistor is operating near the middle of its active region. If there were no input signal, the transistor would have a steady voltage of $+4.0$ V across it, but the voltage across the 2-kΩ load resistor would remain zero because the dc voltage would be blocked by the output coupling capacitor, C_2.

The Small-Signal Equivalent Circuit. We now consider the effect of the input signal. In general, all voltages and currents in the circuit will have two components, a dc component and a time-varying component, which we shall call the signal component. For example, the base current would be

$$i_B(t) = I_B + i_b(t)$$

where $I_B = 60 \ \mu\text{A}$ and $i_b(t)$ is the signal component. What is the equivalent circuit that the signals "see," in the sense that Fig. 6.63b is the circuit seen by the dc voltage and current?

Beginning at the input and moving toward the output, we shall now justify the circuit shown in Fig. 6.65 as the appropriate small-signal equivalent circuit. The input source and its resistance, R_s, are now coupled to the amplifier with a short circuit representing the negligible impedance of C_1 to the signal. The base bias resistor R_1 appears as expected, but R_2 now connects to the *signal ground* because of the low impedance of the power supply to the signal. That is, the power supply circuit, not shown but represented by the V_{CC} symbol, would be connected to the circuit ground by a filter capacitor, as shown for example in Fig. 6.12. This filter capacitor not only filters the power supply output but also provides for the signal a low-impedance path to ground.

The input to the transistor is represented by a resistor, r_π, and the output by a

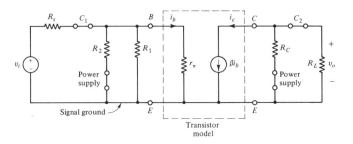

Figure 6.65 Small-signal equivalent circuit.

dependent current source. Representing the base–emitter by a constant 0.7 V, as shown in Fig. 6.52, is adequate for the dc solution but is not accurate for the signal: r_π is a relatively low resistance in the range 300 to 1000 Ω which may be determined from theory (see Problem P6.19), from measured transistor input characteristics, or from published specifications of the transistor. The dependent current source, $i_c = \beta i_b$, represents the current gain of the transistor.

In the output circuit, the collector bias resistor, R_C, is shown connected to the signal ground for the reason given above for R_2. Finally, the collector is connected to the load resistor, R_L, through the output coupling capacitor, which acts as a short circuit to the signal.

Small-Signal Voltage Gain.   The small-signal equivalent circuit in Fig. 6.65 may be analyzed for the voltage gain, $A_v = v_o/v_i$, by the standard methods of circuit theory. Our analysis begins at the input with the calculation of the base signal current. From the base current we can calculate the collector signal current, then the output voltage.

The base current is easily calculated if the input source is converted to a Norton circuit, as shown in Fig. 6.66. We thus have a four-way current divider. The base current is

$$i_b = \frac{v_i}{R_s} \times \frac{R_s \| R_2 \| R_1 \| r_\pi}{r_\pi} \tag{6.15}$$

For the values in Fig. 6.64 and $r_\pi = 740\ \Omega$, Eq. (6.15) leads to $i_b = v_i/1.89\ \text{k}\Omega$.

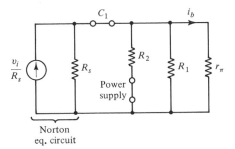

Power
supply

Norton
eq. circuit

Figure 6.66   Analysis of the input circuit is simplified through use of a source transformation.

The collector current is βi_b and the output voltage is thus

$$v_o = -\beta i_b (R_c \| R_L) = -100 \frac{v_i}{1.89\ \text{k}\Omega} (1\ \text{k}\Omega \| 2\ \text{k}\Omega)$$

Thus the voltage gain of the amplifier is $A_v = v_o/v_i = -35.3$. As stated above, the minus sign represents inversion of the signal.

Summary.   The calculation of the small-signal gain of the amplifier requires analysis of a dc equivalent circuit and a small-signal equivalent circuit. In the latter, the transistor is represented by its input resistance and a dependent current source. Because of the unilateral property of the transistor, the analysis proceeds in a straightforward manner from input to output.

PROBLEMS

Practice Problems

SECTIONS 6.2.1–6.2.2

P6.1. Figure P6.1 shows a sinusoidal voltage source, a diode, and a load resistor, R.

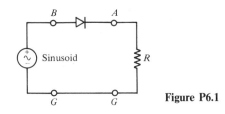

Figure P6.1

    (a) A dc voltmeter measures 10 V when connected between A and G. What would an ac voltmeter measure between B and G? Assume an ideal diode.

    (b) If a very large capacitor were connected between A and G, what then would the dc meter read? (Assume that $RC \gg T$, where T is the period of the sinusoid.)

P6.2. Using the unfiltered half-wave rectifier circuit shown in Fig. 6.4, design a power supply to produce 50 mA of dc current into a 500-Ω load. Give the circuit and the value which an ac voltmeter would indicate for the input voltage.

P6.3. (a) Derive the equations for the dc voltage and ripple voltage for the full-wave rectifier shown in Fig. 6.6a with a capacitor across the load to filter out most of the ripple. Assume that the time constant of the RC part of the circuit is large compared with the period of the input sinusoid. Have f continue to represent the frequency of the input.

    (b) Using your results of part (a), design a 60-Hz filtered power supply to provide 50 V dc, 100 mA dc to a load, with a ripple voltage of 0.5 V. You must specify the rms value of the input voltage and the value of the capacitor to be used. Assume ideal diodes. *Hint:* The dc voltage and current define an equivalent resistance for the load.

P6.4. Draw the circuit of a filtered half-wave rectifier which will deliver 20 dc V and 20 dc mA to a resistive load. Specify the value of all components, including the equivalent load resistor. Assume an ideal diode. The peak-to-peak ripple should not exceed 0.5 V. Also specify the rms input voltage (60 Hz).

P6.5. (a) Draw the circuit of a half-wave rectifier, with a generator, a diode, and a 10-kΩ load.

(b) Consider now that the generator puts out the square wave shown in Fig. P6.5. What would be the dc current in the load?

(c) Now add a 10-μF capacitor across the load. What would in this case be the current in the load?

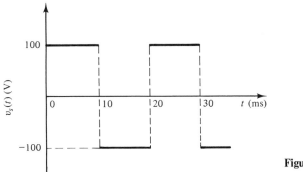

Figure P6.5

SECTIONS 6.3.1–6.3.4

P6.6. (a) Consider that you have a *pn*-junction diode whose characteristic is well described by Eq. (6.7). Plot the i–v characteristic in the region $0 < i_D < 100$ mA, assuming $I_o = 10^{-10}$ A and an ideality factor of 1.7. Plot on a linear scale in both i_D and v_D.

(b) If the diode in part (a) were placed in series with a 12-V source and a 200-Ω resistor, with the battery polarity such as to forward bias the diode, what current would flow? You may use trial and error or an analytic technique.

(c) What current would flow in part (b) if the battery were reversed so as to reverse bias the diode?

P6.7. Figure P6.7 shows a piece of silicon which is doped with donor and acceptor atoms to form a *pn*-junction diode. Assume an ideality factor of 2.0.

(a) Where does there exist: (1) more carrier electrons than holes; (2) more holes than carrier electrons; and (3) relatively few carriers?

(b) If you measured $i_D = 10$ mA of current for $v_D = 0.75$ V, what would you expect for $v_D = 0.80$ V? Assume a temperature of 300 K.

(c) What current would you expect for $v_D = -0.3$ V.

Ans: 26 mA and -5.5 nA

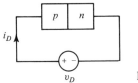

Figure P6.7

P6.8. (a) The usual symbol for a *pn*-junction diode is shown in Fig. P6.8 together with a corresponding hunk of silicon. Indicate which half is *p*-type material and which *n*-type material. Which contains the donor and which the acceptor atoms?

(b) Indicate with an arrow the direction in which holes cross the junction due to diffusion (thermal motion) when the diode is ON.

(c) With the diode reverse biased, the current is determined to be 10^{-11} A. How much voltage would it take to turn ON the diode to a current of 1 mA? Assume an ideality factor of 1.7.

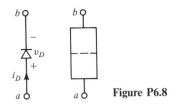

Figure P6.8

SECTION 6.4.1

P6.9. Using the techniques developed in Section 6.4.1, design a zener regulator to supply power to a 6-V 30-W spotlight which must operate from an auto electrical system whose voltage varies between 10.0 and 15.6 V. Specify the resistance, the resistor power rating, and the zener voltage and power rating. Make sure that the regulator does not fail if the spotlight burns out. Assume a knee current of 100 mA.

P6.10. The circuit shown in Fig. P6.10 shows an ac source connected to a 10-V zener diode through a 1-kΩ resistor.

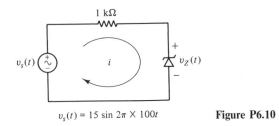

$v_s(t) = 15 \sin 2\pi \times 100t$ **Figure P6.10**

(a) Find the maximum positive current in the reference direction shown.

(b) Find the maximum negative current in the reference direction shown.

(c) Sketch the voltage across the zerner diode, $v_Z(t)$.

SECTION 6.4.2

P6.11. If the gate of the SCR in Fig. 6.35 is limited to 30-mA current, what value of resistance should be placed in series with the thyristor to limit the gate current to this value. Assume that the thyristor fires at 30 V and that the equivalent resistance of the gate is 15 Ω.

P6.12. Normally, the firing time in SCR operation, t_f, shown in Fig. 6.42, is described in terms of a "conduction angle," where

$$\theta_c = 180° - 360° \times \frac{t_f}{T}$$

where T is the period of the frequency, usually $\frac{1}{60}$ s. Clearly, θ_c is the amount of phase angle during which the SCR fires each cycle.

(a) For a firing angle of 90°, $V = 120$ V (rms), and $R_L = 100$ Ω, what is the dc current and the total power in the load? Note that the total power would be a measure of the total heating effect on the load and would be more than the dc power.

(b) Repeat part (a) for a firing angle of 120°. Part (a) can be solved without evaluating any integrals, but part (b) requires two integrations.

Ans: (a) 0.270 A, 36.0 W; (b) 0.405 A, 57.9 W

SECTIONS 6.5.1–6.5.3

P6.13. For the transistor output characteristics shown in Fig. P6.13, determine the approximate value of β in the active region. What value of α does this imply? If the collector current is saturated at a value of 5 mA, what is the collector–emitter voltage that results, and what is the base current required to saturate the transistor at this value of collector current?

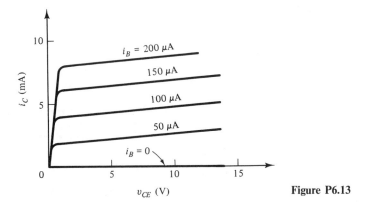

Figure P6.13

SECTION 6.5.4

P6.14. For the transistor amplifier shown in Fig. 6.55, change $V_{CC} = 12$ V and $R_B = 20$ kΩ.

(a) Draw the new load line on the transistor characteristics in Fig. 6.60.

(b) Find the input and output voltages for the following base currents: 0, 20, 40, 60, 80, and 100 μA.

(c) From part (b), plot v_o versus v_i.

(d) Find the incremental gain from the slope, $A_v = \Delta v_o / \Delta v_i$.

P6.15. A simple transistor circuit is shown in Fig. P6.15. The transistor output characteristics are shown in Fig. P6.13.

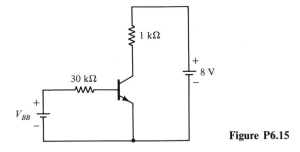

Figure P6.15

(a) Draw the load line on the characteristics. Find the required current into the base to give a collector–emitter voltage of 4 V.

(b) What value of base voltage, V_{BB}, is required to give this amount of base current? Assume 0.7 V between base and emitter.

(c) What value of V_{BB} is required to saturate the transistor for this circuit? What is the collector current at saturation?

P6.16. A transistor amplifier/switch circuit in Fig. P6.16a has a load line as shown on the transistor characteristics in Fig. P6.16b. Assume that $v_{BE} = 0.7$ V.

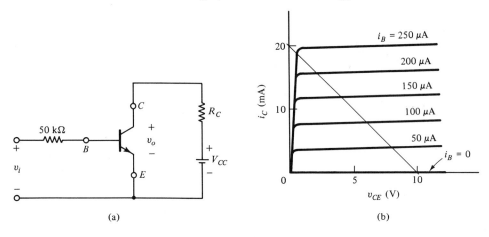

(a)   (b)

Figure P6.16

(a) What are V_{CC}, R_C, and β for the transistor?

(b) What value of v_i is required to saturate the transistor?

(c) What is the output voltage with the transistor saturated?

P6.17. A transistor circuit is shown in Fig. P6.17. Transistor characteristics are those shown in Fig. P6.13. Assume 0.7 V for the base–emitter junction.

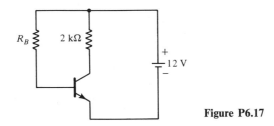

Figure **P6.17**

(a) Draw the load line. What value of R_B will give a dc collector current of 3 mA?
(b) Find the voltage between collector and emitter for this value of R_B.

P6.18. For the transistor amplifier shown in Fig. P6.18:

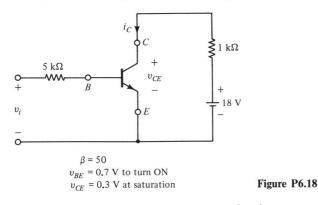

$\beta = 50$
$v_{BE} = 0.7$ V to turn ON
$v_{CE} = 0.3$ V at saturation

Figure **P6.18**

(a) What is the collector–emitter voltage (v_{CE}) if the transistor is cutoff?
(b) What is the collector current (i_C) if $v_{CE} = 10$ V?
(c) What is the minimum base current required to saturate the transistor?
(d) What is the minimum value of v_i to place the transistor in the active region?

SECTION 6.5.5

P6.19. The input characteristic of a transistor is that of a *pn* junction, Eq. (6.7). For a transistor, we would associate the diode current with the base current, i_B, and the diode voltage with the base–emitter voltage, v_{BE}. The appropriate input resistance for the transistor in a small-signal equivalent circuit would be the slope of the input characteristic at the dc base current level: $r_\pi = dv_{BE}/di_B$ at $i_B = I_B$. Using Eq. (6.7) and ignoring the $+1$ term in the ln term, derive an expression for the input resistance to the transistor base.

Ans: $r_\pi = \eta 26$ mV$/I_B$ at 300 K. Also, evaluate with the result r_π for the circuit in Fig. 6.64 and confirm that a value of 740 Ω is reasonable.

P6.20. Figure P6.20 shows a transistor amplifier, biased to operate as a small-signal amplifier.

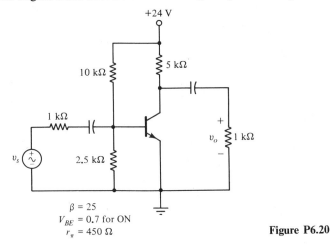

$\beta = 25$
$V_{BE} = 0.7$ for ON
$r_\pi = 450\ \Omega$

Figure P6.20

(a) Draw the dc bias circuit and solve for the base and collector dc currents and the dc voltage from collector to emitter.

(b) Draw the small-signal equivalent circuit and solve for the voltage gain of the amplifier, v_o/v_s. Consider the capacitors as short circuits at the signal frequency. Let $r_\pi = 450\ \Omega$.

(c) Find the input impedance of the amplifier as seen by the input generator. This does not include the 1-kΩ source resistance.

Application Problems

A6.1. The circuit of a battery charger is shown in Fig. A6.1a. The battery model consists of a battery voltage varying from 11.5 V (fully discharged) to 13.6 V (fully charged). Find the rms value of the secondary voltage to fully charge the battery to 13.6 V. For this voltage, what would be the maximum and time average current which would flow into a fully discharged battery. *Hint:* Current will flow only while the instantaneous source voltage exceeds the battery voltage by the 0.7 V required to turn on the diode. Thus, for a fully discharged battery, current will flow only when the instantaneous source voltage exceeds 11.5 + 0.7 V, as indicated in Fig. A6.1b.

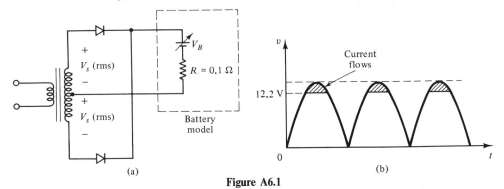

Figure A6.1

A6.2. In Lower Slobovia the ac line voltage varies from 110 V to 130 V, 50 Hz. Your first job as an engineer in the Slobmobile factory is to design a power supply to supply 50 mA at 12 V for testing auto radios. The circuit you select is given in Fig. A6.2. Specify C and R, the power ratings for R and the zener diode, and the peak inverse voltage for the rectifier diode. Use 10 mA for the knee current on the 12-V zener.

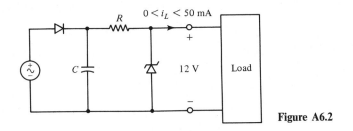

Figure A6.2

A6.3. A three-phase system with a neutral is used in a rectifier configuration as shown in Fig. A6.3. The time-domain line-to-neutral voltages are shown.

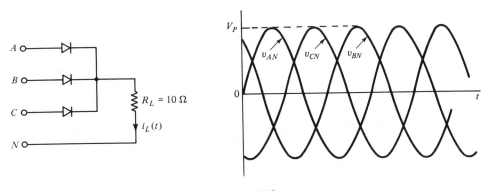

Figure A6.3

(a) If a voltmeter measures 100 V between A and B, what is the peak value of the voltage, V_p?

(b) Sketch the current in the load, $i_L(t)$.

(c) Find the dc current through the load.

A6.4. The circuit shown in Fig. A6.4 consists of a power supply, base input circuit (voltage divider), and transistor amplifier–switch. Your goal is to calculate v_o. The problem is separated into several parts to aid your progress.

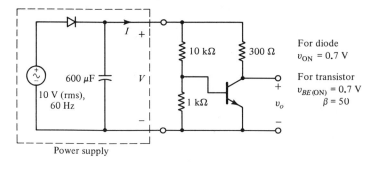

Figure A6.4

(a) The power supply has a box around it. Assume that the ripple is small. The output voltage V will lower as the output current I increases because the capacitor voltage will decrease faster between rechargings. The output is given in Eq. (6.4), where R_L would the output voltage divided by the output current. From this information, and that in Fig. A6.4, determine the Thévenin equivalent circuit for the power supply.

(b) The power supply will furnish current to the base circuit via the voltage divider and also to the transistor collector circuit. Because the output resistance of the power supply is nonzero [part (a)], these two circuits will interact, but for the present we will ignore this interaction. With this assumption, determine the Thévenin equivalent circuit to the base, similar to that in Fig. 6.63b.

(c) Now determine v_o, still ignoring the interaction of the base and collector circuits.

(d) Evaluate the assumption made about the power supply. Would the results of part (c) change if that assumption were denied?

A6.5. The circuit shown in Fig. A6.5 is a zener-regulated power supply.

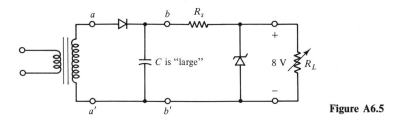

Figure A6.5

(a) If an ac voltmeter measures 10 V rms between a and a', what would a dc voltmeter measure between b and b'? Assume a 0.7-V drop across the ON diode.

(b) What series resistor, R_s, would give 5 mA of current in the 8-V zener when 100 mA of current flows in the load?

(c) If the load were disconnected, what would be the power into the zener?

(d) What would be a suitably "large" value for C, assuming 60 Hz?

A6.6 The circuit shown in Fig. A6.6 will operate as an ac/dc voltmeter, depending on the switch setting. The meter movement responds to dc, 0 to 100 μA full scale.

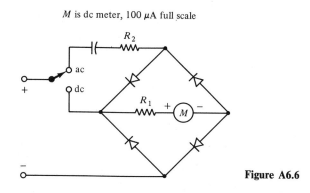

M is dc meter, 100 μA full scale

Figure A6.6

(a) Explain how the circuit works on dc. Mention what would happen if the input polarity were reversed.

(b) Explain how the circuit works on ac. What would happen in this case if the input leads were reversed?

(c) Determine R_1 so that the dc meter indicates full scale with 100 V dc input.

(d) Determine R_2 for the meter to read full scale with 100 V ac (rms) input. Note that the meter movement registers full scale with 100 μA of dc current through it. Assume ideal diodes and that the capacitor is large.

A6.7. An amplifier circuit is shown in Fig. A6.7a. Included in the circuit are an ac ammeter, which responds only to the signal component of the collector current, and a dc voltmeter, which responds only to the dc collector–emitter voltage. The transistor characteristics are shown in Fig. A6.7b. The base circuit resistors, R_1 and R_2, are adjusted such that the dc voltmeter reads 7.5 V.

Figure A6.7

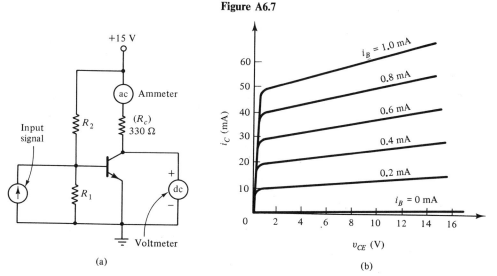

(a)

(b)

(a) What is the dc base current into the transistor?

(b) If the base signal current is increased to its maximum value, stopping short of distortion, what would the ac ammeter read?

(c) Give the effect (increase, decrease, no change, or indefinite) on the two meter readings in response to the following changes (one at a time, then restored to the original condition before the next change)

(1) Decrease base dc current.

(2) Decrease base signal current.

(3) Increase power supply voltage.

(4) Decrease R_c (the 330-Ω resistor).

A6.8. Commercial airplanes carry flashlights which have a red light that flashes occasionally to show that the batteries are still good (and also allow one to find the flashlight in the dark). A circuit that accomplishes this is shown in Fig. A6.8. Assume that the thyristor fires at 3 V. The light-emitting diode (LED) requires at least 10 mA at 0.7 V to glow. Find R_1 and R_2 so that the LED glows for 2.5 ms every 15 s.

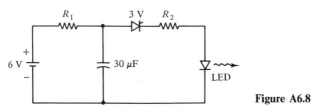

Figure A6.8

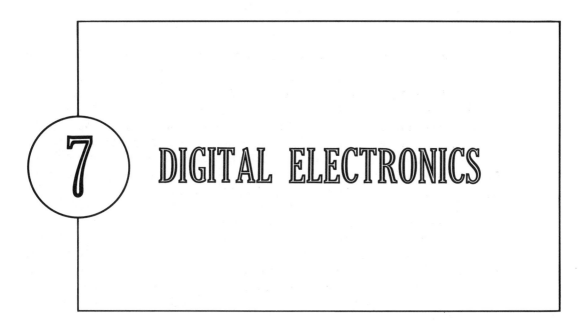

7 DIGITAL ELECTRONICS

7.1 THE DIGITAL IDEA

7.1.1 What Is a Digital Signal?

A Historical Example. "Listen my children and you shall hear/Of the midnight ride of Paul Revere . . . ." According to Longfellow's poem, Paul Revere was sent riding through the New England countryside by a signal from the bell tower of the Old North Church in Boston. "One if by land and two if by sea." That is, one light was to be displayed if the British forces were advancing toward Concord by the road from Boston, and two lights were to be displayed if they were crossing the Mystic River to take an indirect route.

The message received by the Patriot was coded in digital form. We would say today that two "bits" of information were conveyed by the code.* The first light signaled that the British were advancing by either means; no light indicated that they were not yet advancing. The second light indicated by what route they were coming. Because only two routes of advance were envisioned, this second bit of information could be interpreted as indicating one of the two routes.

Information can be communicated in digital form if the message is capable of being defined by a series of yes/no statements. There can be only two states of each variable used in conveying the information. Reducing information to a series of yes/no statements might appear to be a severe limitation on this method, but the method is in fact quite powerful. Numbers can be represented in base 2 and the alphabet by a digital code. Indeed, any situation with a finite number of outcomes can be reduced to a digital code. Specifically, n digital bits can represent 2^n states or possible outcomes. Digital communi-

---

*Strictly speaking, two bits could indicate four possible messages and would require distinguishable lights, say, one red and one white.

cation takes a well-defined code known to the parties at both ends, as in our historical example.

Analysis of the Revere Communication Code.   In order to fix further the idea of digital information, we shall define two digital variables which describe the Paul Revere communication system. Let *B* describe whether the British are coming, and *L* describe the route by which they are coming, provided that they are coming. The mathematical variables, *B* and *L*, are unusual mathematical variables because each can have only two values. We may call those two values by any names we wish: yes/no, true/false, one/zero, high/low, even black/white. When this type of mathematics was used primarily for analysis of philosophical arguments through symbolic logic, the values of the variables were called true or false, according to the validity of the logical propositions being represented. Recently, the names one/zero have come to be preferred by engineers and programmers dealing with digital codes. These names have the obvious advantage of fitting with the binary (base 2) number system for representation of numerical information, but these names occasionally lend confusion to the discussion of digital systems. Nevertheless, we shall use 1/0 or one/zero as our two possible states of the digital variables *B* and *L*. Thus the definitions of *B* and *L* are

$$B = 1 \quad \text{if the British are coming}$$
$$B = 0 \quad \text{if the British are not coming}$$
$$L = 1 \quad \text{if coming by sea}$$
$$L = 0 \quad \text{if coming by land}$$

The first light (B) uniquely determines if he should ride. When the first light appears, he mounts his horse. But he cannot leave until the second light (L) appears (or fails to appear), for the second light reveals the route of the British and hence defines in part the message to be announced.

Representing Digital Information Electrically.   In digital electronics, digital variables are represented by logic levels. At any given time, a voltage is expected to have one value or another, or more precisely to lie within one region or another. In a typical system, a voltage between 0 and 0.8 V would be considered a digital zero, a voltage above 2 V would be considered a digital one, and anything between 0.8 and 2 V would be forbidden; that is, if the voltage fell within this range, you would know that the digital equipment needs repair. These definitions are shown in Fig. 7.1.

As an illustration of a digital circuit, we shall consider the amplifier–switch we analyzed in Chapter 6 as a NOT circuit. The output of a NOT circuit is the digital complement, or the opposite, of the input. First we shall represent the definition of the NOT circuit with a *truth table*. This is shown in Fig. 7.2: *A* represents the input, which may be either 1 or 0; *B* represents the output, which may also be 1 or 0 but depends upon the input. The NOT, or logical complement, operation is indicated algebraically by the equation under the truth table.

We shall now define logic levels for the amplifier–switch we studied in Chapter 6 such that it performs the NOT function. The input–output characteristic of the circuit is

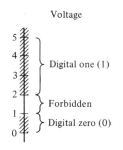

A	B
0	1
1	0

$B = \text{NOT } A = A'$

Figure 7.1 Ranges of voltage represent digital ones and zeros.

Figure 7.2 NOT binary function.

repeated in Fig. 7.3. Clearly, we wish 10 V to be in the region for a 1 and 0.7 V to be in the region for a 0. That is, if the input were 10 V (digital 1), the output should be less than 0.7 V (digital 0), and vice versa. Hence we might consider making the region for a digital 0 to be from 0 to 1 V, and the region for a digital 1 to be, say, from 8 to 10 V. This will work but leaves insufficient range for a working digital system. That is, it is desirable to broaden the range of values in the regions for 1 and 0 to allow for variations in transistors, power supply voltage, noise that might get mixed in with the signal, and the like. In the present case, we can by trial and error determine that the region 0 to 1.5 V as a digital 0 works well with 5 to 10 V as a digital 1; with these definitions the circuit operates as a NOT circuit, that is, it performs the logical complement. These logic levels are shown in Fig. 7.4.

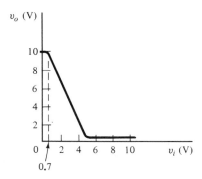

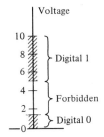

Figure 7.3 Amplifier-switch input-output characteristic.

Figure 7.4 Digital definitions for amplifier-switch.

7.1.2 Digital Representation of Information

An Elevator Door Controller.   Having explained the nature of digital signals, how digital information is represented electrically, and how transistor circuits have the possibility of performing digital operations, we turn to a more complete example showing how to represent a situation in digital form. Our purpose is to introduce the AND and OR digital functions and to illustrate further the language and mathematics of the digital approach.

The door on a typical elevator has a timer on it which closes the door automatically if no one enters the elevator and pushes a button for another floor. It also has an "electric eye" to prevent the door from closing on a passenger. Let us represent a command to the door-closing mechanism with the binary variable D ($D = 1$ if the door is to close). The state of the door (D) will be controlled by three binary variables: T represents the state of the timer ($T = 1$ means that the timer is running, time has not yet elapsed); B represents the results of someone's pushing a button for another floor ($B = 1$ means that a button has been pushed); and S represents the state of the safety device ($S = 1$ means that someone is in the door). We see that D is the dependent variable and is a function of three independent variables (T, B, and S). Keep in mind that these are all binary variables and hence can be only 1 or 0.

$$D = f(T, B, S)$$

How might such a mathematical function be described? One useful method for describing a binary function is a truth table, as shown in Fig. 7.5. Here we have enumerated all possible combinations of the independent variables and shown the appropriate value of the dependent variable. In general, when there are n independent variables, each having two possible states, there will be 2^n possible combinations (2^3 in this case), which may be enumerated to define the function. The truth table offers a systematic form for displaying such an enumeration. The name *truth table* originated historically from the application of this type of representation to the systematic investigation of logical arguments. Because of this association, digital circuits are often called *logic circuits.*

T	B	S	D
0	0	0	1
0	0	1	0
0	1	0	1
0	1	1	0
1	0	0	0
1	0	1	0
1	1	0	1
1	1	1	0

Figure 7.5 Truth table for elevator door controller.

A Truth-Table Representation.   Figure 7.5 presents the truth table for the elevator door function. Let us see how we determined the 1's and 0's in Fig. 7.5. The 1's and 0's in the first three columns resulted from a systematic counting of the eight possibilities, or *states.* We call it "counting" because the pattern we have used constitute counting in the base 2 number system. However you think of it, the pattern is clear: we alternated the 1's and 0's fastest for S, slower for B, and slowest for T, thus covering all possible combinations. These represent the values of our independent variables. For filling out the 1's and 0's in the last column, we looked first at the S column, which represents the safety switch. We do not want the door to close when the safety switch indicates that someone stands in the door ($S = 1$), so we put 0 in the D column ($D = 0$ means do not close the door) for every 1 in the S column. This accounts for four of the eight states. The other four states depend on the button and the timer. If $S = 0$ (nothing blocking the door), the door should close if either the button is pushed ($B = 1$) or the timer expires ($T = 0$). We examine the remaining four states and put a 1 in column D if there is a 1 in the B column or a 0 in the T column. Using this rule, we find three combinations leading to the closing of the door. The first, all 0's, represents the timer running out to close of the door. The second represents a button being pushed and the timer running out simultaneously, and the third represents a button being pushed before the timer runs out. Of course, in all three cases the safety mechanism allows the door to close.

The NOT Function.   The truth-table method is a brute-force way for describing a binary function. The same information can be represented algebraically through the AND, OR, and NOT binary functions. Consider first the NOT function, the logical complement, described in the truth table of Fig. 7.2. The NOT function is involved in this problem because NOT S allows the door to close, and NOT T prompts the closing of the door by the timer. The NOT function is represented algebraically by a prime added to the variable or expression to be NOTed: S' means NOT S. We need the NOT when a 0 is to trigger the OR or AND combinations because these trigger on a 1.

The OR Function.   The OR binary function is defined in Fig. 7.6. The dependent variable, $C = A$ OR B, is 1 when either A or B (or both) is (are) 1. This is thus the *inclusive* OR because it includes the case where both A and B are 1. The OR function is involved in our elevator problem in describing the combined effect of the timer and the button. We wish the door to get the signal to close when the timer elapses ($T = 0$) OR the button is pushed ($B = 1$) or both. The way to express this algebraically is T' OR B. The truth table for this function is shown in Fig. 7.7. In constructing the truth table in Fig. 7.7, we added a NOT T column. Then we put a 1 in the last column wherever there was a 1 in either of the previous two columns because these are the variables we are ORing.

$C = A$ OR B

A	B	C
0	0	0
0	1	1
1	0	1
1	1	1

Figure 7.6   OR function.

T' OR B

T	T'	B	
0	1	0	1
0	1	1	1
1	0	0	0
1	0	1	1

Figure 7.7   Binary function T' OR B.

The AND Function.   Next we need to account for the safety switch. The AND function is required because we must express the simultaneous occurrence of an impulse to close the door and the lack of an obstacle in the door. The truth table for the AND function appears in Fig. 7.8. Here we get a 1 only when both A and B are 1. To complete the truth table for our door-closing variable (D), we must AND S' with the last column in Fig. 7.7 in order to cover all the possibilities. Thus we may state the door-closing function as

$$D = (T' \text{ OR } B) \text{ AND } (S') \tag{7.1}$$

$C = A$ AND B

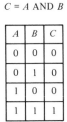

A	B	C
0	0	0
0	1	0
1	0	0
1	1	1

Figure 7.8   AND function.

When interpreted as a binary or logical function, Eq. (7.1) states algebraically the same information as the truth table in Fig. 7.5, which we worked out by considering all possible combinations. We have now introduced a method of representing information in digital form and we have defined some basic logic relationships. We turn next to describing how electrical circuits can perform logical operations such as AND and OR.

7.2 THE ELECTRONICS OF DIGITAL SIGNALS

7.2.1 The NOT Circuit

Circuit Improvements.   In Section 7.1.1 we showed that the transistor amplifier–switch circuit performs the digital NOT function, provided that we define a digital 1 as any voltage between 5 and 10 V and a digital 0 as any voltage between 0 and 1.5 V. We propose here three modifications of the circuit to improve its performance as a NOT circuit. Specifically, we will add two diodes to the base circuit, lower the resistance in that circuit to 10 kΩ, and lower the power supply voltage to 5 V, all shown in Fig. 7.9.

Analysis of the Modified Circuit.   Let us investigate the input–output characteristic of the modified circuit to appreciate the improvements brought by these changes. In our investigation we shall assume the base emitter pn junction and the diodes are OFF when the voltage across them is less than 0.7 V. After they are turned ON and current begins to flow, their voltage will remain at 0.7 V. This is the simple model of the pn junction shown in Fig. 6.52. We shall now derive the input–output characteristic of the circuit in Fig. 7.9. Let us increase the input voltage, beginning at zero volts. No current will flow into the base of the transistor until both diodes and the base–emitter junction turn

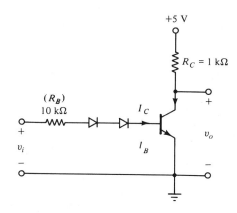

Figure 7.9 Improved NOT circuit.

ON. This requires about 3×0.7 or 2.1 V at the input. Once the voltage at the input rises to 2.1 V, therefore, the three diodes will turn ON, and the transistor will move out of cutoff. Once this occurs, Kirchhoff's voltage law in the base–emitter loop takes the form

$$i_B = \frac{v_i - 2.1}{R_B} = \frac{v_i - 2.1}{10 \text{ k}\Omega} \tag{7.2}$$

The input current into the base will control the flow of current in the collector–emitter loop, which is the output part of the circuit. While the transistor is cutoff, no collector current will flow and the output voltage will remain at +5 V. As the base current begins to flow, however, the transistor will move into the active region and the collector (output) voltage will begin to fall. This moves the operating point from cutoff toward saturation as shown on the load line in Fig. 7.10. Because we changed the power supply voltage, the load line now goes from +5 V on the voltage axis (V_{CC}) to 5/1 kΩ = 5 mA on the current axis. Because the transistor beta (β) is about 100, a base current of about 5 mA/100 = 50 μA will be required for saturation; hence we have put the 50-μA characteristic on Fig. 7.10. Base currents between 0 and 50 μA put the transistor in the active region. Equation (7.2) requires, therefore, that input voltages between 2.1 and 2.6 V correspond to the active region. After the input voltage exceeds 2.6 V, the transistor will be saturated; that is, the collector voltage (v_o) will have fallen to about 0.4 V and further

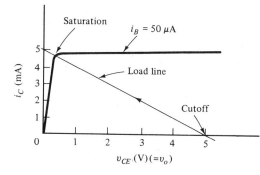

Figure 7.10 Load line for improved circuit.

increases in base current cause little change in the output. Thus the input–output characteristic of the modified amplifier–switch will be as shown in Fig. 7.11.

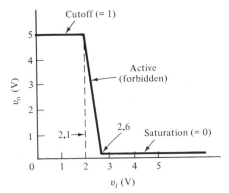

Figure 7.11 Input–output characteristic of improved NOT circuit.

The Benefits of Changing the Circuit.   If you compare the characteristic in Fig. 7.11 with that in Fig. 7.3, you will note several differences. The slope is greater (higher gain) in the active region, a result of decreasing the resistance in the base circuit. We want higher gain so that the transistor passes through the active region (the forbidden region in digital operation) with a smaller range of input voltage and hence passes through with greater speed. Next we notice that the modified circuit remains in cutoff until the input reaches over 2 V, in contrast to 0.7 V for the unmodified circuit. This results from placing the two diodes in the base circuit. Their function is to increase the range for a digital 0 and to enhance symmetry between the allowed ranges of the digital 0 and the digital 1. Finally, we note that the output voltage for a 1 is decreased from +10 V to +5 V. We lowered the value of V_{CC} for two reasons: to save diodes and to save power. We added two diodes to raise the threshold of the active region to roughly half of the +5 V. If we had to add enough diodes to raise the active region up to one-half of 10 V, we would have had to use five or six diodes; thus we save diodes by lowering the power supply voltage. We also save power by lowering the voltage since the saturation current is lowered correspondingly. The power used by the circuit when saturated is approximately V_{CC} times the saturation current; hence we use one-fourth the power with the smaller V_{CC}.

Logic Levels for the Modified NOT Circuit.   Appropriate digital levels for the modified amplifier–switch are shown in Fig. 7.12. The range 0 to 1.0 V defines a digital 0, 4.0 to 5.0 V defines a digital 1, and the range from 1.0 to 4.0 is forbidden. We desire these broad, symmetrical regions for several reasons. Digital equipment is reliable and inexpensive because we do not have to be careful about the exact voltage levels, as long as they lie in the ranges for a 0 or 1. Furthermore, broad regions for the logic levels immunize the digital circuits to noise and false signals as far as possible: you do not want your digital clock resetting itself due to noise on the power line. We will use these voltage ranges as our logic levels for the remainder of this section; that is, when we write 1 (meaning a digital 1, a symbol and not a number) we mean a voltage between 4.0 and 5.0 V, and when we write 0 we mean a voltage between 0 and 1.0 V.

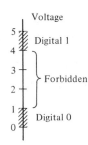

Figure 7.12   Logic levels for improved NOT circuit.

7.2.2 Simple Gates

The NOR Gate.   In common language a gate is a device for admitting (or excluding) someone to (or from) a fenced region. In digital systems, gates are circuits that pass or inhibit signals moving through a logic circuit. We will now examine the NOR gate, a circuit that combines the OR function with the NOT function. A simple NOR circuit is shown in Fig. 7.13. The inputs are A and B, and the output is C. We shall justify the truth table shown beside the circuit. Because there are two inputs, there will be $2^n (n = 2)$ or 4 possible input combinations for A and B, which we have listed systematically in the truth table. The first of these (00) corresponds to having voltages below 1.0 V at both inputs. Although the four *pn* junctions (three diodes and the base–emitter junction) are slightly forward biased, insufficient voltage is present to turn them ON and in particular the transistor base–emitter junction will not turn ON; hence the transistor will remain cut off. This cutoff condition causes the output to be $+5$ V, a digital 1; hence we place 1 in the C column, first row. The next row has a digital 1 at B and a digital 0 at A. The voltage at B will turn ON diodes D_2, D_3, and the base–emitter junction, the last leading to saturation and a digital 0 at the output. Notice that the voltage at P is at least $4 - 0.7 = 3.3$ V and

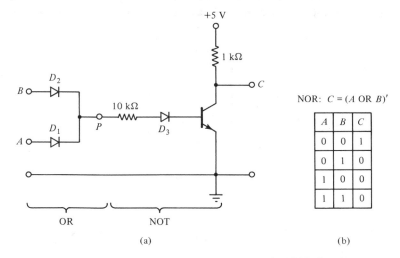

NOR: $C = (A \text{ OR } B)'$

A	B	C
0	0	1
0	1	0
1	0	0
1	1	0

Figure 7.13   (a) NOR circuit; (b) truth table for NOR function.

the voltage at A is at most 1.0 V; hence D_1 is OFF. This diode, acting as an open circuit in its OFF condition, prevents the signal at B from coupling back into the source of A. The last two rows in the truth table follow from similar considerations: clearly, if either (or both) of the inputs is (are) a 1, the transistor will turn ON, current will flow in the saturated transistor, and the output voltage will drop into the range for a digital 0. The diodes perform the OR operation and the transistor gives the NOT. Incidentally, we are not limited to two inputs: we can have three, four, or more inputs coming into the OR part of the circuit.

Whatever goes on at the output C is isolated by the transistor from affecting the inputs. The transistor firms up the output of the diode OR gate and isolates the input circuit from the output circuit. In the process of offering these benefits, the transistor inverts the digital signal and we thus pick up the NOT operation. If we require an OR circuit, we would have to put C into a NOT circuit; this would NOT the NOR to give the OR operation.

The NAND Gate.   The NAND circuit shown in Fig. 7.14 combines the AND and the NOT operations. Ignore for the moment the two inputs with their diodes and think of the circuit as a NOT circuit with the input (to the 10-kΩ resistor) connected to the $+5$-V power supply. Without action at the A and B inputs, the circuit would be a NOT circuit with the input locked at a digital 1 and the output locked therefore at 0. Now let us consider the effect of the inputs. The first row of the truth table has both A and B at low voltage, below 1 V. This will cause both D_1 and D_2 to turn ON and current will flow through the 10-kΩ resistor, through the diodes, and to ground through whatever is controlling A and B. The voltage at P will be at most $1 + 0.7 = 1.7$ V, not enough to turn ON diodes D_3 and D_4 and the base–emitter junction. Hence the transistor will be cutoff and the output voltage will be $+5$ V, a digital 1. The next state (01) leads to similar operation, the only difference being that the voltage at B is now at least 4 V and hence D_2 is OFF. But the voltage at P remains no higher than 1.7 V and the output remains at 1. Only if both A

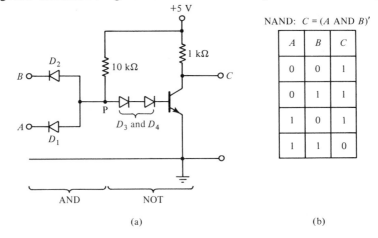

(a)   (b)

Figure 7.14   (a) NAND circuit; (b) truth table for NAND function.

and B are a digital 1 does the current through the 10-kΩ resistor return to the transistor base, turn ON the base–emitter junction, saturate the transistor, and drop the output voltage to a digital 0. The input diodes perform the logical AND function and the transistor the NOT function, giving the entire circuit a NAND function. If we wish the AND function, we can invert C with a NOT circuit. Note that if we add other inputs in parallel with A and B, all would have to be at a digital 1 to give a digital 0 at the output.

The NOR, NAND, and NOT circuits are the basic building blocks from which digital systems are constructed. We shall consider how large systems are formed after we have refined our mathematical language.

7.3 THE MATHEMATICS OF DIGITAL ELECTRONICS

7.3.1 Introduction

The Need for a Mathematical Language.   We showed earlier how to express a digital function with a truth table. We termed this a brute-force method for, in a truth table, we list all possible states of the independent variables (the inputs) and list the corresponding values of the dependent variables (the outputs). This way of expressing a digital function has several limitations. The truth table offers little guidance about how to accomplish the required logical operations with digital circuits in order to produce the desired output. That is, the truth table yields few hints about how to "realize" the required logical function. In our example about the elevator door, for example, we were able to translate the problem description into a logical function by common sense rather than through examination of the truth table. The mathematics we shall develop in this section not only suggests a realization of the logical function but also permits manipulation of the function into different forms, thus offering alternative realizations. Also, we shall demonstrate ways to simplify logical expressions, thus permitting simpler realizations.

Origins of Boolean Algebra.   The mathematics of two-valued variables is called Boolean algebra after George Boole, an English mathematician who first investigated this type of mathematics. Boole was interested in symbolic logic, the formal examination of logical arguments to establish their soundness (or expose their fallacies). As we have already remarked, this early application of the mathematics has influenced the terminology of digital electronics.

A Warning.   Whereas the ideas, theorems, and applications of Boolean mathematics are relatively simple, say, as compared to those of calculus or spherical trigonometry, the nomenclature and language can be confusing. One difficulty is that this mathematics uses some of the same symbols as ordinary algebra, but with different meanings. Thus $A + B = C$ is a meaningful equation in both systems but has totally different meaning and is read differently as a binary expression. For example, the equation $1 + 1 = 1$ is correct in Boolean mathematics, but is incorrect in ordinary algebra. Another difficulty is that common English words such as "and" are given a technical meaning. Until you become accustomed to this new usage, many of the statements about Boolean variables sound like double talk. We shall point out some of these peculiarities as we encounter them in the following introduction to Boolean algebra.

7.3.2 Common Boolean Theorems

Boolean Variables.   We have already stated that a Boolean (or digital, or binary) variable can have two values, which we call 1 and 0. The values are *defined* to satisfy the definitions of OR, AND, and NOT in Fig. 7.15. We note that some of these definitions are invalid if 1 and 0 are considered numbers, but are valid definitions for 1 and 0 as symbolic representations of the two possible states of a binary variable. We note that the + sign is used for OR, the "·" symbol for AND, and the prime for NOT or logical complement.

OR	AND	NOT
$1 + 1 = 1$	$1 \cdot 1 = 1$	$1' = 0$
$1 + 0 = 1$	$1 \cdot 0 = 0$	$0' = 1$
$0 + 1 = 1$	$0 \cdot 1 = 0$	
$0 + 0 = 0$	$0 \cdot 0 = 0$	

Figure 7.15   Basic definitions of OR, AND, and NOT functions.

One-Variable Theorems.   The theorems involving one variable, here A, are shown in Fig. 7.16. All of these may be verified by testing the validity of the expression for both states of A and comparing with the definitions in Fig. 7.15. For example, the bottom expression under AND is verified by the truth table shown in Fig. 7.17.

OR	AND	NOT
$1 + A = 1$	$1 \cdot A = A$	$(A')' = A$
$0 + A = A$	$0 \cdot A = 0$	
$A + A = A$	$A \cdot A = A$	
$A + A' = 1$	$A \cdot A' = 0$	

Figure 7.16   Boolean theorems for one variable.

A	$A \cdot A'$	DEF
1	$1 \cdot 1'$	$1 \cdot 0 = 0$
0	$0 \cdot 0'$	$0 \cdot 1 = 0$

Figure 7.17   Proof of a theorem with a truth table.

Two or Three Variables.   Some useful theorems and properties involving two or three binary variables are shown in Fig. 7.18. Again we see that many of these are deceptively similar to the familiar properties of algebra. Also note that we show the expressions involving AND operations both with and without the "·" symbol. Writing the AND without any symbol is common, even though the confusion with ordinary multiplication is compounded.

All of the relationships in Fig. 7.18 may be verified by examination of all possible states and then applying the earlier properties and definitions. Because the absorption

Commutation:	$A + B = B + A$; $A \cdot B = B \cdot A$; $AB = BA$
Association:	$A + (B + C) = (A + B) + C$
	$A \cdot (B \cdot C) = (A \cdot B) \cdot C$; $A(BC) = (AB)C$
Absorption:	$A + (A \cdot B) = A$; $A \cdot (A + B) = A$; $A(A + B) = A$
Distribution:	$A \cdot (B + C) = (A \cdot B + A \cdot C)$; $A(B + C) = AB + AC$
	$A + (B \cdot C) = (A + B) \cdot (A + C)$; $A + BC = (A + B)(A + C)$
DeMorgan's Theorems:	$(A + B)' = A' \cdot B'$; $(A + B)' = A'B'$
	$(A \cdot B)' = A' + B'$; $(AB)' = A' + B'$

Figure 7.18   Some useful theorems and properties involving two or three binary variables.

rules appear the strangest, we select the first of these for verification in Fig. 7.19. Note that the last column is identical to the column for A, showing that the theorem is valid. Examination of the truth table shows how B gets "absorbed" by A: if $A = 1$, B does not matter; and if $A = 0$, B does not matter. Hence B is absorbed. The proof of the second absorption rule we will reserve for a practice problem at the end of the chapter.

A	B	AB	$A + AB$
0	0	0	0
0	1	0	0
1	0	0	1
1	1	1	1

Figure 7.19   Truth-table proof of an absorption theorem.

De Morgan's Theorems.   De Morgan's theorems reveal how to distribute the NOT over variables which are AND'ed or OR'ed. These relationships are not obvious and are sufficiently important to merit proof, as given in Fig. 7.20 for the first theorem. Notice that the fourth and seventh columns are identical, thus proving the theorem. Proof of the second of the theorems we will save for a homework problem. The form of De Morgan's theorems is that on the left side of the identities, the NOT covers two variables which are either AND'ed or OR'ed, and on the right side the NOT has been distributed to the individual variables. Both rules are summarized by the following statement: A NOT can be distributed in a logical expression involving two variables provided that ANDs are changed to ORs, and vice versa. This rule can be applied to expressions involving more than two variables, provided that care is taken with the grouping of variables.

The importance of De Morgan's theorems in digital electronics issues from the inversion of the output signal relative to the input(s) by the transistor. We saw before in

A	B	$A + B$	$(A + B)'$	A'	B'	$A' \cdot B'$
0	0	0	1	1	1	1
0	1	1	0	1	0	0
1	0	1	0	0	1	0
1	1	1	0	0	0	0

Figure 7.20   Truth-table proof of one of De Morgan's theorems.

examining NOR and NAND gates that the diodes at the input perform the OR or AND logic and the transistor firms up the decision, isolates the input from the output, and inverts the signal in the process, thus adding the NOT unavoidably to the logical operation. This "NOT" makes De Morgan's theorem useful and important in digital electronics.

We will explain the usefulness of De Morgan's theorems in digital electronics after we have introduced symbols for the logical gates. For now, we shall illustrate the use of the various theorems in simplifying logical expressions. In Eq. (7.3) we show that a logical expression which appears to depend on two input variables is in fact constant and remains in the 0 state, regardless of the values of A and B.

$$[(A \cdot B)' + A]' = (A' + B' + A)' = (1 + B')' = 1' = 0 \tag{7.3}$$

In Eq. (7.4) we simplify an expression and discover that the second term adds nothing new to the first. Notice that we used the absorption rule to go from the second form to the third.

$$(A \cdot B)' + (A' + B)' = A' + B' + A'' \cdot B' = A' + B' = (A \cdot B)' \tag{7.4}$$

EXCLUSIVE OR and the Equality Function.   The OR operation is, as we pointed out, an *inclusive* OR, meaning that it includes the case where both variables are 1. It is also useful to define an "EXCLUSIVE OR" function, that is, a logical function which takes the value 1 if either variable is 1 but takes the value 0 if both variables are either 1 or 0. The truth table for such a function is given in Fig. 7.21. The third column contains the definition of the EXCLUSIVE OR; note that the symbol for this operation is a + inside a circle. The bottom row of the table shows the exclusion of the state where both inputs are 1. The EXCLUSIVE OR can be thought of as indicating inequality between the variables, for the output 1 of the EXCLUSIVE OR indicates that A and B are unequal. It follows that the complement of the EXCLUSIVE OR would indicate equality, as the last column of Fig. 7.21 shows.

A	B	$A \oplus B$	$(A \oplus B)'$
0	0	0	1
0	1	1	0
1	0	1	0
1	1	0	1

Figure 7.21   EXCLUSIVE OR function.

7.4 ASYNCHRONOUS DIGITAL SYSTEMS

7.4.1 Logic Symbols and Logic Families

Logic Symbols.   Digital systems consist of vast numbers of NAND, NOR, and NOT gates, plus memory and timing circuits which we will discuss later, all interconnected to perform some useful task, such as count and display time, measure a voltage, or perform arithmetic operations. If we were to draw a circuit diagram for such a system, including all the resistors, diodes, transistors, and interconnections, we would face an overwhelming task, and an unnecessary one. The task would be unnecessary because anyone who read the circuit diagram would in their mind group the components together into standard circuits and think in terms of the "system" functions of the individual gates. For this reason, we design and draw digital circuits with standard logic symbols, as shown in Fig. 7.22.* The small circle at the output indicates the inversion of the signal. Thus without the small circle, the triangle would represent an amplifier (or buffer) with a gain of unity, and the second symbol would indicate an OR gate. As shown above, however, the common circuits are those which invert. These logic symbols show only the input and output connections. The actual gates, when wired into a digital circuit, would have power supply (V_{CC}) and grounding connections as well. Figure 7.23 shows the connections for a quadruple, two-input NAND gate. Notice that the input power is applied between pins 14 and 7, with 7 grounded.

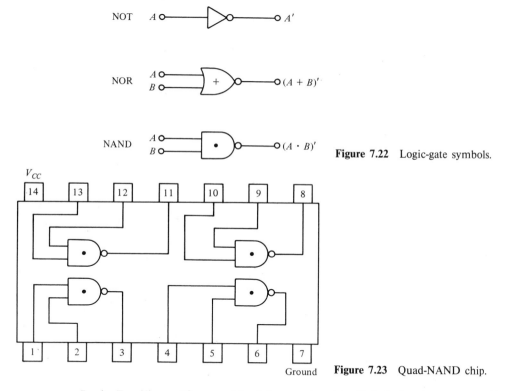

Figure 7.22   Logic-gate symbols.

Figure 7.23   Quad-NAND chip.

Logic Families.   If you wished to construct a digital circuit, you would not assemble a pile of diodes, resistors, and transistors and proceed to wire them together, first into standard gate circuits, then into larger functions. You would purchase the gates already fabricated on an integrated circuit (IC), and packaged within a plastic capsule, as suggested by Fig. 7.24. This commercial IC "chip" includes four NAND gates packaged

*The + (OR) and · (AND) markings in the gates are optional. The shape of the gate symbol defines its function.

Figure 7.24 Physical appearance of a logic chip.

together. An important part of the design would be to select a particular "logic family," depending on the nature and working environment of your eventual product. The logic families are composed of a large selection of compatible circuits which can be connected together to make digital systems. The logic families differ in the details of the circuits used to perform the logical operations. Here are some of the logic families.

1. Diode-transistor logic (DTL) circuits are similar to the circuits in Figs. 7.13 and 7.14. These circuits are now obsolete. We used this simple type of logic only to illustrate the principles of logic gates.

2. High threshold logic (HTL) circuits are similar to DTL gates but include a zener diode in place of the two series diodes in Figs. 7.13 and 7.14, and use a larger power supply voltage, V_{CC}. The zener diode raises the threshold level for switching the transistor and hence separates the voltage regions for a 1 and 0 by a large margin, say 10 V. This logic family is useful where electrical noise is a problem, to prevent moderate noise signals, which might leak into the circuit, from affecting the circuit as valid digital signals. If, for example, your circuit must operate adjacent to a large dc motor or an arc welding machine, you would use HTL circuits.

3. Transistor-transistor logic (TTL) circuits use special-purpose transistors in place of the diodes. These circuits are widely used because they switch rapidly, require modest power to operate, and are inexpensive. The circuits used in this logic family are considerably more complicated that the DTL circuits.

4. Complementary metal-oxide semiconductor (CMOS) logic circuits use field-effect transistors (FETs), which differ from the type of transistors we presented in Chapter 6. These circuits require very little power to operate and are used where low power consumption is an important requirement, as in battery-operated calculators.

5. Emitter-coupled logic (ECL) gates switch very fast and are used in high-speed circuits such as high-frequency counters.

There are even more logic families than these. The design of logic circuits is highly sophisticated, and specialists in this area must become intimately familiar with all the possible products and logic families which are available at a given time.

7.4.2 Realization of Logic Functions

The NOT Function.   Often when a NOT circuit is required, the designer will make one out of a NOR or a NAND circuit. Figure 7.25 shows the two ways to make a NOT (or inverter) out of a NAND circuit. These realizations are based upon the first and third rows under "AND" in Fig. 7.16. The input to be fixed at a digital 1 in the lower realization would be attached to the V_{CC} power supply through a resistor. Similarly, if one required an input to be fixed at 0, this input would be grounded. Grounding an input can be used to realize the NOT function with a NOR gate, which we will leave for a practice problem at the end of the chapter.

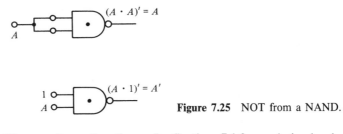

Figure 7.25   NOT from a NAND.

Realizing the Elevator Door Function.   In Section 7.1.2 we derived a logical expression for closing an elevator door ($D = 1$ means close) based on a timer ($T = 1$ means the timer is still running, do not close the door), a button switch ($B = 1$ means someone has pushed the button for another floor, close the door), and a safety device ($S = 1$ means that someone is blocking the door, do not close the door). The logical expression, recast into the notation we have developed, is given in Fig. 7.26. We first shall realize the function with NAND and/or NOR gates. The realization in Fig. 7.26 utilizes six gates: one NOR, one NAND, and four NOTs, which were accomplished with NANDs. This realization is based on direct translation of the logical expression into logic-gate symbols.

We may accomplish a simpler realization by manipulating the expression for D

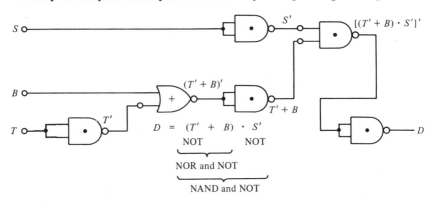

Figure 7.26   Straightforward realization of the elevator door function using NAND and NOR gates.

into a more convenient form. Equation (7.5) shows such a manipulation.

$$D = [[(T' + B) \cdot S']']' = [(T' + B)' + S'']' = [(T' + B)' + S]' \qquad (7.5)$$

In Eq. (7.5) the first form is the same expression for D as in Fig. 7.26, except that we NOTed it twice. We do this because we want the final result for D to be primed, that is, to be the NOT of something. This is required if we are to realize the final operation with a NOR or NAND gate. The second form results from using De Morgan's theorem to distribute one of the NOTs to the individual terms. The third form is the same as the second, except that we have removed the double complement from S. This final form proves convenient for realization with NOR gates. Figure 7.27 shows the realization. Note that we made a NOT out of NOR similar to the way we realized a NOT with a NAND earlier. De Morgan's theorem thus leads to a simpler realization.

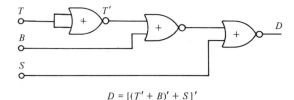

$$D = [(T' + B)' + S]'$$

Figure 7.27 Simple realization of the elevator door function using NOR gates.

7.4.3 Binary Arithmetic

Number Bases. High-speed arithmetic is one of the spectacular achievements of digital electronics. Computers perform arithmetic with logic circuits through the representation of numbers as *binary numbers*, that is, in base 2 form. In this section we present the bare outlines of how binary arithmetic can be accomplished by digital circuits.

The development of efficient arithmetic methods was retarded for centuries by the lack of a convenient system and notation for the representation of numbers. You earned the equivalent of a Ph.D. in ancient Egypt if you managed to work out the answer to something like 426×678. Their hindrance was that the multiplication tables were endless, and if you do not believe us try to multiply two two-digit numbers with Roman numerals.

The breakthrough came, of course, when the 10 Arabic* numerals were used to write numbers in base 10, that is, allowing the repetition of numerals, with the position representing powers of ten. For example, the number 806.1 means

$$806.1_{10} = 8 \times 10^2 + 0 \times 10^1 + 6 \times 10^0 + 1 \times 10^{-1}$$

The advantage of such a system is that the addition and multiplication tables assume manageable size; and the rules for applying these properties follow simple patterns.

The triumph of the base 10 number system was so successful that, until recently, few could write and perform arithmetic in some other base, say, base 5. This is no longer true because "modern math" in the secondary school system includes arithmetic in

---

*Actually, these symbols are thought to have been first used in ancient India and introduced into Western society through Islamic culture.

nondecimal number bases. You presumably required little explanation of the binary counting already used in Fig. 7.5. In that table, the first row represents zero, the second one, and the seventh, for example, represents six in binary form:

$$6_{10} = 110_2 = 1 \times 2^2 + 1 \times 2^1 + 0 \times 2^0$$

Thus it takes three binary digits, or *bits*, to represent six in base 2, or binary, form. In general, n bits can represent numbers 0 to $2^n - 1$ and, if we wish to consider signed numbers (plus or minus), we require another bit to represent the sign. An ordered grouping of binary information, a group of bits, is called a *word*. Thus a digital computer would represent a number as a word of, say, 32 bits, and hence be limited to numbers smaller than about 1 billion. From programming courses, you realize that we are speaking here of an integer format (or straight binary) for the numbers; we can, of course, use a floating (exponential) form to represent a wider range of numbers.

Binary-Coded Decimal (BCD).   Thus far we have presented pure decimal and pure binary representation of numbers. A hybrid system, binary-coded decimal (BCD), is frequently used in calculators and digital instrumentation. With BCD, the decimal format of the numbers is preserved, but each digit is represented in binary form. Because we must represent 10 decimal possibilities $(0, 1, 2, \ldots, 9)$, we require 4 bits of digital information for each decimal digit to be represented. Figure 7.28 offers an example, representing 906_{10} as a 12-bit BCD word. From Fig. 7.28 we deduce that n bits (where n is a multiple of four) can represent numbers up to $10^{n/4} - 1$ in BCD form. This represents a reduction from what we can represent with pure binary, but facilitates input and output interactions between system and operator. Because we think in decimal, for example, we want 10 keys on our calculators for entering numbers and we also require outputs in decimal. The internal manipulation of the binary information is complicated somewhat by the BCD form, but this is the problem of calculator designers, who obviously are up to the challenge.

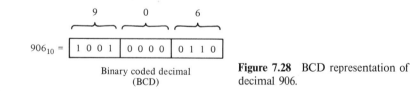

Binary coded decimal (BCD)

Figure 7.28   BCD representation of decimal 906.

The Hexadecimal System.   With the emergence of microcomputers, which work with words of 8 and 16 bits, the hexadecimal system has gained importance. Hexadecimal is base 16 and requires six "new" number symbols in addition to the 10 Arabic numerals. For convenience in using standard printers the letters A through F are used for 10 through 15, respectively. Thus in hexadecimal the number A8F represents

$$A8F_{16} = 10 \times 16^2 + 8 \times 16^1 + 15 = 2703_{10}$$

The numbers zero through sixteen are represented in decimal, hexadecimal, and binary in Table 7.1.

TABLE 7.1   NUMBERS IN DECIMAL,
HEXADECIMAL, AND BINARY

Decimal	Hexadecimal	Binary
0	0	0000
1	1	0001
2	2	0010
3	3	0011
4	4	0100
5	5	0101
6	6	0110
7	7	0111
8	8	1000
9	9	1001
10	A	1010
11	B	1011
12	C	1100
13	D	1101
14	E	1110
15	F	1111
16	10	10000

7.4.4 Digital Arithmetic Circuits

A BCD Adder.   Let us design a logic circuit to add two decimal digits represented in BCD form. The problem is symbolized by

$$A + B = S \Rightarrow A_4A_3A_2A_1 + B_4B_3B_2B_1 = S_4S_3S_2S_1 \tag{7.6}$$

where in the second form the A's and B's with the subscripts are binary variables representing the 4 bits required for the BCD representation.*

First Stage of the Adder.   Addition in both decimal and binary is shown below.

$$
\begin{array}{cc}
 & 1 \\
9_{10} & 1001_2 \\
\underline{5_{10}} & \underline{0101_2} \\
14_{10} & 1110_2
\end{array}
$$

We have shown the carry for the binary addition in the traditional manner to emphasize that we must consider carries in the design of our circuit. That is, when we add in the lowest-order bits, 1 and 1, we obtain 10_2, which is written as a 0 with a carry of 1. Thus our binary circuit which adds the lowest order bit must produce two binary outputs: the lowest-order bit of the sum and the carry to be added with the next-higher-order bits. Figure 7.29 indicates what the first stage of addition must accomplish. It must take two binary inputs, A_1 and B_1, and produce two outputs, the lowest-order bit of the sum

---

*In the development which follows, you will see that we need an additional carry bit to handle all possibilities.

(S_1) and the carry (C_1) to the next-higher stage of addition. Examination of the truth table in Fig. 7.30 reveals that the carry bit is the AND of the inputs and the sum bit is the EXCLUSIVE OR of the two inputs,

$$C_1 = A_1B_1 \qquad \text{and} \qquad S_1 = A_1 \oplus B_1 \qquad (7.7)$$

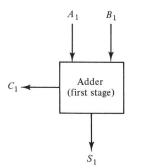

Inputs		Outputs	
A_1	B_1	C_1	S_1
0	0	0	0
0	1	0	1
1	0	0	1
1	1	1	0

Figure 7.29   The first stage of the adder has two inputs and two ouputs.

Figure 7.30   Truth-table representation of the adder inputs and outputs.

We thus need a realization for the EXCLUSIVE OR operation. We may express the EXCLUSIVE OR in terms of OR and AND functions through either of the following equivalent forms,

$$A_1 \oplus B_1 = A_1B_1' + A_1'B_1 = (A_1 + B_1)(A_1B_1)'$$

The first form expresses the two ways the function can be 1, namely, either (A_1 is 1 AND B_1 is 0) OR (A_1 is 0 AND B_1 is 1). The second form uses the ordinary OR and then removes the case where both A_1 and B_1 are 1 by ANDing the ordinary OR with $(A_1B_1)'$. We can manipulate either expression into a form suitable for realization with NAND and NOR gates, but in this case the second expression should be chosen because it involves the AND of the two inputs, which we need for the carry bit. Using the technique illustrated on page 282, we express the EXCLUSIVE OR in the form

$$A_1 \oplus B_1 = [[(A_1 + B_1)(A_1B_1)']']' = [(A_1 + B_1)' + (A_1B_1)'']' \qquad (7.8)$$

We will use only NAND and NOR gates in our realization and there is hence no benefit in canceling the double NOT on the A_1B_1 term. Figure 7.31 shows the realization of the lowest bit adder, and represents in detail what is indicated by Fig. 7.29.

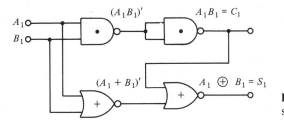

Figure 7.31   Realization for the first stage of the adder.

Higher Stages of the Adder. The second and higher stages of the adder must consider the carry from the next lower stage. Thus these stages of addition must function with three inputs and two outputs, as indicated by Fig. 7.32 for the second stage. In Fig. 7.33 we show the truth table relating the two outputs to the three inputs. We will not work out the complete realization for the higher-order stages of the BCD adder, beyond working out the Boolean expressions for the outputs.

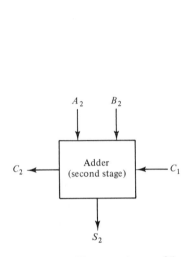

Inputs			Outputs	
A_2	B_2	C_1	C_2	S_2
0	0	0	0	0
0	0	1	0	1
0	1	0	0	1
0	1	1	1	0
1	0	0	0	1
1	0	1	1	0
1	1	0	1	0
1	1	1	1	1

Figure 7.32 The second stage of the adder has three inputs and two outputs.

Figure 7.33 Truth-table representation of the second-stage inputs and outputs.

Examination of the truth table shows there to be four ways to achieve a 1 in the output bit, S_2. These may be expressed as the ORs of a group of ANDs. This form is called a *sum of products*, although that name seems inappropriate because neither sums nor products are defined in Boolean algebra. In this form, the second output bit of the sum would be

$$S_2 = A_2'B_2'C_1 + A_2'B_2C_1' + A_2B_2'C_1' + A_2B_2C_1 \tag{7.9}$$

Similarly, the carry bit from the second stage can be expressed in a similar form:

$$C_2 = A_2'B_2C_1 + A_2B_2'C_1 + A_2B_2C_1' + A_2B_2C_1 \tag{7.10}$$

The theorems of Boolean algebra permit simplification of this expression. Notice the last two terms. They can be combined as shown in Eq. (7.11). Thus we can replace two terms by a simpler term.

$$A_2B_2C_1' + A_2B_2C_1 = A_2B_2(C_1' + C_1) = A_2B_2(1) = A_2B_2 \tag{7.11}$$

Notice also that we could have used the same trick by combining the last term with either of the first two terms. Too bad we do not have more terms like that last term in Eq. (7.10) to use in simplifying the first two terms. Fortunately, the theorems of Boolean algebra allow us to add the desired terms. Notice the third property of the OR in Fig. 7.16:

$A = A + A + A + \cdots$, where we have expanded the expression to as many A's as we require. Using this property we can insert two additional $A_2B_2C_1$ terms in Eq. (7.10), then combine them with each of the first three terms in the manner shown in Eq. (7.11), and hence reduce Eq. (7.10) to

$$C_2 = B_2C_1 + A_2C_1 + A_2B_2 \tag{7.12}$$

Given this success in simplifying Eq. (7.10), you might try to simplify Eq. (7.9) in a similar way, but your effort would be unsuccessful because Eq. (7.9) is already in its simplest form. Techniques exist for identifying terms which may be combined but we shall not carry this discussion in that direction. Clearly, we could complete the realization of the BCD adder with NAND or NOR gates by manipulating these expressions for the sum and carry bits of the higher-order stages of the adder into suitable forms, as we did in Eq. (7.8).

7.5 SEQUENTIAL DIGITAL SYSTEMS

The logic circuits in Section 7.4.4 are called *combinational* circuits because the output responds immediately to the inputs and there is no memory. When memory is a part of a logic circuit, the system is called *sequential* because its output depends on the inputs *plus* its history. In this section we show how memory is developed in logic circuits and how memory elements increase greatly the possible applications of logic circuits.

7.5.1 The Bistable Circuit

The basic memory circuit is the bistable circuit, which we examine in this section. In Fig. 7.34 we show two amplifier–switch circuits in cascade, the output of the first (T_1) providing the input to the second (T_2). This is called a two-stage amplifier, for amplification takes place in two distinct stages. Each of these stages is identical to the original amplifier–switch we analyzed in Section 6.5.3 to obtain the input–output characteristic shown in Fig. 6.61. Placing the two stages of amplification in cascade requires that we consider the output of the first stage as the input of the second stage. In Fig. 7.35 we have shown the overall input–output characteristic of the amplifier. We derived this characteristic from Fig. 6.61 by increasing the input voltage starting with zero volts. With zero volts

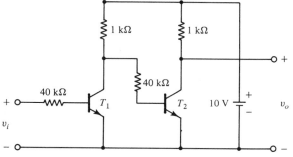

Figure 7.34 Two-stage amplifier.

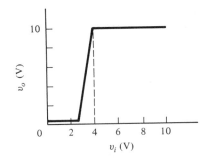

Figure 7.35   Input–output characteristic of two-stage amplifier.

input, the first transistor (T_1) will be cut off and the second transistor (T_2) will be saturated. As the input voltage rises, the first transistor will leave cutoff as the input voltage rises above 0.7 V, but the second transistor will remain saturated until the output of the first stage drops to about 4 V. This output of 4 V requires an input of about 3 V (see Fig. 6.61); hence the input must increase to approximately 3 V before T_2 comes out of saturation and the output voltage of the entire two-stage amplifier begins to rise. In this region, both transistors are in the active region and the output rises rapidly. The second transistor will reach cutoff when the output of the first transistor falls below 0.7 V, which will occur when the input of T_1 exceeds about 4 V. Thus both transistors are in the active region for the range of input voltages between 3 and 4 V.

What will happen if we connect the output of the two-stage amplifier to its input? Figure 7.36a shows the circuit redrawn with this connection made and with T_1 turned around to emphasize the symmetry of the resulting circuit. We have also added inputs, which we shall discuss presently. Mathematically, this connection requires $v_i = v_o$, which defines a straight line passing through the origin and having a slope of unity. In Fig. 7.36b we have drawn this line on the amplifier characteristic, which also has to be satisfied. This straight line is not a load line, but the same reasoning applies here as we followed in thinking about load lines: to satisfy both characteristics, the solution must lie at their intersection(s). Consequently, we have marked and labeled the three intersections on the graph. The two intersections labeled S_1 and S_2 are stable solutions, but the intersection labeled U is unstable. In Fig. 7.36c we suggest a mechanical analog: the lever will have stable equilibria when resting against either wall but with a frictionless pivot the balanced position will be unstable and will not occur in practice. The stable position marked S_1 occurs with transistor T_2 saturated and transistor T_1 cutoff, and the stable position marked S_2 has transistor T_2 cutoff and transistor T_1 saturated. The circuit will remain in one of these stable states forever unless an external signal is applied to force it to the other stable state, just as the lever in Fig. 7.36c will lean against one wall unless an external force moves it to the other wall. By applying sufficient positive voltage to the input of the transistor which is cut off, we can switch the state of the circuit. The diodes are placed in the inputs to ensure that the state of the circuit will not affect the input drivers.

The circuit shown in Fig. 7.36a is called a *bistable* (or *latch*) *circuit* and it can serve as electronic memory. When you depress a button on your calculator for example, the signal sets latch circuits in the calculator to retain the keyed information after you release

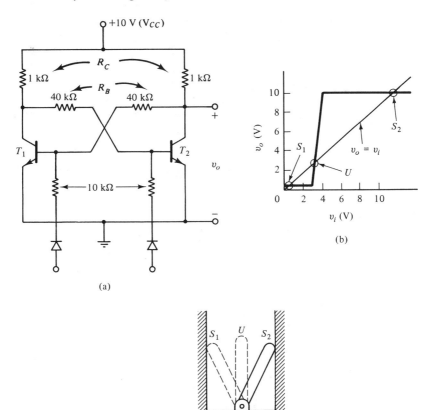

(a)

(b)

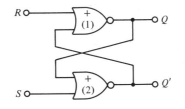

(c)

Figure 7.36 (a) Two-stage amplifier with output connected to input; (b) the circuit has two stable and one unstable operating point; (c) mechanical analog.

the button. The information thus retained is then available for processing after all numerical information is entered.

7.5.2 Flip-Flops

The R-S Flip-Flop.   We can realize the latch function with standard logic gates. Figure 7.37 shows a latch constructed from two NOR gates. The output of each NOR provides one of the inputs for the other NOR. The other inputs are labeled S (for SET)

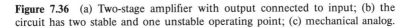

Figure 7.37 Latch from NOR gates.

and R (for RESET). The outputs are labeled Q and Q' because the latch provides complementary outputs. This circuit, called an *R-S flip-flop*, is similar in its operation to the bistable in Fig. 7.36a.

Let us consider that R and S are both 0 but that Q' is 1. In this case one of the inputs to NOR$_1$ is 1 and hence its output (Q) is 0. This is consistent with the assumption that Q' is 1, which implies that both inputs to NOR$_2$ are 0. By symmetry, the latch will also be stable with $Q = 1$ and $Q' = 0$.

If now S momentarily becomes 1, the output of NOR$_2$ (Q') will drop to 0, resulting in NOR$_1$ having 0 at both inputs. This will force Q to 1, and this 1 will keep Q' at 0 after S returns to the 0 state. Thus Q is SET by a 1 at S. Similarly, we can RESET the latch $(Q = 0)$ with a 1 at R.

The truth table for the R-S flip-flop is shown in Fig. 7.38 together with the logic symbol for an R-S flip-flop. The logic symbol covers the various types of realizations for the R-S flip-flop, as well as the variety of logic families which might be used. Seldom would individual gates be used for constructing a flip-flop; rather the logic designer would use one of the special types of flip-flops packaged on a single chip. In the following section we discuss the more popular types of flip-flops.

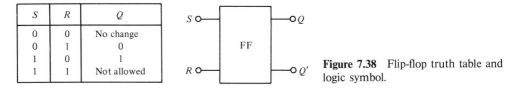

S	R	Q
0	0	No change
0	1	0
1	0	1
1	1	Not allowed

Figure 7.38 Flip-flop truth table and logic symbol.

Gated and Clocked Flip-Flops.

The R-S flip-flop requires a number of refinements to achieve its full potential for memory and digital signal processing. One problem is that the R-S flip-flop responds to its input signals at R and S immediately and at all times. Timing problems can occur when logic signals which are supposed to arrive at the same time actually arrive at slightly different times due to separate delays. Such timing problems can create short, unwanted pulses called "glitches."

The gated flip-flop in Fig. 7.39a will respond to the R or S inputs only when a gating signal arrives at the G(gate) input. Note that here we have built the flip-flop out of NAND gates. In this form, the forbidden state at the inputs to the cross-coupled NANDs is 00, which corresponds to 11 at the R and S inputs, as before. This flip-flop also has Present (Pr) and Clear (Cr) inputs which set the latch independent of the gate. These are active when in the 0 state, as indicated by the circle at their inputs on the logic symbol (in Fig. 7.39b). The truth table in Fig. (7.39c) now lists the output state after the gating pulse (Q_{n+1}) as a function of the R and S inputs and the state prior to the gating pulse (Q_n). Specifically, with $RS = 00$, the gating signal produces no change in the output state $(Q_{n+1} = Q_n)$. Notice that by using an inverter on the gate input, we could have the gating occur at $G = 0$. This would be indicated by a circle at the gating input of the flip-flop symbol in Fig. 7.39b.

The R and S input are thus active when the signal at the gate input is 1. Normally,

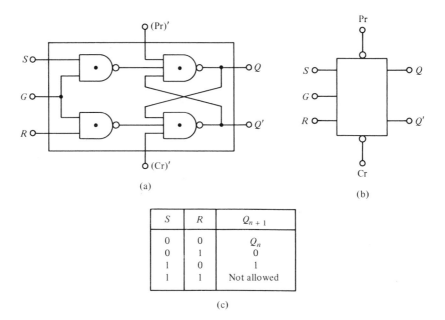

(a)

(b)

S	R	Q_{n+1}
0	0	Q_n
0	1	0
1	0	1
1	1	Not allowed

(c)

Figure 7.39   (a) Gated flip-flop; (b) logic symbol; (c) truth table.

such timing, or synchronizing, signals are distributed throughout a digital system by clock pulses, as shown in Fig. 7.40. The symmetrical clock signal provides two times each period when switching may be accomplished [i.e., when $Ck \Rightarrow 1$ and when $(Ck)' \Rightarrow 1$].

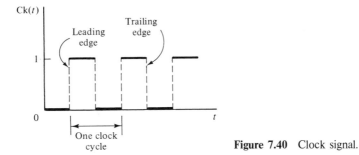

Figure 7.40   Clock signal.

The gating time of the inputs can be further reduced by differentiating the clock signal in an *RC* circuit as shown in Fig. 7.41 and applying the result to the input gate. This *edge triggering* is accomplished by building small coupling capacitors into the input of the integrated circuit. The circuit can be designed to trigger at the leading or trailing edge of the clock. The symbol for an edge triggered flip-flop is shown in Fig. 7.42 for both leading and lagging edge triggering. The distinguishing mark for edge triggering is a triangle at the clock input. Triggering at the edges of the waveforms limits the time during which the inputs are active and thus serves to eliminate glitches. By using circuits

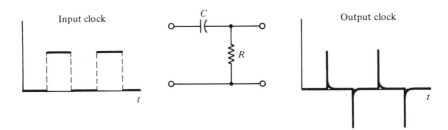

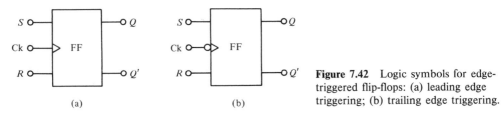

Figure 7.41 The *RC* circuit differentiates the clock signal to limit the time during which state changes can occur.

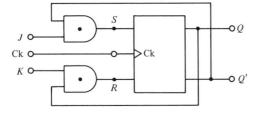

Figure 7.42 Logic symbols for edge-triggered flip-flops: (a) leading edge triggering; (b) trailing edge triggering.

that trigger at either the leading or training edges, the designer can pass signals in a circuit at two times in each clock cycle.

The J-K Flip-Flop. Another problem with the basic *R-S* flip-flop is the forbidden state at the input. This can be eliminated by ANDing the inputs with the output of the flip-flop, thus blocking one of the inputs, as shown in Fig. 7.43. The added gates here have the effect of inhibiting the 1 input to the gate whose output is 1. Therefore, the input (*J* or *K*) which is passed will always change the state of the output. Thus the truth table for the *J-K* flip-flop (Fig. 7.44) is the same as the truth table in Fig. 7.39, except that we indicate a change of output state $(Q_{n+1} = Q'_n)$ for the hitherto forbidden input state. The *J-K* flip-flop thus gives us, in addition to a latched memory of the input, the capacity to "toggle"* when both inputs are 1. This toggle feature reveals why we must use edge trig-

Figure 7.43 *J-K* flip-flop.

J	K	Q_{n+1}
0	0	Q_n
0	1	0
1	0	1
1	1	Q'_n

Figure 7.44 Truth table for *J-K* flip-flop. Q_n represents the stage before the clock pulse and Q_{n+1} the stage after the clock pulse.

*A lever-actuated switch, like the ordinary light on-off switch, is called a *toggle switch*. Thus *to toggle* means to switch from one state to another.

gering for this flip-flop; for if the clock pulse were extended in time, the state would oscillate back and forth and the eventual output would be indeterminent. The toggle mode of the J-K flip-flop is useful in counters and frequency dividers.

The D- and T-Type Flip-Flops.   The J-K (or R-S) flip-flop can be converted to a D-type flip-flop (D for delay) by connecting an inverter between the inputs as shown in Fig. 7.45. This effectively eliminates the forbidden state for the R-S flip-flop and has the effect of delaying the output by one clock cycle, as shown by Fig. 7.46. Although the input signal and the clock appear here to change at the same time, the clock transition occurs *before* the input transition marked with the arrow on the leading edge. Thus the output is assured to have the value which was present at the input during the *previous* clock cycle (i.e., the output is delayed one clock cycle).

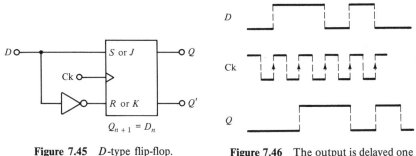

Figure 7.45 D-type flip-flop.

Figure 7.46   The output is delayed one clock cycle.

Tying the J and K inputs together produces a T-type flip-flop (T for toggle). The T-type flip-flop toggles with the clock pulse when its input is 1 and does not toggle when its input is 0. This is useful, as stated above, for counters and divide-by-2 applications. The logic symbol and truth table for the T-type flip-flop is shown in Fig. 7.47.

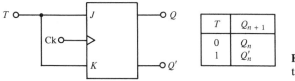

T	Q_{n+1}
0	Q_n
1	Q_n'

Figure 7.47 T-type flip-flop with truth table.

The Master–Slave Flip-Flop.   Timing problems in logic circuits can be further eliminated through the use of the master–slave configuration shown in Fig. 7.48. In this configuration, two J-K flip-flops are used: the first receives the input state at the trailing edge of the clock pulse and passes that state to the "slave" flip-flop at the leading edge of the next clock pulse. Notice that the clock pulse also gates the output of the master flip-flop to ensure that signals will be passed only at the correct time.

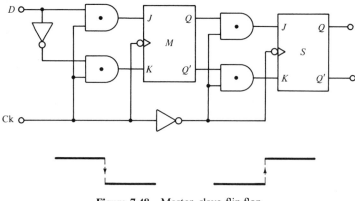

Figure 7.48   Master–slave flip-flop.

Summary.   In this section we have presented the *R-S* flip-flop as the basic memory element in logic circuits. A variety of refinements were given, leading to several types of flip-flops. In all cases, we have shown only one way to realize the characteristic of the different flip-flops. In the various logic families, these circuits could be realized through many variations, depending on the properties of the specific family.

In the next section we show some common applications of flip-flops. Our purpose is to indicate broadly how these system components can be used.

7.5.3 Flip-Flop Applications

Frequency Dividers.   The clock frequency can be halved with a *J-K* flip-flop by connecting the *J* and *K* inputs to 1 and letting the clock toggle the output. Since the output changes states with each clock cycle, the output frequency will be half the clock frequency. If this output is then used as a "clock" for a second flip-flop, a second division by 2 is performed, and so forth. Figure 7.49 shows four stages of division, which would produce an output frequency one-sixteenth of the input clock frequency. Division by a factor which is not an exact power of 2 can be accomplished by using an AND gate to detect the appropriate state. Figure 7.50 shows a divide-by-3 counter. If the two flip-flops start off cleared (both Q's $= 0$), then at the end of the third clock cycle both Q's would be

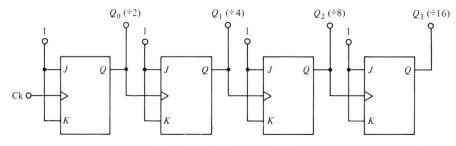

Figure 7.49   Frequency divider.

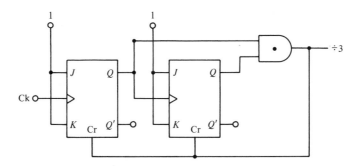

Figure 7.50   Divide-by-3 circuit.

1, which would produce a 1 at the AND output. This 1 clears the flip-flops and acts as output.

Counters.   As implied in the discussion above, a string of *J*-*K* flip-flops connected as in Fig. 7.49 acts as a binary counter, the output being $Q_3 Q_2 Q_1 Q_0$. If the flip-flops are cleared in the beginning, after the first input pulse the output state would be 0001, after the second 0010, and so on, up to the sixteenth, which would in effect reset the counter with 0000. Note that the chain counts input pulses in binary even if the pulses are unevenly spaced.

Figure 7.51 shows how to convert the binary counter to a decade counter. The AND gate detects a count of 10, clears the counter, and provides an input to the next stage. The $Q_3 Q_2 Q_1 Q_0$ output from the stage could be transferred to a BCD-to-decimal display while the counter is counting another sample of the input.

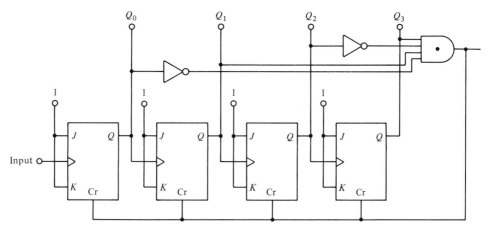

Figure 7.51   One stage of a decade counter.

The types of counters that we have presented above are called *ripple counters* because the flip-flop transitions move in sequence from left to right through the counter. By contrast, a synchronous counter uses a simultaneous clock pulse at all flip-flops and controls the counting operation with external gates. Also possible are up/down counters, which increase or decrease the count depending on an input signal.

The digital designer has available frequency dividers, counters, and many other useful system functions on LSI (large-scale integration) chips. Such LSI circuits manage the internal connections and provide only the necessary external inputs and outputs.

Registers. A register is a series of flip-flops arranged for organized storage or processing of binary information. Before describing several types of registers, we must introduce some concepts that relate to the use of registers in computers.

Information is represented in a computer by groups of 1's and 0's called *words.* Thus a word in a 8-bit microprocessor might be 01101101, or 6D in hexadecimal. An 8-bit word is also called a *byte* and is a convenient unit for digital information. One byte can represent two BCD digits or an ASCII (American Standard Code for Information Interchange) code alphanumeric symbol. Current microprocessors work with words of 4, 8, or 16 bits, whereas larger computers work with words of 32 or more bits. A register in a computer with 8-bit words would require eight flip-flops to store or process simultaneously the 8 bits of information.

Words of information are moved around in a computer or other digital system on a *bus.* As in a city bus line where passengers can enter or leave at a variety of points along the way, so in a computer bus the words can originate at any of several registers or arrive at any of several destination registers. The bus itself consists of the required number of wires* connecting all potential source registers with all potential destination registers.

Figure 7.52 shows a bus of four wires connected to a destination register of four D-type flip-flops. At the leading edge of the LOAD signal, the information on the bus will be stored in the register. The information does not come down the bus and get off at the

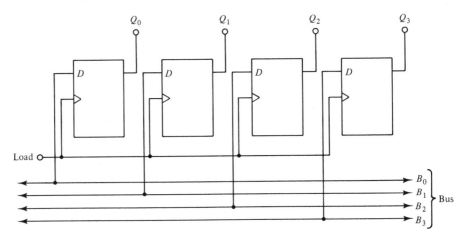

Figure 7.52 Loading a register from a bus.

*Actually, conducting paths is more appropriate, for no wires are used for internal information transfer in a microprocessor.

register (that is pushing the analogy too far); rather, the information appears simultaneously all along the bus and may be loaded simultaneously into several registers. If the register is part of a computer memory, the register will have an address consisting of one or more words, and this address is decoded to give the LOAD signal at the appropriate location in memory.

Whereas many registers can be loaded simultaneously from the bus, it should be clear that only one register can put information on the bus at one time. We need a way to connect the outputs of all source registers to the bus, such that only one register can transfer its output word to the bus at any time. An ordinary gate will not accomplish this, for its output must be either 1 or 0 and hence connecting the outputs of all source registers would result in a tug of war. A *three-state gate*, shown in Fig. 7.53, is required. In the absence of an ENABLE signal, the output of the gate approximates an open circuit and the gate is disconnected from the bus. When the ENABLE signal is present, the gate is connected to the bus and the output is 1 or 0 according to the input. Thus we can connect all source registers to the bus with three-state gates and ENABLE one register at a time to transfer data to the bus.

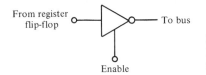

Enable

Figure 7.53   Three-state gate. When ENABLE is 0, the output is disconnected from the bus.

The Shift Register.   The data storage register described in the preceding section receives and delivers a word simultaneously: it transfers information with parallel input and parallel output. Sometimes digital information must be sent over one wire, as when a telephone circuit is used. In this case bits are sent in time sequence, or serial, form. When digital information must be received in serial form, a shift register may be used to accept the serial information and convert it to parallel form.

The D-type flip-flops in Fig. 7.54 will act as a shift register. Recall that the D flip-flop input immediately before the clock pulse shifts to its output after the clock pulse. The input to D_0 will appear at Q_0 after one clock pulse, at Q_1 after two, at Q_2 after three, and at Q_3 after four clock pulses. Thus the 4-bit sequence (word) input at D_0 will fill the register $Q_3Q_2Q_1Q_0$ after four clock pulses. The word can then be put on a parallel bus with three-state gates.

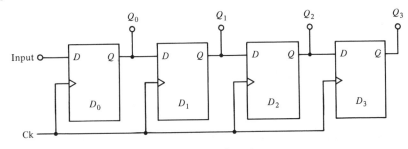

Figure 7.54   Shift register.

The shift register takes a serial input and produces parallel output when operated in the manner of Fig. 7.54. The same register can accept a parallel word at the Preset inputs to the flip-flops. The word can then be driven out Q_3 by the clock to produce serial output.

The shift register in Fig. 7.54 shifts its bit pattern one unit to the right with each clock pulse. A more versatile shift register results when the flip-flop inputs are connected with gates to their neighbors on left or right. Such a register can shift its bit pattern to the right or left, depending on which set of gates are ENABLED. This facility is useful in arithmetic operations. Multiplication by 10 in the decimal number system can be accomplished by shifting the decimal point one digit to the right. Similarly, multiplication by 2 in binary can be accomplished by moving the binary point one place to the right. This can be effected in hardware by shifting all bits one place to the left in a shift register, and clearly multiplication by 2^n requires n left shifts. A combination of shifting and adding intermediate results is required for multiplication by numbers which are not exact powers of 2. In like manner, division is accomplished by shifting bits to the right in a shift register.

Summary.   In this section we have shown how flip-flops are used to store digital information. Groups of flip-flops called registers can receive or deliver words of information in parallel with a bus, or information can be stored and delivered in serial form. The shift register is useful in binary arithmetic operations. These are the principal components of computers, to which we now turn.

7.6 COMPUTERS

7.6.1 Computers Are Important

Electrical engineers have produced some passing fads, but computers are here to stay. These versatile devices increasingly influence modern business and pleasure; seers predict an even broader place for computers in future society. The computer epitomizes many of the themes of this book. In Chapter 1 we stressed the speed with which electrical phenomena occur: computer magic arises, for the most part, out of the speed with which they operate. We stressed in Chapters 1 and 6 that microscopic matter is electrical in nature: from manipulation of the electrical properties of matter come the tiny transistor switches and electrical connections that physically constitute computer circuits. The digital idea finds, of course, its fullest expression in computers.

Computers have changed drastically since their development three decades ago. The original computers were big and expensive; only large institutions could justify their purchase for demanding computational tasks. Costs were in hundreds of thousands of dollars, if not millions. Of course, large (mainframe) computers still command an important place in the computer market. Then came the minicomputer, about the size of a large suitcase. The electronics of minicomputers utilize LSI digital circuits. These computers made possible the automation of process control and the processing of data in real time. Costs were in tens of thousands of dollars.

Now we have microcomputers. The low cost of the basic computer chip is revealed by its use in toys which retail for less than $25. Small personal computers including memory, keyboard, and elementary software sell for hundreds of dollars (you provide a TV set for a display). The computer on a chip finds more and more applications: smart typewriters, cash registers, appliances, and sewing machines; in the automobile, in electronic instruments, at the video arcade, in the nursery. Perhaps the personal computer itself best demonstrates the potential of microprocessor technology.

Computers are complicated systems, not easily explained. Aside from the fundamental principles of the computer itself, a myriad of related topics could be discussed: interfacing the computer to peripheral devices such as keyboards, CRTs, and printers; software development, including programming languages, editors, assemblers, and compliers; information representation matters such as fixed- and floating-point representation of numbers, data structures, codes for representing text; potential applications such as numerical calculation, word processing, accounting systems, real-time process control, and time sharing. To add to the complexity, the world of computers has developed its own esoteric and colorful language.

We have listed above topics that, for the most part, we are not going to cover in this section. Our brief introduction to this complex topic will answer two basic questions: What is a computer, and how does it work?

We urge the reader to seize opportunities to work with microcomputers. Only thus can one recognize when a professional task requires a computer. Computers will not solve all your technical problems and may not solve any of them; but you cannot be technically literate without gauging the power and limitations of these information-processing marvels.

7.6.2 What Is a Computer?

The four basic elements of a computer are memory, an arithmetic-logic circuit, a control circuit, and input–output. These elements are present when you use a hand-held calculator. In that case you supply the program by pushing the various buttons that sequence the calculations. When the calculator has the capability of storing a "program" of keystrokes, it is a computer, albeit a limited, slow, and highly specialized computer.

The Central Processing Unit (CPU).   Figure 7.55 shows the system configuration of a typical computer. The central processing unit (CPU) contains an arithmetic-logic unit (ALU), control circuits, and several registers for storage and general manipulation of words of data. (We shall hereafter speak as if our computer is a 8-bit microprocessor.) The CPU is the brains of the computer and, at our level of understanding, the most complex and mysterious. Although we have touched on hardware implementation of binary arithmetic and data storage, the complexity of the CPU places it far beyond the level of this introduction.

The CPU has an instruction register which is sequentially loaded with words from the program in memory. These instructions, mere strings of 1's and 0's, are decoded and executed by the CPU: words are brought from memory, placed in designated registers, processed according to the instruction, then stored in memory at a specified location or

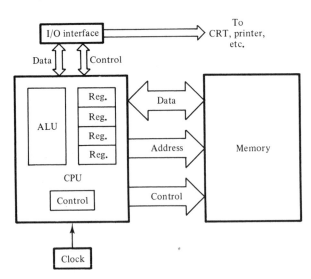

Figure 7.55 Elements of a computer.

perhaps transferred to an output device. The best way to review the diverse operations of the CPU is to examine the instruction set of a microprocessor. Within the CPU are a number of registers to hold data and addresses of special places in memory relating to the program or to blocks of data. A program-counter register is incremented after each instruction is "fetched" from memory into the instruction register so that the CPU knows where to get the next instruction. The CPU communicates with memory with a two-way (8-bit) data bus, an address bus (16 bits for a 64K* memory) controls the location in memory to furnish or receive data or program words, and a control bus to coordinate timing and order.

Memory. Computer memory may be of several types. A random access memory (RAM) is an array of memory registers with which data may be exchanged. The memory access is random because any memory location is equally accessible for reading or writing of data or program instructions. Normally, the program would not write into the portion of the memory containing program instructions, but the entire program must initially be written (loaded) into memory. Most semiconductor memories are "volatile" because they lose the stored information when the computer is turned off. For this reason, and generally to store large amounts of information, disks or magnetic tapes are used for storage of digital information.

Also important are read-only memories (ROMs), which contain information (usually programs) that can be read but not modified by the computer. One way to acquire programs is to purchase a commercial ROM containing software for your computer. Programmable ROMs (PROMs) allow the user to store information by burning microscopic fuses in the ROM. An erasable PROM (EPROM) is a PROM that is capable of being reprogrammed after its information has been erased by ultraviolet light.

*In computer talk, K means 2^{10}, or 1024. Thus a 16-bit address can specify one location out of 2^{16} or 64 K or 65,536 words.

Input and Output (I/O).   Although the computer is self-contained for its internal calculations and data manipulations, interaction with the outside world must occur. Often computer–person interaction is provided through "terminals" with keyboard and display and with printers and plotters. Such devices translate between people-oriented symbols such as alphanumeric text or graphical representation of information and computer-oriented representation as bits (1's and 0's) of information.

Computers interact with external systems such as robots, electronic instruments, and manufacturing processes. Such interactions often involve digital-to-analog (D/A) and analog-to-digital (A/D) conversion. Communication between the computer and external devices occurs over an external data bus, which is separate from the computer's internal data bus. Several protocols exist for announcing which component has information to transfer to the computer, for keeping two devices from "talking" at once, and for assuring that the target component "heard what was said."

7.6.3 Programming

Programming Languages.   Programming is the art of translating a problem into words of 1's and 0's which the computer CPU executes to solve the problem. Figure 7.56 summarizes the communication problem: we think in terms of language, mathematical notation, accounting conventions, and so on, and the computer "thinks" in 1's and 0's. Programming languages provide the bridge between human thought processes and the binary words that control computer operations. Computer languages which are deliberately close to human thought processes or notations are called high-level languages: FORTRAN, BASIC, LISP, ALGOL, C, FORTH, and PASCAL are examples of high-level languages. Low-level languages are called *assembly languages* and these are oriented toward the instructions that the specific computer can execute. The binary words are loaded into computer memory and executed as machine or object code. One line of code in a high-level language can produce many lines of machine code, whereas one line of assembly code, being close to the computer operations, will produce one or at most a few

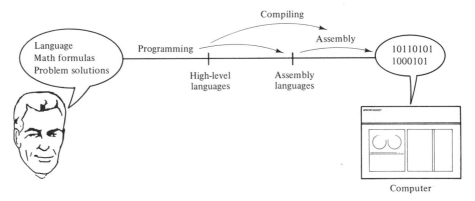

Figure 7.56 Communicating with a computer.

lines of machine code. For the microprocessor, the assembly language is also called the *op code* for the machine.

Compilers, Assemblers, Loaders, and Interpreters.   A compiler is a program that accepts as input a program in a high-level language (usually as a string of ASCII characters) and produces as an output either an assembly language program or a machine code program. The compiler is thus a computer program which performs language translation, from one computer language to another language closer to what the computer can use. An assembler performs the same function, using the assembly code as input (source code) and producing output (object code) in machine code.

Compilers and assemblers are large programs which function on a specific computer. Such programs would thus be developed and furnished by the computer manufacturer (often on a ROM chip) or perhaps by an institution providing computer services. The object code that is produced usually is stored on disk or magnetic tape or in memory at a different location from where it will eventually reside. After the compiler or assembler does its work, it often would be erased from memory to make room for the program machine code and whatever data the program requires. A small program called a *loader* will load the machine code into memory for execution. When the program consists of several separate parts (main program, subroutines, functions), the loader will load these into consecutive memory and, in effect, tell each where all the others are so that they can communicate. When one begins with a completely empty memory, a small program called a *bootstrap* can be loaded into memory by means of a keyboard or front panel switches. The bootstrap program can load the loader, which loads the program and signals the beginning of execution.

Some high-level languages, such as BASIC, are interpreted instead of compiled. This means that the program is converted into machine language and executed line by line. This is inefficient in computer time but efficient for finding certain program errors because mistakes can be corrected as they occur without having to reexecute the program up to the point where the error has occurred. This is especially appropriate when the programmer is working on a small computer.

In the past, programs were usually punched on cards or perhaps paper tapes and read into the computer with a card (or paper tape) reader. Modern systems use editors that allow typing of alphanumeric text directly into the computer and offer the programmer easy modification through a variety of editing commands. Indeed the storage, editing, and manipulation of text has opened new computer applications for the production of printed matter. Such "text processing" computer systems offer many advantages over traditional methods involving writing by hand and revising a number of times in order to produce final copy.

7.6.4 Postscript

To end this section on computers, we wish to discuss two final matters: one practical, the other philosophical. The practical matter concerns the speed with which computers operate. This speed is the real secret to their usefulness.

The clock speed on a microprocessor is approximately 1 MHz. Typical instructions take four to six clock cycles; hence a typical instruction might be executed in 4 to 5 μs or at a rate of about 200,000 instructions per second. Although a single instruction in a microcomputer does not accomplish much—an addition, a word stored in memory, a change in the program counter—the rapid execution of many instructions can create a lot of action. In arcade space warfare, for example, the CRT picture must be refreshed, trajectories must be calculated, joysticks and triggers responded to, scores computed, big winners stored to impress later generations, and so on. These operations require the execution of many instructions, but still the machine can outthink most of us.

The philosophical matter concerns the hierarchy of order and meaning involved in explaining the workings of a computer. We shall start with the apparently random motion of the electrons in a semiconductor chip. If we could follow the motion of a single electron, we would discover two features of its movement. One is that its motions are, to all appearance, totally random and meaningless. The other is that its behavior obeys a statistical regularity which is describable by the laws of quantum mechanics. Were we able to solve the relevant equations, we would have a complete statistical explanation of the phenomenon at the level of quantum mechanics; no further assumptions would be required to account for what is happening. As the bumper sticker claims: "Everything is physics."

At another level, however, the apparently random movement of the electrons reveals a pattern. Upon and within the semiconductor chip are formed patterns representing definite devices: transistors, diodes, resistors, capacitors, and conducting paths. A device-oriented person, looking at the semiconductor chip, would assert: "These are semiconductor devices; that's all there is to it." This is a correct description at the level of electronic devices.

But the semiconductor devices are organized into electronic circuits—gates, to be specific. If the circuit theorist were looking for logic gates, he would be correct to identify the semiconductor chip as nothing but an electronic circuit containing a large number of gates.

At a yet higher level, the gates could be described as flip-flops, registers, ALU, memory, and buses. That is, the circuit can be described as well, and as completely, by the system function of the gates, considered as organized units having inputs and outputs. The system-oriented person who recognizes the existence of these larger functions sees an order that escapes one who understands the circuit at the level of gates, devices, or electron motion.

Finally, we reach the level of computer. Through the interaction of memory, ALU registers, and buses, the overall purpose of the semiconductor chip as a computer is realized. Thus the observer who sees the chip as a computer could justly say that the others, however right and complete at their levels of perception, missed the overall purpose.

Enter the programmer, who knows that an unprogrammed computer is an unfulfilled computer. The computer comes to life only when executing a program designed toward a purpose external to the computer. Apart from the program, the silicon might as well still be on the beach.

There are yet higher levels of order and purpose: in the goals of the organization that purchased the computer and hired the programmer; in society at large; in the history of humankind; perhaps in higher realms of meaning known only to God.

So what is the point? The point is that although it it possible to have an adequate and even exhaustive explanation of a phenomenon or process at one level of meaning, this explanation cannot rule out explanation and meaning at higher levels. Specifically, many claim that higher levels of meaning are excluded by scientific description of the world in terms of "natural law." Thus everything (we are assured) can be explained in terms of physics, chemistry, genetics, evolution, psychology, sociology, and so on. This argument, as our semiconductor chip shows, is fallacious, for description of a particle, or a device, or a circuit, or a system at one level does not rule out the possibility of order and purpose at a higher level. Complete understanding must offer an explanation at *every* level of meaning, including the philosophical and, we hasten to add, the theological.

PROBLEMS

Practice Problems

SECTION 7.1.1

P7.1. Susie is sure to accept a date $(D = 1)$, provided that she likes the guy $(L = 1)$ and has no test the next day $(T = 0)$. But if it's Henry $(H = 1)$, she will go even if she has a test.

    (a) Make a truth table relating the dependent variable, D, to the independent variables: L, T, and H.

    (b) Express the Susie dating function, $D(L, T, H)$, in terms of NOTs, ANDs, and ORs.

    Comment: In this situation, there are some states which are impossible. Specifically, no case where $L = 0$ and $H = 1$ can occur, because clearly Susie likes Henry. States such as these are called *don't-care* states and are indicated by a "×" in the D column of the truth table rather than a 1 or 0. The logical function is still valid; it is just that certain combinations of the independent variables never occur in practice. If we were designing a digital system to aid Susie with her social decisions, we would not care what the system indicates on those states which are impossible, because presumably they would never occur.

P7.2. When in the 1980 NBA playoffs the Philadelphia 76ers basketball team led the Boston Celtics three games to one, the 76ers could have won the best of seven series by winning any of the fifth, sixth, or seventh games. (You may recall that they lost all three games.) Let G_5, G_6, and G_7 be digital variables to describe the outcome of those games, with $G_5 = 1$ if the 76ers win the fifth game and $G_5 = 0$ if Boston wins, and so forth for G_6 and G_7. For the sake of completing the truth table, we will award the remaining games to the 76ers should they win the series before the seventh game. The dependent variable is $P = 1$ if the 76ers win the series and $P = 0$ if the Celtics win. Make a truth table showing the relationship between G_5, G_6, and G_7 as input (independent) variables and P as output (dependent) variable. This situation has some don't-care states; see the comment in Problem P7.1.

SECTIONS 7.2.1–7.2.2

P7.3. For the NOT gate in Fig. 7.9 to operate properly, the transistor must be saturated when the input is a digital 1. Assume that the transistor characteristics are those in Fig. 6.54 and that the minimum voltage input for a 1 is 4.0 V. Assume that the *pn* junctions have a voltage of 0.7 V when ON.
(a) What is the largest value of R_B (10 KΩ in Fig. 7.9 but now a variable) which allows the transistor to remain saturated for an input 1?
(b) Keeping R_B at 10 kΩ, what is the smallest value of the transistor β that will keep the transistor saturated? (Now the transistor characteristics are different from those in Fig. 6.54.) Assume that $V_{CE(sat)} = 0.5$ V.

P7.4. The circuit shown in Fig. P7.4 performs a logical function on the inputs A and B to produce an output C. Consider a digital 0 as a voltage between 0 and 0.5 V and a digital 1 as between 4 and 5 V. Assume that 0.7 V is required to turn ON the base–emitter junction. Assume that $V_{CE(sat)} = 0.3$ V.

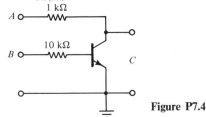

Figure P7.4

(a) Make a truth table for the circuit, assuming that the β of the transistor is 200. This requires a careful analysis of the voltages and currents in the transistor circuit.
(b) Is the circuit a NOT, NAND, NOR, OR, AND, or none of these?

SECTIONS 7.3.1–7.3.2

P7.5. Make a truth table with two inputs, A and B, and the outputs OR, NOR, AND, NAND, EXCLUSIVE OR, and the equality function.

P7.6. With a truth table verify the second of the absorption rules in Fig. 7.18.

P7.7 Use the theorems of Section 7.3.2 to simplify the following digital functions:
(a) $[A' + (B + A)B']'$
(b) $A + [B(1 + A')]'A$
(c) $(ABC)' + A'B'(C + A)$

SECTIONS 7.4.1–7.4.4

P7.8. Show two ways for realizing a NOT function with a NOR gate.

P7.9. Give a realization of the Susie dating function of Problem P7.1 with NOR and NAND functions. Do this directly with the function as you derived it and then use De Morgan's theorem to place the function into a more favorable form and try for a simpler realization.

SECTION 7.5.1

P7.10. For the bistable circuit in Fig 7.36 to operate as described, sufficient current must flow into the base of the ON transistor to permit saturation (i.e., $i_B \geq \beta i_C$). For the resistor values shown, this imposes a minimum value of β.

    (a) What is the minimum value of β for the circuit in Fig. 7.36 to function as a bistable? Assume 0.7 V for the ON base–emitter voltage and 0.3 V for $V_{CE(\text{sat})}$.

    (b) Find the formula for the minimum value of β in terms of the power supply voltage (V_{CC}), base resistor (R_B), and collector resistor (R_C).

    (c) For the circuit in Fig. 7.36, and for a β of 60, what voltage at the input to the OFF transistor is required to switch the circuit? Assume 0.7 V to turn ON the diode and transistor *pn* junctions. *Note:* The bistable will switch if the current into the ON transistor is dropped below the value required to saturate the transistor. This implies a certain voltage at the collector of the OFF (but coming ON) transistor. This implies in turn a certain collector current, and so on.

SECTION 7.5.2

P7.11. Design an *R-S* flip-flop circuit with NAND gates rather than NOR gates, as shown in Fig. 7.37. Determine the inputs (1 or 0) required to SET and RESET the flip-flop.

P7.12. Consider a 1-MHz square-wave clock signal with voltage levels of 0 and 5 V, as shown in Fig. P7.12. Find the RC time constant of the circuit in Fig. 7.41 to differentiate this signal to produce a gating time of 30 ns. Assume that the gate is open when the voltage is above 2.0 V.

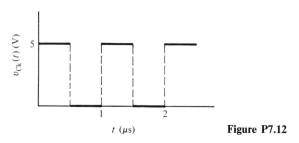

$v_{CK}(t)$ (V)

t (μs)   **Figure P7.12**

P7.13. Construct a truth table for the *J-K* flip-flop with J, K, Q, and Q' as the inputs and R and S as intermediate outputs. Use the truth table to confirm the truth table in Fig. 7.44. You must eliminate all rows which contradict the definitions for the gates or the R-S flip-flop.

SECTION 7.5.3

P7.14. Using T-type flip-flops and an AND gate, design a circuit to divide the clock frequency by 5.

P7.15. Repeat Problem P7.14, except use a NOR gate rather than an AND gate.

P7.16. Figure P7.16 shows four T-type flip-flops clocked for synchronous counting. Using AND gates, design logic such that at the clock leading edge, FF_1 changes states when $Q_0 = 1$, FF_2 changes states when $Q_0 Q_1 = 1$, and FF_3 changes states when $Q_0 Q_1 Q_2 = 1$. What is the function of this circuit?

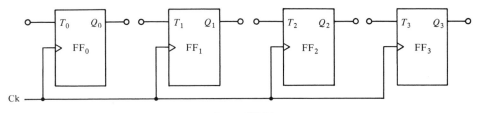

Figure P7.16

P7.17. The shift register in Fig. P7.17 illustrates one way that a left/right shift register can be configured.

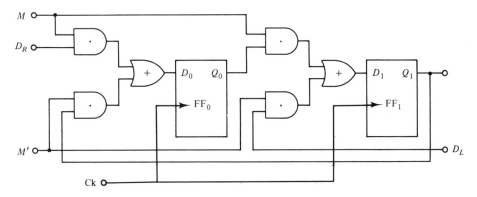

Figure P7.17

(a) Show that with $M = 1$, the flip-flops are controlled from the left: D_R to D_0 and Q_0 to D_1.

(b) Show that with $M = 0$, the flip-flops are controlled from the right: D_L to D_1 and Q_1 to D_0.

(c) Using techniques similar to those on page 282, replace the AND and OR logic with NAND-gate logic.

ANALOG
ELECTRONICS

8.1 INTRODUCTION

8.1.1 The Contrast between Analog and Digital Electronics

We have already explored how transistors and diodes are used as switching devices to process information which is represented in digital form. Digital electronics uses transistors as electrically controlled switches: transistors are either saturated or cut off. The active region is used only in transition from one state to the other.

By contrast, analog electronics depends on the active region of transistors and other types of amplifiers. The Greek roots of "analog" mean "in due ratio," signifying in this usage that information is encoded into an electrical signal which is proportional to the quantity being represented.

In Fig. 8.1 our information is some sort of music, originating physically in the excitation and resonances of a musical instrument. The radiated sound consists in the ordered movement of air molecules and is best understood as acoustic waves. These pro-

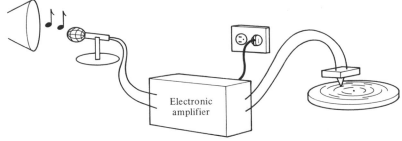

Electronic
amplifier

Figure 8.1 Analog system.

duce motion in the diaphram of a microphone, which in turn produces an electrical signal. The variations in the electrical signal are a proportional representation of the sound waves. The electrical signal is amplified electronically, with an increase in signal power occurring at the expense of the input ac power to the amplifier. The amplifier output drives a recording head and produces a wavy groove on a disk. If the entire system is good, every acoustic variation of the air will be recorded on the disk and, when the record is played back through a similar system and the signal reradiated as sound energy by a loudspeaker, the resulting sound should faithfully reproduce the original music.

Electronic systems based on analog principles form an important class of electronic devices. Radio and TV broadcasting are common examples of analog systems, as are many electrical instruments used in monitoring deflection (strain gages, for example), motion (tachometers), and temperature (thermocouples). Many electrical instruments—voltmeters, ohmmeters, ammeters, and oscilloscopes—utilize analog techniques, at least in part.

Analog computers existed before digital computers were developed. In an analog computer, the unknowns in a differential equation are modeled with electrical signals. Such signals are integrated, scaled, and summed electrically to yield solutions with modest effort compared with analytical or numerical techniques.

8.1.2 The Contents of This Chapter

Analog techniques employ the frequency-domain viewpoint extensively. We begin by expanding our concept of the frequency domain to include periodic, nonperiodic, and random signals. We will see that most analog signals and processes can be represented in the frequency domain. We shall introduce the concept of a *spectrum*, that is, the representation of a signal as the simultaneous existence of many frequencies. Bandwidth (the width of a spectrum) in the frequency domain will be related to information rate in the time domain.

This expanded concept of the frequency domain also helps us distinguish the effects of linear and nonlinear analog devices. Linear circuits are shown to be capable of "filtering" out unwanted frequency components. By contrast, new frequencies can be created by nonlinear devices such as diodes and transistors. This property allows us to shift analog signals in the frequency domain through AM and FM modulation techniques, which are widely used in public and private communication systems. As an example we shall describe the operation of an AM radio.

Next we study the concept of *feedback*, a technique by which gain in analog systems is exchanged for other desirable qualities such as linearity or wider bandwidth. Without feedback, analog systems such as audio amplifiers or TV receivers would at best offer poor performance. Understanding of the benefits of feedback provides the foundation for appreciating the many uses of operational amplifiers in analog electronics.

Operational amplifiers (op amps, for short) provide basic building blocks for analog circuits in the same way that NOR and NAND gates are basic building blocks for digital circuits. We will present some of the more common applications of op amps, concluding with their use in analog computers.

8.2 THE FREQUENCY-DOMAIN REPRESENTATION OF SIGNALS

8.2.1 Frequency-Domain Concepts

We first introduced the concept of the frequency domain in Chapter 4 for the analysis of ac circuits. There we showed that sinusoidal sources can be represented by complex numbers through

$$v(t) = V_p \cos(\omega t + \theta) = \text{Re } \underline{\mathbf{V}} e^{j\omega t} \tag{8.1}$$

where the phasor voltage, $\underline{\mathbf{V}} = V_p \underline{/\theta}$, is a complex number representing the sinusoidal function in the frequency domain. When only one frequency is present, the frequency normally is not stated explicitly. The time-domain and frequency-domain representations for a sinusoidal signal appear in Fig. 8.2.

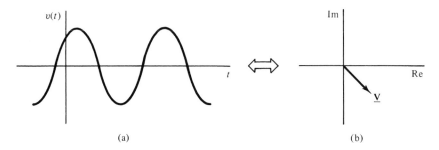

(a)   (b)

Figure 8.2   Sinusoid in (a) the time domain and (b) the frequency domain.

8.2.2 Frequency-Domain Representation of Periodic Functions

An Example.   Our frequency-domain representation of signals is enriched, however, when we consider periodic signals which are nonsinusoidal. As a beginning, we will consider the function

$$
\begin{aligned}
v(t) &= \frac{V}{2} + \frac{2V}{\pi}\left(\cos \omega_0 t - \frac{1}{3} \cos 3\omega_0 t\right) \\
&= \text{Re}\left[\frac{V}{2} + \frac{2V}{\pi} e^{j\omega_0 t} + \frac{2V}{3\pi}\underline{/180°}\, e^{j3\omega_0 t}\right]
\end{aligned}
\tag{8.2}
$$

Leaving aside for the moment the significance of this particular mathematical function, let us consider the time-domain representation in Fig. 8.3. In the time domain, we see nothing unusual except that we now have a more complicated function than a simple sinusoidal function. The dc component of the voltage is easily identified in Eq. (8.2) as $V/2$. The time-varying part of the function consists of two sinusoidal terms. Because the frequency of the second term is an integral multiple of the frequency of the first term, the

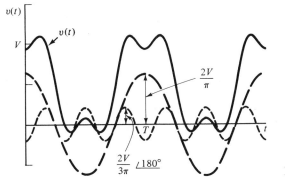

Figure 8.3 Time function.

frequencies are said to be *harmonically related*. The term "harmonic" (from the Greek, meaning "to fit together") is used because in music the tones which share harmonics fit together to form pleasant-sounding chords. Because the two sinusoids are harmonically related, they form a stable pattern and thus repeat with the period of the lower frequency, $T = 2\pi/\omega_0$. To summarize, we would describe this function in the time domain as having three harmonic components, a dc component (zero frequency), a fundamental (or first harmonic), and a third harmonic. The dc component is described by its magnitude, and the two sinusoidal components are described by their amplitudes and phases.

The Spectrum.   Because we now have three frequencies involved, we no longer can describe the signal in the frequency domain by a point in the complex plane. Frequency becomes important in a new way and cannot be relegated to the memory or the margin of the page. A common way to handle this new complexity is shown in Fig. 8.4. We have made frequency (ω) the independent variable and shown the three harmonics as lines at their respective frequencies. The magnitude (for the dc component) and phasors (for the ac components) have been placed beside the various frequency components, drawn roughly to scale.

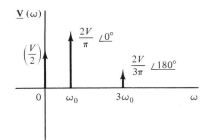

Figure 8.4   Spectrum consists of three frequencies.

Figure 8.4 is called a *spectrum*, more specifically, a voltage spectrum. The word "spectrum" (from the Latin, meaning "to behold") is derived by association with the visible spectrum, that is, the frequencies of light that we can see. The voltage spectrum in Fig. 8.4 is the frequency-domain representation of the time-domain voltage in Fig. 8.3. This expanded concept of the frequency domain enriches greatly the concept presented in Chapter 4. Figures 8.3 and 8.4 portray the complementary nature of the two domains: time and frequency are shown as complementary variables. Subsequent exploration of the frequency domain will be based on this expanded concept.

Fourier Series.   In 1822, a French artillery officer immortalized his name among electrical engineers yet unborn by proving that any periodic function can be expressed as a series of harmonically related sinusoids. Fourier's theorem expressed mathematically what musicians had long known by ear, that a steady tone consists of many harmonically related frequencies. The mathematical expression of Fourier's theorem is

$$f(t) = C_0 + C_1 \cos(\omega_0 t + \theta_1) + C_2 \cos(2\omega_0 t + \theta_2) + \cdots$$
$$+ C_n \cos(n\omega_0 t + \theta_n) + \cdots \tag{8.3}$$

where $\omega_0 = 2\pi/T$, $T =$ the period of $f(t)$. Fourier's theorem shows that any periodic function can be expressed as a spectrum consisting of dc, a fundamental frequency, and in general all the harmonic frequencies, that is, all frequencies that are integral multiples of the fundamental.

Fourier's theorem gives a mathematical explanation for many musical phenomena. Musicians have known by ear that the overtones of musical instruments are important. The piano sounds different from the harpsichord, for example, because the harpsichord strings are excited in a way which creates more sound energy at the higher harmonics than piano strings produce. In the language of the frequency domain, the various musical instruments sound different because they produce different acoustic spectra.

The musical scale is based on the harmonic relationship between the frequencies of the notes. Two tones are said to be consonant, to sound good together, when they are harmonically related, that is, when their harmonic structures are compatible. For example, in the pure scale the frequency of G would be $\frac{3}{2}$ that of C. It follows that the third, sixth, ninth, . . . harmonics of C coincide with the second, fourth, sixth, . . . of G. This shared harmonic structure defines all major fifth intervals in the pure musical scale and is representative of the harmonic relationships that are foundational to musical theory.

Our digression on music was to show that frequency-domain concepts are really commonplace in certain areas of experience. For our purposes Fourier's theorem, as expressed by Eq. (8.3), shows that all periodic functions can be described by harmonic spectra. We shall bypass the mathematical procedures for computing the amplitudes and phases of the harmonics of a periodic function because such a development would lead us away from our primary purpose, the exploration of our expanded concept of the frequency domain. Fourier series are studied in many courses in differential and partial differential equations and in engineering courses on signal analysis.

Two Spectra.   We will give two examples of Fourier series. The first is the Fourier series representation of the periodic series of pulses shown in Fig. 8.5a. The Fourier series for the square wave in Fig. 8.5a is

$$v(t) = \frac{V}{2} + \frac{2V}{\pi}\cos(\omega_0 t) - \frac{2V}{3\pi}\cos(3\omega_0 t) + \frac{2V}{5\pi}\cos(5\omega_0 t) + \cdots \tag{8.4}$$

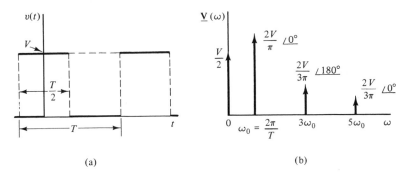

Figure 8.5 (a) Periodic time function; (b) voltage spectrum.

We have shown only the first four components of the frequency-domain representation in Fig. 8.5b. Comparison of Eqs. (8.2) and (8.4) will show that the first three frequency components of the square wave are those which are plotted in Fig. 8.3. Indeed, you should see some resemblance between the series of pulses and the sum of the dc, the fundamental, and the third harmonics in Fig. 8.3. If the fifth harmonic were added, it would steepen the slope at the sides and flatten the top and bottom. With the addition of each higher harmonic the series would approximate more closely the periodic pulses.

A second example of the Fourier series of a periodic function is shown in Fig. 8.6a, the half-wave rectified sinusoid. The Fourier series is

$$v(t) = V_p \left(\frac{1}{\pi} + \frac{1}{2} \sin \omega_0 t - \frac{2}{3\pi} \cos 2\omega_0 t + \frac{2}{15\pi} \cos 4\omega_0 t - \cdots \right) \qquad (8.5)$$

The spectrum (Fig. 8.6b) consists mainly in the even harmonics, unlike that in Fig. 8.5b, which contains the odd harmonics only. These are but two common spectra, which we introduce as examples to be explored in this and later sections.

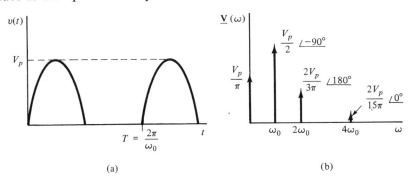

Figure 8.6 (a) Half-wave rectified sinusoid; (b) voltage spectrum.

Voltage Spectra and Power Spectra.   The spectra shown in Figs. 8.5b and 8.6b are called *voltage spectra* because they correspond to the amplitudes and phases of the harmonics in their respective signals. In many applications the power in a signal carries great importance; hence we shall extend the concept of a spectrum to include power and energy concepts.

Power is defined as the energy flow per unit time. Figure 8.7 shows a sinusoidal voltage source connected to a resistive load. The power into the resistor, as given on the figure, depends only on the amplitude and the resistance; frequency and phase do not matter. The factor of 2 in the denominator results from time averaging [this result was worked out in Eq. (5.7)] and can be absorbed into the voltage term if we use the rms value of the voltage instead of the peak.

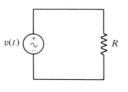

$$v(t) = V_p \cos(\omega t + \theta) \text{ volts}$$
$$P_R = V_p^2 / 2R \text{ watts}$$

Figure 8.7   Average power in a resistor.

The presence of the resistance in the power relationship in Fig. 8.7 poses to us a mild dilemma in developing the concept of the power spectrum of a signal. One possibility is to state explicitly the impedance level of an assumed load (say, 50 Ω) and to compute the power that each harmonic would deliver to such a load. With this approach, watts would be the units of the harmonics in the power spectrum. The second possibility leaves the resistance term unstated. In this case the "power" would be $(V_p)^2/2$, it being understood that to get actual power one has to divide by an appropriate resistance. In this second approach, which is more general, the units of "power" are volts^2. One would be wise to state explicitly these units to stress that the power spectrum is of this second variety. Because of its generality, this second approach is frequently preferred and will be used here.

We will therefore write the power spectrum of a signal in units of volts^2. Table 8.1 states the power spectrum of an arbitrary signal, plus the specific application to the case of the series of pulses in Fig. 8.5. Note that the dc term in the fourth column is not divided by the factor of 2 because the time average of the square of the dc term is equal to the dc value squared. The power spectrum cannot be written in the form of a Fourier series such as Eq. (8.3), but it can be graphed in the manner of Fig. 8.5b. Indeed, a graphical presentation of the last column in Table 8.1 would look like the voltage spectrum plot in Fig. 8.5b, except that the powers of the higher harmonics diminish at a greater rate due to the squaring.

Accounting for All the Power.   The legitimacy of the concept of a power spectrum must rest on a proper accounting for all the power in the frequency domain. Certainly, the concept of a power spectrum would be questionable if the power in the time domain could not be equated with the sum of the powers of the individual harmonics in the frequency domain.

TABLE 8.1   VOLTAGE AND POWER SPECTRA

Harmonic	Frequency	General		Series of pulses	
		Voltage spectrum (V)	Power spectrum (V^2)	Voltage spectrum (V)	Power spectrum (V^2)
0	dc	C_0	C_0^2	$V/2$	$V^2/4$
1	1000	$C_1\underline{/\theta_1}$	$C_1^2/2$	$2V/\pi\,\underline{/0°}$	$2V^2/\pi^2$
2	2000	$C_2\underline{/\theta_2}$	$C_2^2/2$	0	0
3	3000	$C_3\underline{/\theta_3}$	$C_3^2/2$	$2V/3\pi\,\underline{/180°}$	$2V^2/9\pi^2$
4	4000	$C_4\underline{/\theta_4}$	$C_4^2/2$	0	0
5	5000	$C_5\underline{/\theta_5}$	$C_5^2/2$	$2V/5\pi\,\underline{/0°}$	$2V^2/25\pi^2$
⋮	⋮	⋮	⋮	⋮	⋮

We shall now confirm equivalence of power between the two domains. The time-average power in the series of pulses is

$$P_{\text{avg}} = \frac{2}{T}\int_0^{T/2} v^2(t)\,dt = \frac{2}{T}\int_0^{T/4} V^2\,dt = \frac{2V^2}{T} \times \frac{T}{4} = \frac{V^2}{2} \qquad \text{volts}^2 \qquad (8.6)$$

The power in the frequency domain would be the sum of the harmonics in the last column of Table 8.1.

$$P = \frac{V^2}{4} + \frac{2V^2}{\pi^2} + \frac{2V^2}{9\pi^2} + \frac{2V^2}{25\pi^2}$$

$$= V^2\left[\frac{1}{4} + \frac{2}{\pi^2}\left(1 + \frac{1}{9} + \frac{1}{25} + \cdots\right)\right] \qquad \text{volts}^2 \qquad (8.7)$$

The infinite sum in Eq. (8.7) converges rapidly to the value $\pi^2/8$, as you can confirm by looking up the series in math tables or summing terms on your calculator until you get tired. We note that half the power comes from the dc component and half from the ac harmonics. We see that the sum of the power spectrum harmonics equals the power in the time domain. The equal division of power between dc and ac components is not a general result, but the accounting of the power between the time and frequency domains will always come out correct. The verification of this equality for the spectrum of the half-wave rectified sine wave we will leave for a homework problem.

Summary.   We have shown in this section that a periodic signal can be represented by an infinite series of sinusoidal components at frequencies harmonically related to the fundamental frequency of the signal. The voltage spectrum of a periodic signal consists, therefore, of a series of harmonics, each being described by an amplitude and phase. The power spectrum has no phase information and may be described in either watts or volts^2. The power spectrum accounts for the total power as the sum of the powers in the individual harmonics.

8.2.3 Spectra for Nonperiodic Signals

An Interesting Question.   Granted this success in representing periodic signals in the frequency domain through Fourier series, we are emboldened to ask: How far can we carry this line of development? This is an important question, for in electronic systems we must deal with both periodic and nonperiodic signals. Examples of periodic signals are: ac voltages, pulses sent out by radar transmitters, timing signals in a TV signal, clocking signals in a computer, and steady musical tones. Examples of nonperiodic signals are: music, speech, information pulses in a computer, and radar pulse returns from a maneuvering target or a diffuse target, such as a thunderstorm.

In Section 8.2.2 we have shown how to represent periodic signals in the frequency domain through the use of Fourier series. In subsequent sections of this chapter we will explore the operation of filters in the frequency domain. We will show further that nonlinear devices such as diodes can effectively move, or translate, signals around in the frequency domain. Common communication devices, such as radio and TV transmitters and receivers, operate by combining frequency translation, filtering, and amplification. Thus we must press for the extension of frequency-domain concepts to include a wider class of signals.

The Spectrum of a Single Pulse.   How, then, might we explore the possibility of having a spectrum for a nonperiodic signal, such as a pulse which happens only once? One approach, which works in this case, begins with a series of pulses and then increases the period of the series, keeping the width of the individual pulses constant. As we let the period of the repeating pulses get larger and larger, in the limit we would have a single pulse. This limiting process in the time domain will affect the spectrum, and we suppose that the resulting spectrum would correspond to the spectrum of a single pulse. Figure 8.8 suggests such a series of pulses by showing two pulses. The individual pulse widths are τ, which will be kept constant, and the period is T, which will be allowed to increase so as to isolate a single pulse. We shall consider the time origin to be at the center of one of the pulses; hence the signal has even symmetry about the origin. This means that the Fourier series for this series of pulses will contain harmonics in the form of Eq. (8.3), but every harmonic will have a phase angle (θ) of 0 or 180°, because these phases alone possess even symmetry. Because the phase of 180° is equivalent to a negative sign in front

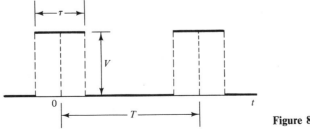

Figure 8.8   Series of pulses.

of the amplitude, the Fourier series for this signal must be of the form

$$v(t) = A_0 + A_1 \cos \omega_0 t + A_2 \cos \omega_0 t + \cdots + A_n \cos n\omega_0 t + \cdots \qquad (8.8)$$

where $\omega_0 = 2\pi/T$ and the A's are amplitudes of the harmonics, which may be negative. The amplitudes can be computed by the standard methods of Fourier series, with the result

$$A_n = \frac{V\tau}{T} \frac{\sin(n\omega_0\tau/2)}{n\omega_0\tau/2} = \frac{V\tau}{T} \frac{\sin(n\pi\tau/T)}{n\pi\tau/T} \qquad (8.9)$$

Recall that the spectrum consists of components at $\omega = 0, \omega_0, 2\omega_0, \dots, n\omega_0, \dots$, with harmonic amplitudes given by Eq. (8.9). We reserve for a homework problem the proof that Eq. (8.9) yields the results presented in Eq. (8.4) when $\tau = T/2$. As a example, we have plotted in Fig. 8.9 the spectrum for the case where $T = 10\tau$. We see in Fig. 8.9 that there are now many harmonics which are significant, unlike the spectrum in Fig. 8.5b, where the harmonic amplitudes fall off rapidly after the first few harmonics. The negative amplitudes indicate a phase of 180° from the tenth to the twentieth harmonics. Thus we note that the spectrum consists of a large number of harmonics spaced apart $\Delta f = 1/T$, where T is the period of the time function.

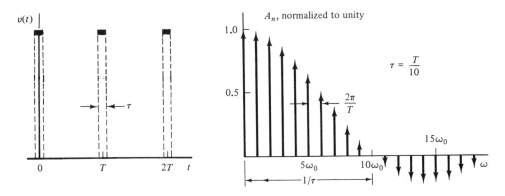

Figure 8.9 Time-domain and frequency-domain representations for a series of pulses with period T and width $\tau = T/10$.

Although for graphing purposes we have normalized the harmonic amplitudes to unity relative to the dc component, Eq. (8.9) reveals that the individual harmonics become small as T is made large. There are two reasons for this decrease. As the period is increased (with τ kept constant), the pulses come less and less frequently and hence the power in the signal decreases accordingly. Specifically, the average power must decrease as $1/T$ due to the spreading of the pulses in time. Compounded with this effect, the power in any individual harmonic must decrease even faster because the power in the series of pulses distributes between an increasingly larger number of harmonics. As Eq. (8.9) shows, the power in, say, the first harmonic $(n = 1)$ decreases as $(1/T)^2$ because power is proportional to the square of the amplitude.

Isolating a Single Pulse.   With the aid of Fig. 8.9 and a measure of imagination, we can now anticipate what will happen to the spectrum as we allow the period, T, to approach infinity. As $T \longrightarrow \infty$, the frequencies of the individual harmonics will crowd closer and closer together and, in the limit, constitute a continuous spectrum. Individual harmonics must disappear because there is no longer any specific period; hence frequency becomes a continuous variable. We conclude that a single pulse contains all frequencies in the frequency domain. The shape of the spectrum, however, will not change as T becomes large. The period, T, affects the spacing of the harmonics, but the shape of the spectrum depends only on τ. Thus the shape of the spectrum will retain the form in Eq. (8.9), which approaches

$$A(\omega) = E_0 \frac{\sin (\omega\tau/2)}{\omega\tau/2} \tag{8.10}$$

where the continuous variable ω replaces the discrete variable $n\omega_0$ and E_0 is a constant. This continuous spectrum is pictured in Fig. 8.10, where we have called the spectrum $S_V(\omega)$ because this is a voltage spectrum. We wish also to investigate the power spectrum, as we did in Table 8.1, but here we must be careful. As implied earlier, the single pulse carries no average power because the energy in the pulse is spread over all time, whereas power implies the continuous flow of energy as a time process. Consequently, we may meaningfully discuss the energy in the pulse but not the power. Our voltage spectrum, when squared, becomes an energy spectrum, not a power spectrum. The energy in the pulse distributes between the infinite number of frequencies making up the spectrum. Each individual frequency carries an infinitesimal quantity of energy, but taken together all the frequencies account for the total energy of the pulse.

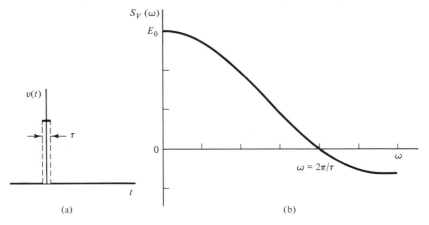

Figure 8.10   (a) Time-domain and (b) frequency-domain representations for a single pulse.

As in the case of periodic power spectra, we have the option of including resistance explicitly or omitting it to leave the definition of an energy spectrum more general. The units of the spectrum given in Eq. (8.10) are volts per hertz. The energy spectrum has the units joule-ohms per hertz, which requires division by a resistance and multiplication by a bandwidth in hertz to have the units for energy.

The Spectrum Amplitude.   Let us apply this interpretation of the square of the voltage spectrum in computing the energy in the pulse in the frequency domain. We can equate this energy to the energy in the time domain in order to evaluate the constant in the spectrum in Eq. (8.10). To have a physical picture, and to get the units to work out to the proper units of energy, let us consider that the pulse has an amplitude of V and is applied to a resistor of value R, as pictured in Fig. 8.11. The total energy in the pulse is the integral of the power, as we stressed in Chapter 3 in deriving the stored energy in inductors and capacitors [see, for example, Eq. (1.13)]. The required calculation is

$$W_R = \int_{-\infty}^{+\infty} vi \, dt = \int_{-\tau/2}^{+\tau/2} \frac{v^2}{R} \, dt = \frac{V^2}{R} \int_{-\tau/2}^{\tau/2} dt = \frac{V^2 \tau}{R} \tag{8.11}$$

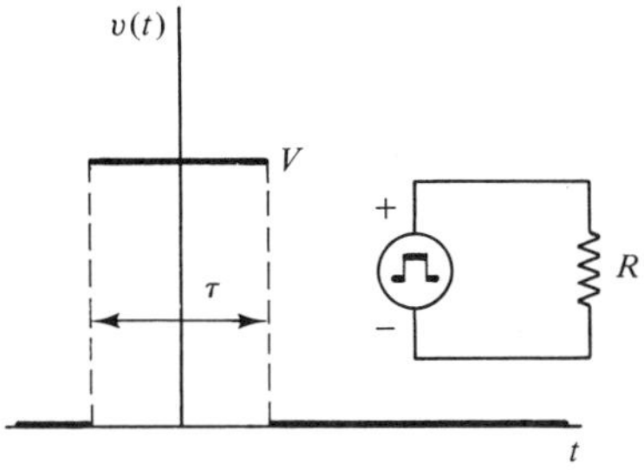

Figure 8.11 Computation of the energy in a single pulse.

The frequency-domain calculation follows the same reasoning as we used in Eq. (8.7), except that here we replace the sum of the contributions of each individual harmonic with the integral of the continuous spectrum. We shall use the rms value of each contributing sinusoid; hence the amplitudes must be divided by $\sqrt{2}$. We no longer must treat the dc term separately because it carries an infinitesimal amount of energy. Equation (8.12) summarizes the required computation.

$$W_R = \frac{1}{R} \int_0^\infty \left[\frac{E_0}{\sqrt{2}} \frac{\sin(\omega\tau/2)}{\omega\tau/2} \right]^2 d\left(\frac{\omega}{2\pi} \right) = \frac{E_0^2}{2R} \left(\frac{1}{\pi\tau} \right) \int_0^\infty \frac{\sin^2 x}{x^2} \, dx = \frac{E_0^2}{4R\tau} \tag{8.12}$$

You will note that we are integrating with respect to $\omega/2\pi$, which is the frequency in hertz. The second form of the integral results from a change in variables from ω to $x = \omega\tau/2$. The definite integral has the value $\pi/2$, leading to the final result. We now equate the two energy calculations, with the result

$$\frac{E_0^2}{4R\tau} = \frac{V^2\tau}{R} \Rightarrow E_0 = 2V\tau \tag{8.13}$$

We have labored greatly to derive this result because of the mathematical limitations we have accepted. Results of this sort can be derived more elegantly using Fourier transform theory.

Bandwidth and Pulse Width.   The spectrum of the single pulse in Eq. (8.10) illustrates an important general relationship between the time and frequency domains. In Fig. 8.12 we show the energy spectrum of the pulse, which is the square of the voltage spectrum in Fig. 8.10. Clearly, most of the energy is contained in the frequency bandwidth below the first null at $2\pi/\tau$. Numerical integration of the energy spectrum to that first null reveals that 90% of the total energy of the single pulse lies in the bandwidth between 0 and the first null in the spectrum. Thus, if we were to pass this spectrum through an ideal lowpass filter which eliminated all frequencies above the first null of the spectrum, most of the energy would get through and the pulse would appear in the time domain without serious change. Similarly, the representation of the periodic pulses (Fig. 8.5a) by the dc, first, and third harmonics contains 95% of the energy in the entire series of pulses, as you can verify from Eq. (8.7). Figure 8.3 verifies that the shape of the pulses is relatively unchanged by the elimination of the higher harmonics. Later in this chapter we will investigate filtering in the frequency domain; for now, we merely wish to establish that there is an effective bandwidth in the frequency domain associated with a pulse.

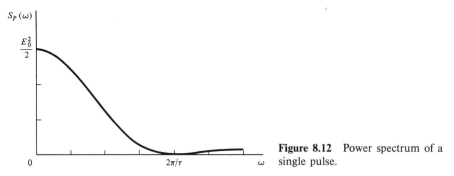

Figure 8.12   Power spectrum of a single pulse.

We may determine this effective bandwidth by setting $\omega\tau$ equal to 2π, the angle in radians where the spectrum first goes to zero. Solving this equation and converting to frequency in hertz, we obtain

$$\omega\tau = 2\pi \Rightarrow \text{BW} = \frac{2\pi}{2\pi\tau} = 1/\tau \tag{8.14}$$

where BW is the significant bandwidth in hertz. Important consequences follow from this approximate relationship between pulse duration in the time domain and bandwidth in the frequency domain. For one, we must provide adequate bandwidth if pulses are to pass through electronic communication or computing systems without serious distortion. If, for example, we have a digital system which utilizes pulses of 1-μs duration, we must provide at least 1×10^6 Hz or 1 MHz of bandwidth to send those pulses to another location for recording or processing.

This reciprocal relationship between pulse length and bandwidth applies also to periodic pulses. The spectrum of the periodic pulses shown in Fig. 8.5 has the same general shape, and hence the same bandwidth, as that of the single pulse. Furthermore, this relationship between time duration and bandwidth does not depend on the exact shape of the pulses. The pulses may be rounded on top or triangular in shape: the same approximate bandwidth would be needed to preserve the identity of the pulses in passing through a communication system.

An Example from Mechanics.   Let us apply these ideas to a mechanical system. Let us imagine that we have before us a large bell such as might be used in a church steeple, and we strike this bell with a metal hammer. We know, of course, that we will "ring" the bell, meaning that the mechanical resonances of the bell will be excited to produce a bell sound. The force between the hammer and the bell would be of short duration, and this short pulse of force will excite the bell structure. Because the energy spectrum of the short pulse is broad in bandwidth, the higher resonances of the bell will be excited. Pick up now a rubber hammer and again strike the bell. We would expect the bell to sound less harsh, more mellow. We can understand this change in sound in terms of the energy spectrum of the force from the rubber hammer. The rubber hammer will remain in contact with the bell much longer because the rubber hammer is more compliant than the metal hammer. The result of this longer pulse of force will be a smaller bandwidth of excitation; consequently, the higher-frequency resonances of the bell would not be greatly excited. These higher resonances, which give the bell a harsh sound, are eliminated; hence a mellow sound results with a rubber hammer.

In Section 8.2.5 we will explore the relationship between bandwidth and information rate in a communication system. There we will show another consequence of the relationship between bandwidth in the frequency domain and duration in the time domain. This relationship also underlies an important result in the next section on random signals.

8.2.4 The Spectra of Random Signals

The Importance of Random Signals.   A random signal is any signal that is unpredictable to us, either because it originates from chance events or because its complexity defies simple analysis. Examples of random signals are broadcast signals over radio and TV channels, digital signals passing from computer to computer, the output of a microphone placed before a symphony orchestra, pulses occurring due to the passing of automobiles over a sensor cable stretched across a street, the amounts of the sales at a cash register, and so on.

Random signals can be carriers of information. We say that a signal contains information when we know something new because we received the signal. This implies that a signal containing the information is unpredictable to us, that is, is random in the sense given above. The examples of random signals given above illustrate the information-carrying capacity of random signals. An exception might be the music, which evokes an esthetic as well as an intellectual response.

The Characterization of Random Signals.   If we are to design electronic circuits to monitor, record, transmit, receive, and process random signals, we must develop appropriate concepts to describe, or characterize, such signals. Probability and statistics are the branches of mathematics that deal with random phenomena. These types of mathematics do not allow us to predict random events but they do allow us to describe the structure of the randomness. Consequently, the application of probability theory to communication systems has yielded many important results, and modern communication theory is based on probability theory.

It is, however, not our aim here to explore the statistical theory of communication. We shall limit our characterization of random signals to that required for the discussion of simple communications systems, such as an AM radio. Our major goal is to show that random signals can be characterized by a power spectrum in the frequency domain. First we shall discuss the power in a random signal; then we shall discuss the time structure and show, in consequence, how the power in the signal is distributed in the frequency domain.

The Power Structure of Random Signals.   We have sketched a random signal in Fig. 8.13. This represents, let us say, an electrical signal monitoring the temperature at a certain point in a chemical plant. We note that the average temperature would be indicated by the average value of the voltage, v_{dc}. We can denote this time average as

$$v_{dc} = \langle v(t) \rangle \qquad (8.15)$$

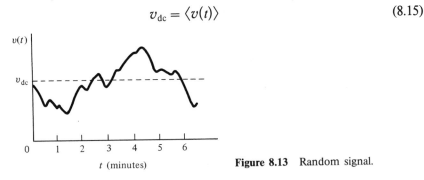

Figure 8.13   Random signal.

where the angle brackets indicate time average. This average could be approximated by averaging over a long period of time

$$v_{dc} = \langle v(t) \rangle = \frac{1}{T} \int_0^T v(t)\, dt \qquad (8.16)$$

Here T would have to be large enough to include many random fluctuations, say, a period of hours for the function in Fig. 8.13. We acknowledge the possibility, of course, that at some future time the plant might shut down and cool off, and hence the average temperature might change. But for normal operation we assume there to be a meaningful average temperature as represented by the dc voltage in Eq. (8.16). The average would be established, therefore, over a suitably long period of normal operation.

We may define the "ac" component of the random signal as the remainder after the average is subtracted:

$$v_{ac}(t) = v(t) - v_{dc} = v(t) - \langle v(t) \rangle \qquad (8.17)$$

By "ac" we mean everything but the dc, and not "alternating current." Thus "ac" means "fluctuating" in this context. Note that, by definition, the time average of the ac component is zero:

$$\langle v_{ac}(t) \rangle = \langle v(t) - v_{dc} \rangle = \langle v(t) \rangle - v_{dc} = v_{dc} - v_{dc} = 0 \qquad (8.18)$$

Because averaging is a linear process, we have distributed the averaging operation to the individual terms in Eq. (8.18).

The time-average power of the random signal holds special interest to us. Leaving out the resistance level for generality, we may define the power, P, to be

$$P = \langle v^2(t) \rangle = \langle [v_{ac}(t) + v_{dc}]^2 \rangle$$
$$= \langle v_{ac}^2(t) + 2v_{ac}(t)v_{dc} + v_{dc}^2 \rangle \qquad (\text{volts})^2$$

Again we may distribute the time average to the sum.

$$P = \langle v_{ac}^2(t) \rangle + 2\langle v_{ac}(t) \rangle v_{dc} + v_{dc}^2$$
$$= \langle v_{ac}^2(t) \rangle + v_{dc}^2 \qquad (\text{volts})^2 \qquad (8.19)$$

The middle term drops out because the time average of the ac component vanishes, as shown in Eq. (8.18). Equation (8.19) indicates that the total power of the signal may be considered as the sum of the powers in the dc and ac components, acting independently.

$$P = P_{ac} + P_{dc}$$

These two types of power produce two distinct components in the power spectrum of the random signal.

The Time Structure and the Power Spectrum.   The time structure of a random function is characterized by a *correlation time*. This is the time, on the average, during which the fluctuations in the signal become independent of the past. In Fig. 8.13, for example, the correlation time is about 1 min. This correlation time suggests how far into the future we might be able to predict the signal. If we know the voltage in Fig. 8.13 at some instant of time, we could predict its value 1 or 2 s later with confidence. But our uncertainty would grow as the prediction time moved into the future, and we can say little if anything about the value of the fluctuations, say, some 3 minutes hence. Thus correlation time characterizes the time structure of a random signal in a crude way.

Probability theory gives a precise definition of the correlation time and furnishes numerical methods for estimating it for a random signal, given that we analyze a sufficiently long sample of the signal. Investigation of such mathematical techniques, however, would lead us far beyond our goal of developing the ideas we need to understand basic communication systems.

Given that we know the correlation time for a signal, we can approximate the signal with a train of pulses, as shown in Fig. 8.14. We recognize that the approximation is poor for details of the fluctuation of the random signal. We introduce it merely as a crude representation of the time structure of the signal.

Let us now consider the spectrum of the random signal. A spectrum is a representation in the frequency domain of a time-domain signal, that is, a representation as a sum of sinusoids. At first glance it would appear to be impossible to represent an unpredictable random signal with stable, predictable sinusoids. This is true in part, for a random

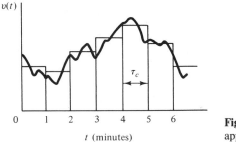

Figure 8.14 Random signal approximated by pulses.

signal canot be represented with a stable sum of sinusoids. The signal contains sinusoids, but the phases of these sinusoids never stabilize but rather vary randomly with time. In effect, an amplitude spectrum exists for random signals but phase is meaningless. Another way to express this fact is as follows: A power spectrum exists for a random signal but a voltage spectrum cannot be meaningfully defined.

The exact shape of the power spectrum of a random signal depends on the details of the nature of the fluctuations, but the meaningful bandwidth depends on the correlation time. If we were to consider the power spectrum of the pulses approximating the random signal in Fig. 8.14, all of them would have a significant bandwidth of $1/\tau_c$ hertz, where τ_c denotes the correlation time. Thus we would require approximately 1 Hz of bandwidth to handle the random signal in Fig. 8.13, but if the correlation time were 1 μs, we would require about 1 MHz of bandwidth.

Thus the power spectrum of a random signal will have one component due to the dc component and one component due to the ac component, as shown in Fig. 8.15. The dc component is represented by a discrete line at zero frequency, but the ac component is represented by a continuous spectrum. The units of this continuous spectrum are volts^2/hertz or volts^2-second. As shown, the area of the continuous power spectrum represents the total power in the random signal fluctuations and must give the same power as the time-domain representation,

$$P_{\text{ac}} = \langle v_{\text{ac}}^2(t) \rangle = \int_0^\infty S_{\text{ac}}(f)\, df \tag{8.20}$$

where $S_{\text{ac}}(f)$ represents the power spectrum of the fluctuations in the random signal. On Fig. 8.15 we have also shown the relationship between significant bandwidth and correlation time, as asserted above.

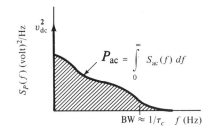

Figure 8.15 Power spectrum of a random signal.

8.2.5 Bandwidth and Information Rate

What Is Information?   Information is a word having meaning in both technical and ordinary language. We might describe of a book as containing information, and we might understand the strange gestures of a baseball coach as offering information to his base runners. In these instances, we think of information as offering knowledge—matters to be observed and acted upon.

The technical definition of information can cover these possibilities but deals primarily with the transmission or storage of symbols with machines. With this theory, we may study the exchange of information between two computers or we may calculate the information stored in a genetic code. This mathematical theory of information is of recent origin, springing from the work of Claude Shannon (1916–). This theory is based on probability concepts and deals with the probability of deciphering a known code which has been transmitted or stored in the presence of random disturbance. Information theory generally supports and enhances the commonsense understanding of information but also allows nonsense to be considered valid information. According to the mathematical theory, for example, the received message, "$3 = 5$," is valid information if that is what the transmitter sent. In other words, the theory measures and describes objective information (messages send and received), not subjective information (the meaning, validity, and importance of such messages).

Information Rate and Bandwidth.   Our purpose here is to show an important relationship between the information rate of a communication system and the bandwidth required by that communication system in the frequency domain. For our purposes, it is unnecessary to distinguish between the common understanding of information and that of the mathematical theory because this relationship is valid in both senses. Our examples will be based on the commonsense understanding of information. The conclusion we reach, however, can be supported with the mathematical theory. We shall consider one analog and one digital example.

First consider spoken language. What is the band of audio frequencies necessary to communicate, say, over a telephone line? Experiments have established that the essential bandwidth of audio frequencies lies between about 800 and 2400 Hz. That is, if someone spoke to you via a telephone system which passed that band of frequencies, you could certainly understand what they said. Although this would be the minimum bandwidth for communication to occur, the phone company actually allocates the band of frequencies between about 400 and 3300 Hz. The additional bandwidth is not required for intelligibility but rather for recognition of the speaker. For satisfactory telephone service we require not only that we can hear and understand the message but also that we be able to recognize the speaker's voice. This recognition factor represents additional information content to the phone message which is very important subjectively even if it does not add greatly to the content of the message.

An AM radio signal uses audio frequencies between 100 and 5000 Hz. This additional bandwidth is required for satisfactory transmission of music. Of course, FM radios sound better because their audio bandwidth is from 50 to 15,000 Hz. The large bandwidth enhances the quality of the music, a subjective measure of its information content.

A TV signal contains audio (voice) and video (picture). The audio bandwidth is similar to that of an FM radio, but the video requires much more bandwidth because the information rate is much higher. Of the 6-MHz bandwidth required for transmission of the entire TV signal, over 99% of the bandwidth is required for the video and synchronization signals, the remainder being used for the audio. Thus we see a clear relationship between bandwidth and information rate in analog systems.

Consider now a digital communication system. You have probably done some interactive computing using a remote terminal connected to a computer over telephone lines. You have no doubt noticed that some computer terminals write a line of text slowly, such that you can read the words as they appear, whereas some write faster than one can read. The difference lies in the information rate between the computer and the terminal. Information rate in this case is described by the *baud rate*, where a *baud* is the unit for a bit per second. Typical baud rates are 110, 300, 1200, and 3000 baud. The lower two rates are appropriate for mechanical printers, which must print the symbols on paper (hardcopy), and the faster rates for CRT terminals, which write the symbols on a TV-type screen.

Let us consider the bandwidth required for a data rate of 1200 baud. Each bit (pulse) will require $\frac{1}{1200}$ s to transmit, so the individual pulses will look as shown in Fig. 8.16. The pulse itself would occupy half the time space and hence the pulse width would be about $\frac{1}{2400}$ s. Using the relation given in Eq. (8.14), we would expect that a bandwidth of about 2400 Hz would be required to send such a pulse without significant distortion; and therefore we would expect that an ordinary phone line would be adequate for connecting the computer with the remote terminal. This is true for rates up to 1200 baud, but the 3000-baud system requires a special phone line. Therefore, for both the analog and digital systems, the higher information rates require larger bandwidth. One of the achievements of the mathematical theory of information was to place this relationship on a sound theoretical basis.

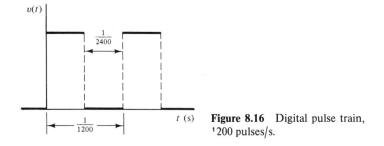

Figure 8.16   Digital pulse train, ¹200 pulses/s.

8.3 FILTERS

8.3.1 Simple RC Filters

The RC Low-Pass Filter.   In Fig. 8.17 we show a simple low-pass filter, so-named because it will pass low frequencies and filter out, or eliminate, high frequencies. We shall now derive the *filter function* of this circuit and show its effect on the spectra of sig-

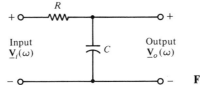

Figure 8.17 Low-pass filter.

nals. The analysis is based on the techniques we developed earlier in dealing with ac circuits. Specifically, we shall use the principle of superposition and the methodology of ac circuits. Consider, for example, that the input to this circuit is a periodic series of pulses of height V volts as shown in Fig. 8.5a. How are we to solve for the output, given that the input is a series of pulses? To analyze this problem in the time domain, we must solve a differential equation. The DE is not impossible, but this method of solution generates little insight into the behavior of the circuit. We will approach the problem in the frequency domain.

Filter Response in the Frequency Domain.   In Fig. 8.18 we show the impedance of the resistor and capacitor and derive the filter function, $\mathbf{F}(\omega)$, by treating the filter as a voltage divider. The characteristics of the low-pass filter can be understood best through considering the effect of frequency on the impedance of the capacitor. Recall that the impedance of a capacitor has a magnitude of $1/\omega C$. At dc the capacitor will act as an open circuit; hence all the input voltage will appear at the output, independent of R. At very low ac frequencies, the impedance of the capacitor will still be very high compared to R. Specifically, as long as $1/\omega C \gg R$, the output voltage will be approximately equal to the input voltage, and thus the filter gain will be approximately unity.

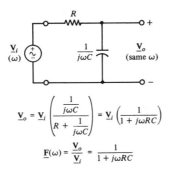

$$\mathbf{V}_o = \mathbf{V}_i \left(\frac{\frac{1}{j\omega C}}{R + \frac{1}{j\omega C}} \right) = \mathbf{V}_i \left(\frac{1}{1 + j\omega RC} \right)$$

$$\mathbf{F}(\omega) = \frac{\mathbf{V}_o}{\mathbf{V}_i} = \frac{1}{1 + j\omega RC}$$

Figure 8.18 Low-pass-filter function derivation.

At very high frequencies, the impedance of the capacitor will approach that of a short circuit and hence little voltage will appear across the capacitor. As long as $R \gg 1/\omega C$, the current will be approximately $\mathbf{V}_i/R$ and the magnitude of the voltage across the capacitor, which is the output voltage, will be $|\mathbf{V}_i|/\omega RC$ and hence will approach very small values as ω increases. Thus the gain of the filter becomes very small at high frequencies. This means that a high-frequency signal would be greatly reduced, whereas low-frequency components would not be reduced by the filter; that is, it will "pass" only components of the signal at low frequencies. The filter characteristics are shown in Fig. 8.19. We have used a log scale to show a large range of frequencies.

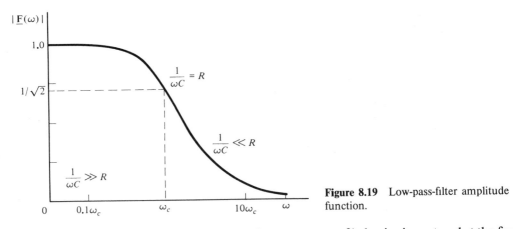

Figure 8.19 Low-pass-filter amplitude function.

The region of transition between these two types of behavior is centered at the frequency where the impedance of the capacitor is equal to that of the resistance:

$$R = \frac{1}{\omega_c C} \Rightarrow \omega_c = \frac{1}{RC} \tag{8.21}$$

where ω_c is called the *cutoff frequency* and characterizes the region in frequency where filtering begins. Notice that the filter function can be expressed in terms of the cutoff frequency:

$$\mathbf{F}(\omega) = \frac{1}{1 + j\omega RC} = \frac{1}{1 + j(\omega/\omega_c)} = \frac{1}{1 + j(f/f_c)} \tag{8.22}$$

At $\omega = \omega_c$ (or $f = f_c$), the filter function has the value $1/(1 + j1)$, which has a magnitude of $1/\sqrt{2} = 0.707$. (This has nothing to do with the rms value of a sine wave.) The significance of a voltage gain of 0.707 lies in the relationship between input and output power. Power is proportional to the square of the voltage; hence at ω_c the output power is reduced by a factor of 2 from what it would have been in the absence of the filter. For this reason, the cutoff frequency (expressed in either rad/s or hertz) is often called the *half-power frequency*.

Phase Effects.   We have plotted only the magnitude of the filter function in Fig. 8.19, $|\mathbf{F}(\omega)|$. The filter function is a complex function with real and imaginary parts. Thus the filter affects both the magnitude and the phase of sinusoids passing through it. For example, at the cutoff frequency the filter function has the value $1/(1 + j1) = 0.707\underline{/-45°}$. This phase shift is an unavoidable by-product of the filtering process, which can cause problems in some applications. In audio systems few problems occur due to phase shift because the ear is largely insensitive to phase. But in video systems the phase shift produced by filtering operations can degrade the appearance of the picture.

Frequency-Domain Analysis.   In the frequency domain we now possess a new tool, Fourier series. We can express the periodic pulses as a sum of sinusoids, each having a different frequency. Each of these represents an ac source and hence leads to a straightforward ac circuit problem, which we can solve using phasors and impedance. Specifically, if we have an input sinusoid of frequency ω and phasor magnitude, $\underline{V}_i$, the output phasor can be determined by considering the circuit as a voltage divider or as a filter. Before exploring the implications of this result, however, let us review the approach which we are following. The basic idea is indicated by Fig. 8.20. This is similar to Fig. 4.4, except that now we are willing to let the input phasor be considered a spectrum of frequencies. Specifically, we must consider the effect of the filter on each input frequency. We know that the circuit, being a linear circuit, will create no new frequencies: each harmonic in the input will produce a harmonic in the output. Put another way, the frequencies in the output will be the same harmonics as exist in the input. The filter will affect each of these frequencies differently because the impedance of the capacitor varies with frequency. This frequency dependence of the circuit is symbolized by the filter function, $\underline{F}(\omega)$. For this circuit the filter function is derived in Figure 8.18 as

$$\underline{F}(\omega) = \frac{1}{1 + j\omega RC} \qquad (8.23)$$

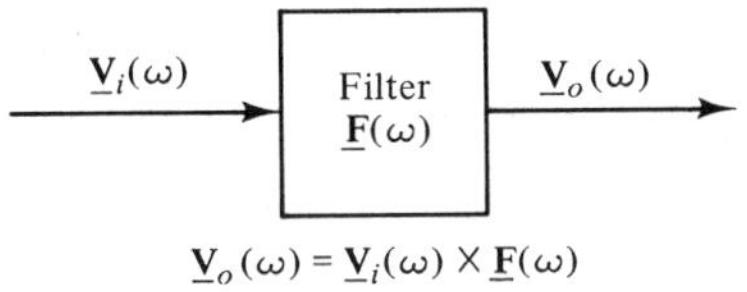

$$\underline{V}_i(\omega) \longrightarrow \boxed{\begin{array}{c} \text{Filter} \\ \underline{F}(\omega) \end{array}} \longrightarrow \underline{V}_o(\omega)$$

$$\underline{V}_o(\omega) = \underline{V}_i(\omega) \times \underline{F}(\omega)$$

Figure 8.20   Filtering a spectrum.

Passing the Signal Through the Filter.   We now can pass the series of pulses in Fig. 8.5a through the filter in Fig. 8.17 by treating separately each frequency component of the signal. Our approach is suggested by Fig. 8.21, where we have indicated that the input pulses can be represented as a series of generators representing the dc component, the fundamental, the third harmonic, the fifth harmonic, and so on. (Recall that the even harmonics are missing from this Fourier series.) Our analysis is based on superposition, adding up at the output the effect of each input sinusoidal source, considered separately. This exchanges the solution of a DE in the time domain for a host of ac circuit solutions

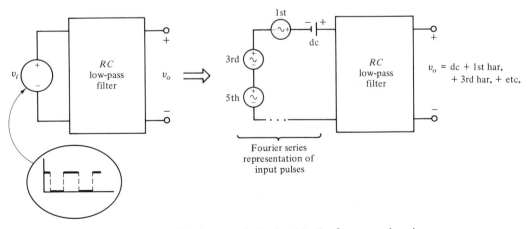

Figure 8.21   Filtering a periodic signal in the frequency domain.

in the frequency domain. Fourier analysis allows us to determine the amplitude and phase of the harmonics of the input signal. This is symbolized as

$$v_i(t) = \text{Re}\left\{V_{i(\text{dc})} + \underline{V}_{i(1)}e^{j\omega_0 t} + \underline{V}_{i(2)}e^{j2\omega_0 t} + \cdots + \underline{V}_{i(n)}e^{jn\omega_0 t} + \cdots\right\} \quad (8.24)$$

where $\underline{V}_{i(n)}$ represents the amplitude and phase of the nth harmonic of the input signal. We must next evaluate the filter function at each harmonic frequency to see how each is affected in amplitude and phase by the filter. This is symbolized by

$$V_{o(\text{dc})} = \underline{F}(\omega = 0) \times V_{i(\text{dc})}$$
$$\underline{V}_{o(1)} = \underline{F}(\omega_0) \times \underline{V}_{i(1)} \quad (8.25)$$
$$\underline{V}_{o(n)} = \underline{F}(n\omega_0) \times \underline{V}_{i(n)}$$

where $\underline{V}_{o(n)}$ represents the amplitude and phase of the nth output harmonic and $\underline{F}(n\omega_c)$ is the filter function evaluated at the frequency of the nth harmonic. Finally, we sum the frequency components of the output and transform back to the time domain, as shown in

$$v_o(t) = \text{Re}\left\{V_{o(\text{dc})} + \underline{V}_{o(1)}e^{j\omega_0 t} + \underline{V}_{o(2)}e^{j2\omega_0 t} + \cdots + \underline{V}_{o(n)}e^{jn\omega_0 t} + \cdots\right\} \quad (8.26)$$

A Numerical Example. We will work through the details for one case. We let the input signal consist of pulses of 1-V amplitude, $\frac{1}{2}$-ms duration, and a repetition period of 1 ms. This produces a fundamental frequency of 1 kHz, a third harmonic at 3 kHz, and so on. For our filter we will use a resistance of 2 kΩ and a capacitor of 0.1 μF. This corresponds to a cutoff frequency of 5000 rad/s, or about 800 Hz. Thus the fundamental of the input signal is approximately equal to the cutoff frequency of the filter and all the higher harmonics well exceed the cutoff frequency. We would therefore anticipate that the higher harmonics would be reduced substantially by the filter compared with the fundamental. The details are given in Table 8.2. Thus the Fourier series of the output

TABLE 8.2   CALCULATION OF OUTPUT HARMONICS

Harmonic number, n	Frequency, $\omega_n = 2\pi(nf_0)$	Input phasor, $V_{i(n)}$	Filter function, $\underline{F}(\omega_n) = \dfrac{1}{1 + j\omega_n RC}$	Output phasor $\underline{V}_{o(n)}$
0	dc	0.5	1	0.5
1	$2\pi(1000)$	$0.637\,\underline{/0°}$	$\dfrac{1}{1 + j1.26} = 0.623\,\underline{/51.6°}$	$0.397\,\underline{/-51.6°}$
3	$2\pi(3000)$	$0.212\,\underline{/-180°}$	$\dfrac{1}{1 + j3.77} = 0.256\,\underline{/-75.1°}$	$0.0544\,\underline{/-255.1°}$
5	$2\pi(5000)$	$0.127\,\underline{/0°}$	$\dfrac{1}{1 + j6.28} = 0.157\,\underline{/-81.0°}$	$0.0200\,\underline{/-81.0°}$
7	$2\pi(7000)$	$0.091\,\underline{/-180°}$	$\dfrac{1}{1 + j8.80} = 0.113\,\underline{/-83.5°}$	$0.0103\,\underline{/-264°}$

voltage is the sum of the harmonics in the last column of Table 8.2, as expressed in

$$v_0(t) = 0.5 + 0.397 \cos(\omega_0 t - 51.6°) + 0.0544 \cos(3\omega_0 t - 255.1°)$$
$$+ 0.0200 \cos(5\omega_0 t - 81.0°) + 0.0103 \cos(7\omega_0 t - 264°) + \cdots \qquad (8.27)$$

This voltage is plotted in Fig. 8.22. We see that the shape of the pulses is radically altered by the filter. The dc component and the fundamental are evident, but the sharp corners of the input pulses are gone. This results from filtering out the higher harmonics, which originate in the steep slope and sharp corners of the input pulses.

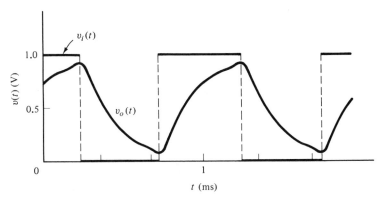

Figure 8.22 Filter input and output in the time domain.

8.3.2 The Bode Plot

The Decibel Scale.   The gain of amplifiers and the loss of filters are frequently specified in *decibels* (dB). This unit refers to a logarithmic measure of the ratio of output power to input power, as defined in

$$\text{gain (dB)} = 10 \log \frac{P_o}{P_i} \qquad (8.28)$$

Because the power gain is proportional to the square of the voltage gain, the voltage gain is defined to be

$$\text{gain (dB)} = 20 \log \left| \frac{\mathbf{V}_o}{\mathbf{V}_i} \right| \qquad (8.29)$$

For example, a voltage gain of 50 would be $20 \log(50) = 34.0$ dB. We note in comparing Eqs. (8.28) and (8.29) that power ratios require a factor of 10 in the dB calculation and voltage ratios a factor of 20.

A loss can also be described by Eq. (8.28) but the "dB gain" is negative for a loss: for example, the gain of a lowpass filter [Eq. (8.22)] at $f = f_c$ was shown to be $1/\sqrt{2}$; Eq. (8.28) yields

$$G_{\text{dB}} = 20 \log |\mathbf{F}(\omega_c)| = 20 \log \frac{1}{\sqrt{2}} = -3.0 \text{ dB}$$

Thus we could say the gain of the filter at its half-power frequency is negative 3.0 dB or, equivalently, that its loss is 3.0 dB.

The dB scale is also useful for expressing the power in a signal. In this context, the power is expressed as a logarithmic ratio between the signal power and an assumed power level, usually 1 W or 1 mW. Thus dBw, dB relative to 1 W, is defined as

$$\text{dBw} = 10 \log \frac{P}{1 \text{ watt}}$$

where P is the power and dBw means that same power relative to 1 W expressed in dB. For example, a satellite transmitter with an output power of 1000 W would have an output power of 30 dBw. This power could also be called 60 dBm, where dBm means dB relative to a milliwatt.

Why dB Is Useful.   There are three reasons why electrical engineers prefer the dB scale for describing gains, losses, and signal levels. Two of these reasons can be demonstrated by an example. Figure 8.23 suggests a communication satellite in stationary orbit around the earth. The satellite transmitter produces 1000 W and operates at a frequency of 5 GHz [λ (wavelength) = 6 cm]. We shall calculate the power collected by an earth antenna. The formula for received power is well known to communication engineers.

$$P_{\text{rec}} = \frac{P_t G_t G_r}{(4\pi D/\lambda)^2} \tag{8.30}$$

where P_{rec} = received power, to be calculated
$\quad$ P_t = transmitted power; assume 1000 W (30.0 dBw)
$\quad$ G_t = transmitting antenna gain; assume 400 ($G_{t(\text{dB})}$ = 26.0 dB)
$\quad$ G_r = receiving antenna gain; assume 5000 ($G_{r(\text{dB})}$ = 37.0 dB)
$\quad$ D/λ = distance between antennas/wavelength of the signal; assume 22,000 miles/6 cm

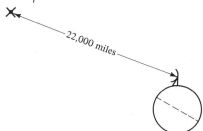

Figure 8.23   Satellite communication system.

With these values, the term in the denominator, which is called the *space loss* (SL), has the value 5.50×10^{19} or 197.4 dB.

If we evaluate Eq. (8.30) in the normal fashion, we get

$$P_{\text{rec}} = \frac{(1000)(400)(5000)}{5.50 \times 10^{19}} = 3.64 \times 10^{-11} \text{ W}$$

(This is a tiny amount of power, but it is sufficient, as evidenced by the growing number of satellite antennas around the countryside.)

To use dB in the calculation of received power, we take the common log of Eq. (8.30) and multiply by 10 (it is a power equation), with the result

$$P_{\text{rec(dB)}} = P_{t\text{(dB)}} + G_{t\text{(dB)}} + G_{r\text{(dB)}} - SL_{\text{(dB)}}$$
$$= 30.0 \text{ dBw} + 26.0 \text{ dB} + 37.0 \text{ dB} - 197.4 \text{ dB} = -104.4 \text{ dBw}$$

The minus sign on the space-loss term occurs because this term is in the denominator. The answers are equivalent: $10 \log (3.64 \times 10^{-11}) = -104.4$ dBw.

This example illustrates two virtues of dB. Because of the compressive nature of the logarithmic function, the numbers involved in the dB calculation are quite moderate compared to the other calculation. This feature of dB measure is particularly useful when plotting quantities that vary greatly in magnitude, as will be illustrated below. The second virtue of dB measure is that the calculation is accomplished by addition and subtraction, rather than by multiplication and division. This feature results from the familiar technique for multiplying numbers by adding their logarithms. In these days of calculators this seems no great benefit, but the custom of using dB became universal among electrical engineers in an earlier day and will no doubt continue in the future.

The Bode Plot.   A third virtue of dB measure involves a special way of plotting the characteristic of a filter, amplifier, or spectrum. The Bode plot (named for H. W. Bode, 1905–   ) is a log power versus log frequency plot. For example, Fig. 8.24 shows the Bode plot of the function

$$\underline{A}(f) = \frac{100}{1 + j(f/10)}$$

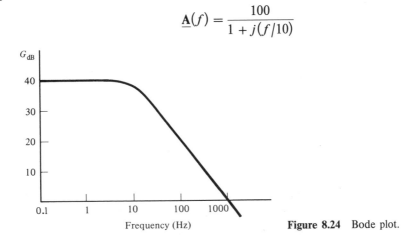

Figure 8.24   Bode plot.

which combines an amplifier (voltage gain of $100 = 40$ dB) with a low-pass filter ($f_c = 10$ Hz). In Fig. 8.24 the vertical axis is proportional to log power expressed in dB. The horizontal axis expresses frequency on a log scale, although normally the frequency (not the logarithm of the frequency) is marked on the graph, as in Fig. 8.24. This is a log plot because equal spacings on the graph are given to equal ratios of frequency (1 : 10, 10 : 100, etc.).

The Bode plot is useful in two ways. As mentioned earlier, the use of a log scale allows the representation of a wide range of both power and frequency. In this case, the vertical scale of 0 to 40 dB represents a range of power gain from 1 to 10^4 (or a voltage gain from 1 to 100). Similarly, in the horizontal scale, we have represented four decades (factors of 10) in frequency. These wide ranges are possible because of the compressive nature of the logarithmic function.

The other advantage of the Bode plot is that a filter characteristic can usually be well represented by straight lines. The straight lines are evident in Fig. 8.24: the characteristic is quite flat for frequencies well below $f_c = 10$, and slopes downward with constant slope for frequencies well above this value. Only in the vicinity of f_c does the exact Bode plot depart significantly from the straight lines. In the following section we examine in more detail the relationship between the (exact) Bode plot and the straight-line approximation, which is called the *asymptotic Bode plot*.

The Bode Plot for a Low-Pass Filter.   As shown in Fig. 8.18 and Eq. (8.22), the low-pass filter in Fig. 8.25 has the characteristic

$$\underline{F}(f) = \frac{1}{1 + j(f/f_c)}$$

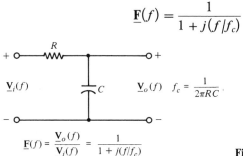

$$\underline{F}(f) = \frac{\underline{V}_o(f)}{\underline{V}_i(f)} = \frac{1}{1 + j(f/f_c)}$$

Figure 8.25 Low-pass filter.

The Bode plot represents the magnitude of the voltage gain and hence requires the absolute value

$$G_{dB}(f) = 20 \log \left| \frac{1}{1 + j(f/f_c)} \right| = 20 \log \frac{1}{\sqrt{1 + (f/f_c)^2}}$$

$$= -10 \log \left[1 + \left(\frac{f}{f_c} \right)^2 \right] \tag{8.31}$$

We may generate the exact Bode plot using normalized frequency f/f_c by numerical calculation, as shown in Fig. 8.26, where we emphasize the tendency of the exact Bode plot to approach a straight line above and below the cutoff frequency, $f/f_c = 1$ on the frequency scale. At the exact cutoff frequency, the filter gain is -3.0 dB, as noted earlier.

We may derive the asymptotic Bode plot by taking the limiting forms of Eq. (8.31). For frequencies well below the cutoff frequency, $f \ll f_c$, the gain is approximately

$$G_{dB} \approx -10 \log (1 + \text{small}) \approx 10 \log (1) = 0 \qquad (f \ll f_c) \tag{8.32}$$

This accounts for the flat section. For frequencies well above the cutoff frequency, the gain approaches

$$G_{dB} \longrightarrow -10 \log \left(\frac{f}{f_c} \right)^2 = -20 \log f + 20 \log f_c \qquad (f \gg f_c) \tag{8.33}$$

f/f_c	$G_{dB}(f)$
0.01	−0.00043
0.1	−0.043
0.5	−0.97
1.0	−3.01
2	−6.99
10	−20.043
100	−40.00043

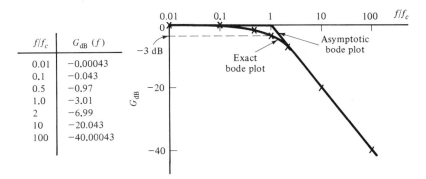

Figure 8.26 Bode plot for the low-pass filter.

Recalling that our horizontal variable is $\log f$, we see that Eq. (8.33) is the form of a straight line, $y = mx + b$, where m is -20, x is $\log f$ and b is $20 \log f_c$. Thus the high-frequency asymptote will be a straight line in this coordinate system. When $\log f = \log f_c$, the y of the straight line has zero value and hence the line passes through the horizontal axis at $f = f_c$. The slope is -20: usually we say that the slope is negative 20 dB/decade because dB of gain and decades of frequency are the units for a Bode plot.

The low- and high-frequency asymptotes of the exact Bode plot combine to form the asymptotic Bode plot. As shown in Fig. 8.26, the exact Bode plot is well represented by this approximation, the largest error being 3.0 dB at the cutoff frequency. For most filter applications, we may use the asymptotic Bode plot to give an adequate picture of the filter characteristics.

The Bode Plot for a High-Pass Filter. Figure 8.27 shows a filter that will block low frequencies and pass high frequencies. The filter characteristics can be determined by considering the circuit as a voltage divider in the frequency domain

$$\underline{\mathbf{F}}(\omega) = \frac{\underline{\mathbf{V}}_o(\omega)}{\underline{\mathbf{V}}_i(\omega)} = \frac{R}{R + 1/j\omega C} = \frac{j\omega RC}{1 + j\omega RC} = \frac{j(f/f_c)}{1 + j(f/f_c)} \tag{8.34}$$

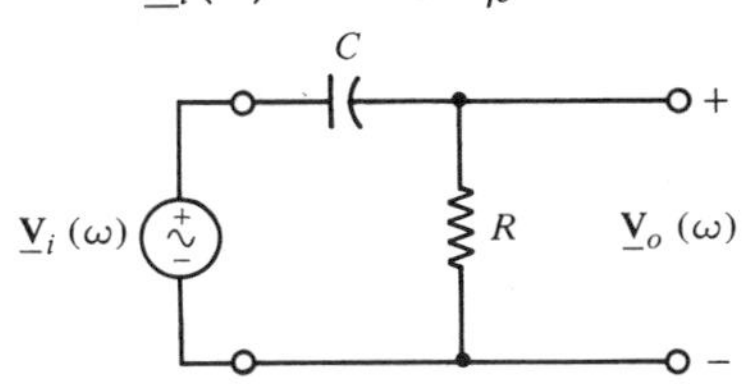

Figure 8.27 High-pass filter.

where $f_c = 1/2\pi RC$ is the cutoff frequency. At low frequencies, $f \ll f_c$, the capacitor has a high impedance and allows little current; hence little voltage will develop across the resistor. Indeed, at zero frequency (dc), the capacitor acts as an open circuit and no voltage appears at the output. At high frequencies, $f \gg f_c$, the capacitor impedance will be low, and essentially all the voltage appears across the resistor. Consequently, the filter passes high frequencies and blocks low frequencies. The transition between these two regimes occurs when the impedance of the capacitor is comparable to that of the resistor, and the cutoff frequency occurs where they are exactly equal, as given by Eq. (8.21).

The asymptotic Bode plot for the filter function given in Eq. (8.34) may be derived by the same method as that which we used for the low-pass filter. The Bode plot is derived from the absolute value of the filter function,

$$G_{dB} = 20 \log \left| \frac{j(f/f_c)}{1 + j(f/f_c)} \right| = 20 \log \frac{f/f_c}{\sqrt{1 + (f/f_c)^2}} \tag{8.35}$$

At frequencies well below the cutoff frequency, the frequency term in the denominator can be neglected and the asymptote becomes

$$G_{dB} \longrightarrow 20 \log \frac{f}{f_c} = 20 \log f - 20 \log f_c$$

Plotted against $\log f$, this is a straight line having a slope of $+20$ dB/decade and passing through the horizontal axis at $f = f_c$ as shown in Fig. 8.28. For high frequencies, the filter function approaches unity, or 0 dB. This asymptote is also shown in Fig. 8.28. The full Bode plot for the filter combines these two asymptotes and makes a smooth transition between them near the cutoff frequency. At the cutoff frequency, the exact Bode plot lies 3 dB below the intersection of the low- and high-frequency asymptotes.

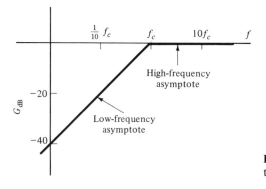

Figure 8.28 Asymptotic Bode plot for the high-pass filter.

Combining Filter Characteristics. In Fig. 8.29 we show a combination of a low-pass filter, an amplifier, and a high-pass filter. We shall examine their combined characteristics. The overall input–output characteristic proves to be the product of the individual characteristics, as shown by

$$\underline{\mathbf{F}}(f) = \frac{\mathbf{V}_4}{\mathbf{V}_1} = \frac{\mathbf{V}_2}{\mathbf{V}_1} \times \frac{\mathbf{V}_3}{\mathbf{V}_2} \times \frac{\mathbf{V}_4}{\mathbf{V}_3} = \underline{\mathbf{F}}_1(f) \times \underline{\mathbf{F}}_2(f) \times \underline{\mathbf{F}}_3(f) \tag{8.36}$$

By expressing the total gain in dB, we convert the product to a sum

$$G_{dB} = 20 \log |\underline{\mathbf{F}}_1 \times \underline{\mathbf{F}}_2 \times \underline{\mathbf{F}}_3| = 20 \log |\underline{\mathbf{F}}_1| + 20 \log |\underline{\mathbf{F}}_2| + 20 \log |\underline{\mathbf{F}}_3|$$

$$= G_{1(dB)}(f) + G_{2(dB)}(f) + G_{3(dB)}(f)$$

Figure 8.30 shows the effect of adding up the individual Bode plots. The amplifier has a constant gain of 20 dB and raises the sum of the filter Bode plots by that amount. The high-pass filter blocks the low frequencies, and the low-pass filter blocks the high frequencies. The resultant is a bandpass filter; the "band" being passed in this case contains

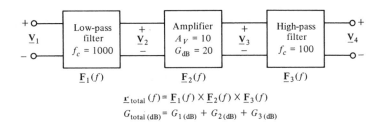

$$\underline{F}_{total}(f) = \underline{F}_1(f) \times \underline{F}_2(f) \times \underline{F}_3(f)$$
$$G_{total\,(dB)} = G_{1\,(dB)} + G_{2\,(dB)} + G_{3\,(dB)}$$

Figure 8.29 Two filters and an amplifier in cascade.

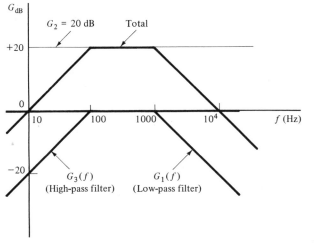

Figure 8.30 Adding Bode plots.

the frequencies between 100 and 1000 Hz. Thus we combine filter effects by adding the Bode plots.

Narrowband RLC Filters.   As shown above, we can produce a filter to pass the frequencies in a certain band by combining high-pass and low-pass *RC* filters. This approach works well only when the passband of frequencies is rather broad. However, most communication systems, such as radios, require filters which pass a narrow band of frequencies. This is normally accomplished with an *RLC* filter such as appears in Fig. 8.31. This particular *RLC* filter produces a parallel resonance between the inductor and capacitor about 1125 kHz, which is near the middle of the AM radio band. The resistor shown in parallel with the inductor is not present in the physical circuit but is placed in

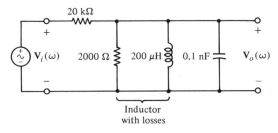

Figure 8.31 *RLC* filter.

the circuit to represent the losses of the inductor. For frequencies far away from the resonant frequency, the impedance of either the inductor or the capacitor becomes small and the filter response drops. At the resonant frequency, the inductance and capacitance cancel out each other and the output is maximum.

Figure 8.32 shows the passband characteristics of this filter in linear and dB scales. It might be noted that this filter has an inherent loss, even at its maximum output, -20.8 dB in this case. Thus such a filter must be used in conjunction with an amplifier in order to compensate for this loss. This is no disadvantage in most communication circuits, for the radio already must provide considerable amplification—this just requires a bit more. We might mention in closing this section that the filter characteristic of the *RLC* filter in Fig. 4.38 is similar to that in Fig. 8.31. The parallel form of the filter is accomplished more easily in electronic circuits with realistic inductors and capacitors, and hence this is the form of narrowband filter commonly found in radios. In the next section, after we have discussed the properties of nonlinear circuits, we will discuss the role of such filters in typical radio circuits.

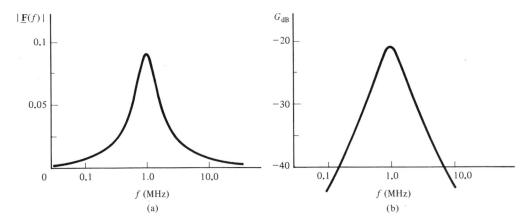

Figure 8.32   Narrow band filter characteristics: (a) linear plot; (b) dB plot.

8.4 NONLINEAR DEVICES IN THE FREQUENCY DOMAIN

8.4.1 General Principles

One of the important principles we used in our analysis of ac circuits was that linear circuits create no new frequencies. Back in Chapter 4 we argued that if you excite a circuit at a certain frequency it will respond only at that frequency. This principle allowed us to simplify greatly our analysis of ac circuits; indeed, the approach using phasors and impedance is founded entirely upon it.

The reverse principle is that nonlinear circuits create new frequencies. We have already discovered that fact in Chapter 6, where we showed that a diode, a nonlinear device, can rectify a signal. When we discussed the spectrum of a half-wave rectified sinusoid [Fig. 8.6b and Eq. (8.5)] we noted that it contains, in addition at the frequency of the input sinusoid, frequency components at dc and all the even harmonics of the input frequency. These new frequencies are created by the nonlinear action of the diode.

The spectrum of a rectified sinusoid is a particular instance of the general principle suggested in Fig. 8.33. When we have a nonlinear circuit with a sinusoidal input, the output of the circuit will in general contain all the harmonics of the input. The amplitudes of the various output harmonics depend on the details of the circuit, but in principle all harmonics can be present in the output of a nonlinear circuit.

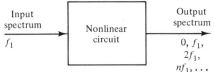

Figure 8.33 Nonlinear circuits create new frequencies.

This property of nonlinear circuits finds direct application in many practical devices. One obvious example is a power supply, where the output frequency component of interest is the zeroth harmonic of the input. Another application occurs in the generation of stable high-frequency sinusoids for precision communication systems or spectroscopic measurement systems. In such applications high-frequency signals of the required stability are often generated by taking the output of a highly stable low-frequency oscillator, say, at 5 MHz, and putting this signal through a chain of frequency multipliers until it reaches the required high frequency, say, 3240 MHz. In such a scheme, the output of the multiplier chain will be harmonically related to the low-frequency source and hence will exhibit the same stability.

On the other hand, the creation of new frequencies may be undesirable. A good audio amplifier, for example, should have low "harmonic distortion," meaning that it should not generate harmonics of the audio signal. The amplifier should also have low "intermodulation distortion." This term refers to the creation of new frequencies through the interaction of separate frequencies in the audio spectrum.

The creation of harmonics in a nonlinear circuit, as suggested in Fig. 8.33, is a special case of a more general property of nonlinear circuits, as suggested by Fig. 8.34. Here we show a nonlinear circuit with two input frequencies. The figure indicates that the output will in general contain all the harmonics of both input frequencies *plus* all the sum and difference frequencies of those harmonics. For example, if we had inputs of 30 and 100 Hertz, the output would contain not only the harmonics of 30 (60, 90, 120, . . .) and the harmonics of 100 (200, 300, . . .), but the output would also contain the various sum and difference frequencies, such as $100 - 30$, $100 + 30$, $200 - 30$, $200 + 30$, $100 - 2 \times 30$, $200 - 2 \times 30$, ... What (you may wonder) are we going to do with all

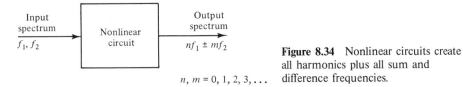

Figure 8.34 Nonlinear circuits create all harmonics plus all sum and difference frequencies.

those frequencies? The remainder of this section addresses this question; the answer, in brief, is that we are going to make radio communication systems out of such frequencies.

Although a multitude of frequencies can appear in the output of such a nonlinear circuit, all these frequencies will not in general be significant. The purpose of the circuit designer is to ensure that only the desired frequencies are strong in the output and that undesired components are minimized. Filters are used to remove unwanted frequency components in the output spectra of nonlinear devices. Only through careful control of the filtering properties of such circuits can the important benefits of nonlinear action be achieved.

The design and operation of nonlinear circuits become complicated by the creation of new frequencies and the filtering of these frequencies to enhance desired and diminish unwanted components. To avoid these complexities, we shall not investigate the details of such circuits. In the following we will examine the applications of a few nonlinear circuits, emphasizing especially their role in communication systems. Our goal is to develop a sufficient basis for discussing the operation of the common AM radio.

8.4.2 Applications of Nonlinear Circuits

Modulation. In communication engineering, *modulation* is the process by which information at a low frequency is coded as part of a high-frequency signal. The higher frequency is called the *carrier* because it carries the information signal. In this section we will discuss the amplitude modulation (AM) scheme which is widely used in commercial broadcasting of radio and TV signals. Initially, we will investigate the nature of amplitude modulation, emphasizing the frequency-domain viewpoint throughout. Then we will discuss why modulation is required in communication systems. Finally, we will present a circuit that accomplishes amplitude modulation.

The audio range covers frequencies from about 30 Hz to about 15 kHz. In a communication system we do not broadcast audio frequencies directly; rather, we use high-frequency carriers, as in the AM radio band from 540 to 1600 kHz.

Let us first consider the nature of amplitude modulation and how modulation affects the spectrum of the carrier. Equation (8.37) gives the equation of a modulated carrier in the time domain.

$$v_{\text{AM}}(t) = \underbrace{V_c(1 + m_a \cos \omega_m t)}_{V_e(t),\text{ the envelope}} \cos \omega_c t \qquad (8.37)$$

Here ω_m is the angular frequency of the modulating sinusoid, ω_c is the frequency of the modulated carrier, and V_c is its amplitude. The parameter m_a is called the *modulation*

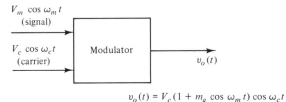

$$v_o(t) = V_c(1 + m_a \cos \omega_m t) \cos \omega_c t$$

Figure 8.35 A modulator circuit has two inputs and one output.

index and indicates the strength or degree of modulation. The modulation index is proportional to the amplitude of the modulating signal, which is indicated by V_m in Fig. 8.35, subject to the restriction that m_a never exceeds unity (i.e., $m_a \leq 1$). The character of the modulated carrier and the role of the modulation index is pictured in Fig. 8.36, where we have shown modulated carriers for $m_a = 0.5$ and $m_a = 1.0$.

Let us examine the role of the various factors characterizing the AM signal. The carrier is the basic sinusoid whose amplitude is being modulated. The carrier frequency controls the spacing of the zero crossings and individual peaks of the composite waveform. If we were to keep everything else the same yet double the carrier frequency, everything would look the same except there would be twice as many zero crossings and peaks.

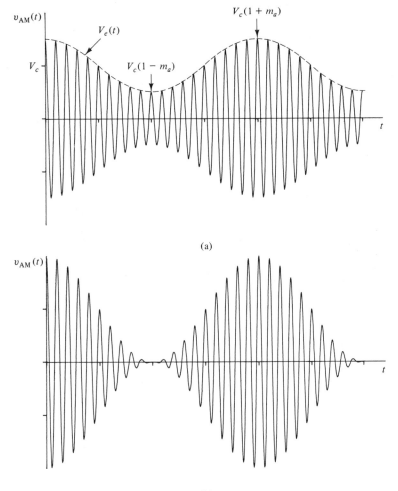

Figure 8.36   Two AM waveforms: (a) $m_a = 0.5$; (b) $m_a = 1.0$.

In other words, the hills and valleys of the amplitude would look the same, but the underlying sinusoid would have a higher frequency.

The amplitude of the carrier, V_c, characterizes the overall strength of the signal. With a broadcast signal, the amplitude of the carrier depends on the power of the station transmitter and the distance to the receiver.

The frequency of the modulating signal, ω_m, determines the distance between the hills and valleys of the envelope, $V_e(t)$, as defined in Eq. (8.37). If, for example, the modulating frequency were doubled, the envelope of the carrier would vary up and down twice as fast.

Finally, the modulation index controls the degree of modulation, the ratio of hills to valleys in the envelope. This parameter is limited to values less than unity, so that the envelope factor in Eq. (8.37) $V_e(t)$, never goes negative. The modulation index is proportional to the amplitude of the modulating signal. The design of the modulator must ensure that the maximum amplitude of the modulating signal can be accommodated without exceeding the allowed value of unity

$$m_a = K V_m \qquad \text{but } m_a \leq 1$$

where K is a constant.

An AM Modulator Circuit. The circuit in Fig. 8.37 will act as a simple modulator circuit. Ignore for the moment the RC filter between a-a' and b-b', and consider the part of the circuit that consists of two generators, a diode, and a resistor. The generator marked $v_c(t)$ is the carrier and has a high frequency. The generator marked $v_m(t)$ is the modulating signal, and it has a lower frequency and an amplitude smaller than the carrier. Consider now the action of the diode on the signal at a-a'. When the instantaneous carrier voltage, $v_c(t)$, exceeds the instantaneous modulating voltage, $v_m(t)$, the diode will turn ON and current will flow clockwise around the circuit. The output voltage in this case will be equal to the instantaneous modulating voltage plus the few tenths of a volt required to turn ON the diode. On the other hand, when the instantaneous carrier voltage is smaller than the instantaneous modulating voltage, the diode will be OFF and no current will flow. During this time, there will be no voltage drop across the resistor and hence the voltage at a-a' will be equal to the carrier voltage. The result is shown in Fig. 8.38.

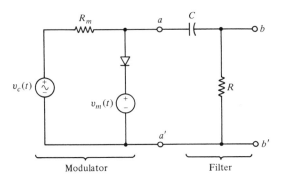

Modulator   Filter   **Figure 8.37** AM modulator circuit.

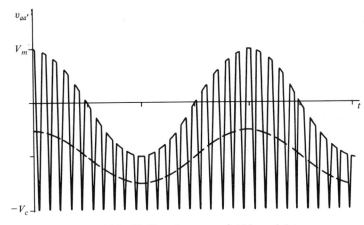

Figure 8.38 Unfiltered output of AM modulator.

The signal at a-a', shown in Fig. 8.38, differs from a pure AM signal in two respects: (1) only the top is modulated, the bottom being flat; and (2) the carrier waveform is not round on top and is thus not a pure sinusoid with varying amplitude. We can solve the first problem by filtering out the dc and the component at ω_m (the dashed line in Fig. 8.38). We can eliminate them with a high-pass filter that passes the carrier while rejecting the modulating frequency. This is the function of the high-pass filter between a-a' and b-b'.

The other problem, that the tops of the modulated carrier are distorted, implies that some higher harmonics of the carrier are also created in the modulator circuit. We could also eliminate them with proper filtering, but this does not concern us here. With proper filtering, the output of the modulator in Fig. 8.38 would look like the ideal AM waveform in Fig. 8.36a.

The Spectrum of an AM Signal.   If your AM radio received the AM signals in Fig. 8.36, and if the modulating frequency were in the audio band, you would hear a steady tone from the radio speaker, and you might suppose that a Civil Defense test was under way. (Actually, the Emergency Broadcast signal uses simultaneous broadcast of tones at 853 and 960 Hz.) In normal broadcasting, the program material would be music and voice signals, not steady tones. The modulated signal would thus be of the form

$$v_{\text{AM}}(t) = V_c[1 + m_a s(t)] \cos \omega_c t \qquad (8.38)$$

where $s(t)$ represents the program material and hence would be a random function. In Eq. (8.38) $s(t)$ must be remain smaller than unity so that the product $m_a s(t)$ never exceeds unity at any time.

The spectrum of an AM signal can be determined through expansion of Eq. (8.37) with the trigonometric identity

$$\cos a \cos b = \tfrac{1}{2}[\cos (b + a) + \cos (b - a)]$$

When we expand Eq. (8.37) and apply the identity letting $a = \omega_m t$ and $b = \omega_c t$, we obtain

$$v_{AM}(t) = V_c \cos \omega_c t + m_a V_c \cos \omega_m t \cos \omega_c t$$

$$= V_c \cos \omega_c t + \frac{m_a V_c}{2} [\cos \underbrace{(\omega_c + \omega_m)t}_{\text{sum freq.}} + \cos \underbrace{(\omega_c - \omega_m)t}_{\text{diff. freq.}}]$$

This expression suggests a spectrum with three frequency components: the carrier at ω_c, an "upper sideband" at $\omega_c + \omega_m$, and a "lower sideband" at $\omega_c - \omega_m$. The spectra before and after modulation are shown in Fig. 8.39. We have broken the frequency scales in Fig. 8.39 to indicate that the carrier frequency is normally much higher than the modulating frequency. For example, the Emergency Broadcast warning tone of 853 Hz, broadcast on a carrier of 1200 kHz, would produce upper and lower sidebands at 1,200,853 and 1,199,147 Hz, respectively.

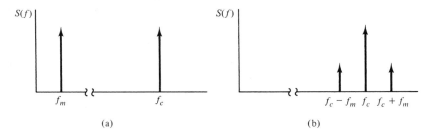

Figure 8.39 Modulation in the frequency domain: (a) before modulation; (b) after modulation.

When the program information consists of a random signal such as music or voice, the individual frequencies comprising the spectrum of the audio signal are shifted to the sidebands. Let us investigate the case of normal AM broadcasting. By law the spectrum of the audio signal is limited to the bandwidth between 100 and 5000 Hz. The effect of modulating with such a signal is indicated by Fig. 8.40. We notice several features of the spectrum of the modulated carrier. Although the carrier still consists of a single frequency, the sidebands now become continuous spectra. The upper sideband is identical in shape with the audio signal, whereas the lower sideband is reversed. The bandwidth of

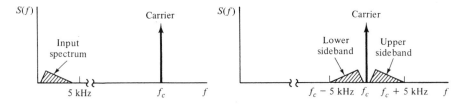

Figure 8.40 Modulation with a spectrum: (a) before modulation; (b) after modulation.

the radio-frequency (RF) spectrum is twice the audio bandwidth, and hence 10 kHz of RF bandwidth is required for each AM station.

If you could examine the spectrum of radio waves received by an AM radio antenna, you would see carriers spaced 10 kHz apart, each bracketed by their upper and lower sidebands. Since the AM broadcast band covers frequencies from 535 to 1605 kHz, there are positions for 106 stations in a single geographic area, although obviously most positions are unoccupied.

Because AM detectors are simple, and hence AM radios are cheap, standard AM modulation is used for commercial broadcasting. This AM system is inefficient in its use of the available bandwidth and the transmitter power. Bandwidth is wasted because, although the lower sideband and carrier are broadcast, all the information is contained in the upper sideband. Consequently, the station power used in transmitting the carrier and lower sideband is wasted because these carry no information not already contained in the upper sideband. For these reasons, sophisticated radio communication systems such as a telephone system uses for long-distance messages employ single-sideband modulation for economy in both power and bandwidth. Single-sideband modulation, as the name suggests, transmits only one of the sidebands over the communication path, and the receiver must generate the carrier and missing sideband to detect the information.

The effect of modulation is to shift the information from the audio band to the RF radio band. We shall now discuss the necessity for modulation; that is, why do we need to modulate? In short, we modulate to solve antenna problems, filtering problems, and confusion problems. The last problem is the easiest to explain: clearly, we cannot allow every radio station to broadcast their signals in the same frequency band because then our radio would receive all stations simultaneously. Hence modulation allows us to assign different portions of the RF spectrum to different radio stations, and they can place their signal neatly into their allotted frequency bands. We choose stations by tuning our radio receivers to different frequencies.

Discussion of the antenna problems solved by modulation would take us beyond the aim of this text. How modulation solves filtering problems we shall consider later in this section, after we have discussed AM detectors and mixers. Having shown how the information in the audio band is shifted to higher frequencies for broadcast, we turn now to the reverse operation: how your radio shifts the information back to the audio band so that you can hear it.

The AM Detector.   We mentioned earlier that standard AM modulation is utilized for commercial broadcasting because of the ease with which the information can be detected. Figure 8.41 shows a simple circuit which will extract the audio information from the AM signal envelope. We studied this circuit earlier (page 216) as a power supply

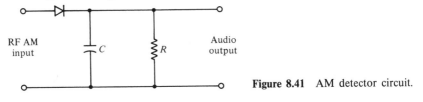

Figure 8.41 AM detector circuit.

circuit. In that application the input is a simple sinusoid. The diode allows the capacitor to charge to the peak voltage of the input, and the resistor represents the load. To work well as a power supply, the circuit requires an RC time constant much larger that the period of the input sinusoid. This long time constant ensures that the output voltage of the power supply will remain essentially at the peak voltage of the input.

To function as an AM detector, this circuit requires an RC time constant intermediate between the period of the carrier and that the audio information, that is, $\omega_m < 1/RC < \omega_c$. With this provision, the detection of the envelope of the carrier takes place as suggested by Fig. 8.42. The output of the detector will follow the peaks of the input and hence will approximate the envelope of the AM signal. For purposes of drawing the waveform, we have drawn the period of the carrier relatively large, resulting in a jagged appearance for the output; but in practice the period of the carrier is so short that the output is essentially smooth. The small amount of jaggedness, which is inevitable, constitutes a small noise component in the output, easily removed with a low-pass filter. Disregarding this noise component, we can see that the effect of the detector in the frequency domain is to shift the information in the sidebands of the carrier back to the audio band, as shown in Fig. 8.43.

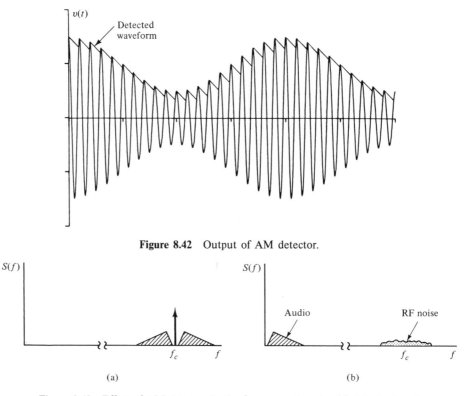

Figure 8.42 Output of AM detector.

(a)   (b)

Figure 8.43 Effect of AM detector in the frequency domain: (a) detector input spectrum; (b) detector output spectrum.

Examining the relationship between the spectra in Figs. 8.40 and 8.43, we see that the detector reverses the modulation process. For this reason the word *demodulation* is often used to describe this process. You might suppose that we now have all the makings of a radio: we can shift the audio information to the sidebands of a carrier for broadcast and we can shift the information back to the audio region for conversion to acoustic waves with loudspeakers. In a sense you suppose correctly, for a simple communication system can be made with direct modulation and demodulation, but the usual radio utilizes another nonlinear process, mixing, to manipulate the spectrum of the received signal.

The Mixer.   A *mixer*, like a modulator, is a nonlinear circuit with two inputs and one output, as suggested by Fig. 8.44. Indeed, modulators and mixers operate on similar principles, yet differ in their effect and function in communication systems. Figure 8.45 shows a simple circuit which will act as a mixer. We assume the LC to be resonance at the intermediate frequence; hence the capacitor's impedance will be so small at RF that the current which flows through the circuit at RF will be controlled by the diode characteristic (Eqs. 6.6 and 6.7, assume $\eta = 1$).

$$i_D = I_0(e^{v_D/V_T} - 1) = I_0\left[1 + \frac{v_D}{V_T} + \frac{1}{2}\left(\frac{v_D}{V_T}\right)^2 + \cdots - 1\right]$$

$$= a_1 v_D + a_2 v_D^2 + \cdots$$

(8.39)

Because the impedance at RF is small, the voltage across the diode will be the sum of the two voltage sources:

$$v_D \approx V_1 \cos \omega_1 t + V_2 \cos \omega_2 t$$

and therefore the power series form of Eq. (8.39) will have many terms involving the two input frequencies, of which we shall investigate but a few.

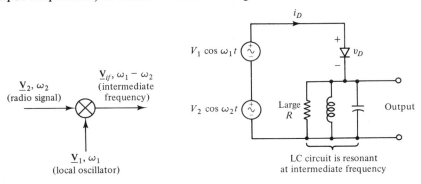

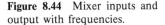

Figure 8.44 Mixer inputs and output with frequencies.

Figure 8.45 Mixer circuit. The LC filter presents a small impedance of the RF but a large impedance at the intermediate frequency.

In Table 8.3 we have shown some of the products that will appear in the mixer output, and we have shown in a separate column the frequency components that will be produced. The output provides an example of the principle suggested in Fig. 8.34,

TABLE 8.3 SOME OF THE FREQUENCIES CREATED IN THE MIXER

	Frequency component
$i_D = a_1(V_1 \cos \omega_1 t + V_2 \cos \omega_2 t) + a_2(V_1 \cos \omega_1 t + V_2 \cos \omega_2 t)^2 + \cdots$	
$\quad = a_1 V_1 \cos \omega_1 t$	ω_1
$\quad + a_1 V_2 \cos \omega_2 t$	ω_2
$\quad + a_2 V_1^2 \cos^2 \omega_1 t \; [= a_2(V_1^2/2)(1 + \cos 2\omega_1 t)]$	DC, $2\omega_1$
$\quad + 2a_2 V_1 V_2 \cos \omega_1 t \cos \omega_2 t \; [= a_2 V_1 V_2 \cos(\omega_1 t + \omega_2 t)$	$\omega_1 + \omega_2$
$\qquad\qquad\qquad\qquad\qquad + a_2 V_1 V_2 \cos(\omega_1 t - \omega_2 t)]$	$\omega_1 - \omega_2$
$\quad + a_2 V_2^2 \cos^2 \omega_2 t \; [= a_2(V_2^2/2)(1 + \cos 2\omega_2 t)]$	DC, $2\omega_2$

namely, that a nonlinear circuit with two input frequencies will in general create all the harmonics of those input frequencies, plus the sums and differences of those harmonics. Had we extended our power series in Eq. (8.39) to higher-order terms and used trigonometric identities to expand these terms, we would have verified the presence of yet higher harmonics, with their sums and differences.

You may wonder if these frequencies really exist. After all, sines and cosines come from triangles; and what do triangles have to do with all this business? If you find yourself a doubter at this stage, you are like many students who have been led down this path of reasoning, which seems so obscure and abstract. Our only defense for the physical existence of these frequencies is to point to your radio, your TV, your telephone. These devices all use mixers, and mixers depend for their operation on the generation of these new frequencies.

Of the various frequencies in the mixer output, only one is desired; the rest are noise to be eliminated with filters. We require the difference frequency, $\omega_1 - \omega_2$, listed in the column in Table 8.3. Because the LC circuit will present a high impedance at the intermediate frequency $(\omega_1 - \omega_2)$, substantial voltage will appear at the output at this frequency. This component represents a shift of the information in the carrier and sidebands to a lower frequency. The effects of mixing and demodulation are shown in Fig. 8.46. The higher frequency (f_1) is called the *local oscillator* (LO) *frequency,* "local oscillator" because this frequency is generated within the radio by an oscillator circuit. The mixer combines this frequency with the carrier and sidebands, acting on each frequency component separately, and shifts the entire RF spectrum to an intermediate frequency (IF). The IF signal is amplified and filtered and is then detected by a normal AM

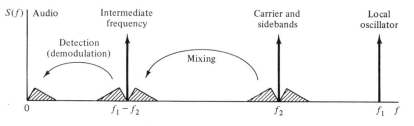

Figure 8.46 Mixing and demodulation in the frequency domain.

detector; the audio signal is thus recovered. This audio signal is further amplified, low-pass filtered, and furnished to a loudspeaker. As an example, consider a radio station broadcasting at 1200 kHz, near the middle of the AM radio band. The standard IF for an AM radio is 455 kHz; hence the local oscillator frequency must be $1200 + 455 = 1655$ kHz. The resulting AM radio circuit is pictured in Fig. 8.47.

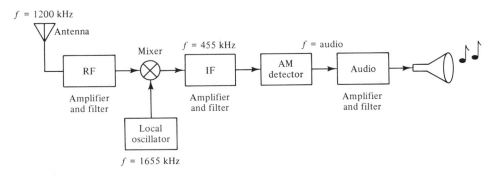

Figure 8.47   Basic AM radio system.

The Reasons for Using an Intermediate Frequency.   We shall now describe in detail the reasons why we need to shift the information to an IF rather than use direct demodulation of the RF signal. As a preliminary we shall review the filtering that we have already encountered in the circuit. We require low-pass filtering at the output of the AM detector to separate the audio signal from the noise at the IF being detected. This filtering is usually incorporated into the audio amplifier circuit and is relatively easy to accomplish because of the wide difference in frequency between the signal spectrum and the noise spectrum.

Similarly, we need filtering at the mixer output in order to eliminate all but one of the frequencies created by the nonlinear action of the mixer diode. This filtering is easily accomplished because there is again a large difference between the desired and the undesired frequency components in the mixer output. As we shall see, this filtering is accomplished by the IF amplifier.

But the primary filtering task of the radio, and the most demanding, we have neglected to mention to this point. We refer to the requirement to filter the RF spectrum to eliminate all the other radio stations which are broadcasting simultaneously with the station of interest. If we were to portray the true effect of mixing in Fig. 8.46, we would have put the signals from many AM radio stations side by side in our RF spectrum, and our radio has to select one and eliminate the rest. How is this filtering to be accomplished?

The simple answer does not work well for practical reasons. We refer to the obvious solution: put a filter in the RF amplifier which will select the desired station and reject the rest. The two reasons why this is impractical are as follows. The requirements on such a filter would strain the limits for practical inductors and capacitors; and the requirement for a tunable filter further complicates the design of an effective RF filter.

Let us consider the second difficulty first. To construct a radio, we need to filter out all the radio stations except one; but in addition, we need to be able to tune the radio to any of the available stations. We have not discussed tunable filters hitherto, nor shall we here except to confess that they are difficult to design. The other problem is that, even without the requirement that the filter be tunable, it would be challenging to design an RF filter selective enough to reject all but one of the available radio stations. The combination of these two difficulties has driven radio designers to the use of intermediate frequencies, because both difficulties are lessened by mixing to an IF.

In Fig. 8.47 we have followed the mixer with an IF amplifier–filter. In the normal AM radio, this amplifier would have high gain only for frequencies within ± 5 kHz of the IF frequency of 455 kHz and hence would act as a filter eliminating all frequencies out of this band of frequencies. The IF amplifier is fixed-tuned, that is, its properties remain the same as you tune the radio; and since its operating frequency is relatively low, it can provide effective filtering. Consequently, not only does the IF amplifier–filter eliminate the undesired frequencies created by the mixer, it eliminates the radio stations adjacent to the desired station. Thus it performs the same function as would an RF filter, but it filters at a lower frequency and does not need to be tunable.

Tuning of the radio is accomplished by varing the local oscillator (LO) frequency, f_1. When you turn the tuning knob on your radio, the primary effect within the radio is to vary the frequency of the LO. We saw before that to receive a radio station broadcasting at 1200 kHz, we had to set the LO frequency to 1655 kHz. Clearly, if we increase the LO to 1665 kHz, the station broadcasting at 1210 kHz would now be mixed into the passband of the IF amplifier–filter, and the station at 1200 kHz would be eliminated. By this technique, one radio station is selected and all the others are rejected.

Image Rejection. To the previous filtering actions of the radio circuit, we must add yet one more. In Fig. 8.47 you will note that the box representing the RF amplifier is also called a filter. This does not contradict our earlier assertion that practical problems prevent us from designing an effective RF filter. This filter is not required select the radio station, but RF filtering is necessary to eliminate the *image band* of the mixer. To explain about the image of the mixer, and why we need one more filter to get rid of it, we must refer back to our explanation of mixer operation. You will note in Table 8.3 that the term relating to the mixer action was of the form $\cos(\omega_1 - \omega_2)t$. This is the logical way to write this term when ω_1 (the LO frequency) is higher than the RF frequency and when there is only one RF frequency coming into the mixer. But if the RF frequency were higher than the LO frequency, we would write the mixer term as $\cos(\omega_2 - \omega_1)t$. So if, for example, our LO were tuned to 1655 kHz and there were an RF input at $1655 + 455 = 2110$ kHz, this higher frequency would also be mixed into the IF passband. We conclude that the mixer would not distinguish between RF frequencies above and below the LO frequency and hence would mix both into the IF amplifier–filter passband. Thus without additional filtering, the radio would receive two stations simultaneously, one above and one below the LO frequency. This second, undesirable frequency is called the *image band* of the mixer, and hence the RF amplifier–filter is required to eliminate any station broadcasting in the image band before it reaches the mixer. Thus the passband of the RF amplifier–filter can be rather broad and often does not have to be tunable, although the RF amplifier–filter is tunable in many radios.

Summary.   The normal radio combines amplification, filtering, and spectrum shifting. Obviously, considerable amplification is required because of the small signals picked up by the antenna, and we have indicated amplification in Fig. 8.47 at RF, IF, and audio frequencies. Spectrum shifting is performed by the mixer, to shift the spectrum of the information from the RF frequency to the IF frequency; and spectrum shifting is also accomplished by the AM detector (or demodulator) to move the information into the audio-frequency band. Filtering is done at RF to get rid of the station broadcasting at the mixer image frequency; filtering is done by the IF amplifier to eliminate unwanted mixer frequencies and adjacent radio stations; and filtering is done at the audio frequencies to eliminate noise created by the detector.

The radio provides a strong example of the importance of the frequency-domain viewpoint in electrical engineering. Of the three essential processes involved in radio circuits, two (filtering and spectrum shifting) are processes which are natural to the frequency domain.

We mention in closing this section that there are obviously other schemes for effecting modulation, with their corresponding detection techniques. Frequency modulation (FM) provides a well-known example. Numerous additional modulation techniques are also used, particularly techniques well adapted to the transmission of digital information. But all communication systems, from the common phone system to the most advanced space communication network, utilize amplification, spectrum shifting, and filtering.

8.5 THE FEEDBACK CONCEPT

8.5.1 An Example

An Amplifier with Feedback.   In Fig. 8.48 we show an amplifier that has been modified from straight amplification by having a sample of the output brought back to interact with the input. A main amplifier is represented by the top box: its input (v_i) and its output (v_o) are related by the main amplifier gain A, as indicated in the box. We assume A to be real and positive. In the main amplifier, the signal goes from left to right.

A feedback circuit, represented by the bottom box, consists in this example of two resistors arranged as a voltage divider. In the feedback circuit, the signal goes from right to left, that is, from the output of the main amplifier back to its input. The input to the feedback circuit is v_o and its output is v_f.

A voltage source, v_{in}, is located at the input to the entire "amplifier with feedback," that is, the entire system, feedback and all. The output of the feedback circuit, v_f, is part of the input loop, which also contains the voltage source and the input to the main amplifier.

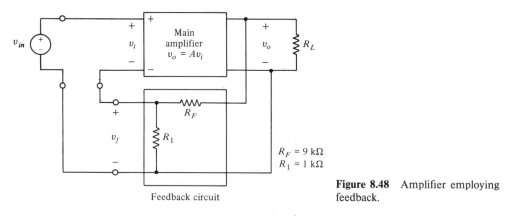

Figure 8.48 Amplifier employing feedback.

Feedback circuit

$R_F = 9 \text{ k}\Omega$
$R_1 = 1 \text{ k}\Omega$

Analysis of Amplifier Gain.   Our goal in this section is to calculate the gain of the amplifier with feedback, A_f. The gain with feedback is the ratio of the output to the source voltage, as in

$$A_f = \frac{v_o}{v_{in}} \tag{8.40}$$

We can write three equations describing the system. These are:

Definition of the main amplifier:

$$v_o = A v_i \tag{8.41}$$

Voltage divider for the feedback circuit:

$$v_f = \frac{R_1}{R_1 + R_F} v_o \tag{8.42}$$

KVL around the input loop:

$$-v_{in} + v_i + v_f = 0 \tag{8.43}$$

Equations (8.41) to (8.43) give us three equations relating four voltages. We can eliminate two voltages between the equations, reducing to one equation in two voltages. We require the ratio of v_o to v_{in} for computing the gain in Eq. (8.40), so we eliminate v_f and v_i. Some straightforward algebra results in

$$A_f = \frac{A(R_1 + R_F)}{R_1 + R_F + R_1 A} = \frac{A}{1 + [R_1/(R_1 + R_F)]A} \tag{8.44}$$

Equation (8.44) is representative of a more general relationship which we will develop presently. We conclude from the equation that the gain with feedback, A_f, is smaller than the gain without feedback, A. Having lost gain, what have we achieved? The answer is that we have improved the reliability, linearity, sensitivity, and bandwidth of the amplifier. We will verify these benefits later in this section. First we investigate how feedback works.

Signal Levels.   We will examine the signal levels for a typical amplifier with feedback. The gain of our main amplifier will be $A = 200$ and the feedback circuit will consist of $R_F = 9 \text{ k}\Omega$ in series with $R_1 = 1 \text{ k}\Omega$. According to Eq. (8.44), the gain of the amplifier with feedback would be

$$A_f = \frac{200}{1 + [1\text{ k}\Omega/(1\text{ k}\Omega + 9\text{ k}\Omega)](200)} = 9.52 \tag{8.45}$$

We shall assume that the output voltage is 10 V and calculate the signal levels throughout the circuit. With 10 V at its output, the input voltage to the main amplifier, v_i, must be $10/200 = 0.05$ V. The input voltage to the entire amplifier would be $10/9.52 = 1.050$ V. The output of the feedback circuit, v_f,

$$v_f = \frac{R_1}{R_1 + R_F}v_o = \frac{1\text{ k}\Omega}{1\text{ k}\Omega + 9\text{ k}\Omega}(10) = 1.000\text{ V}$$

Hence we confirm that KVL is satisfied in the input loop, as it must be if our derivation is valid.

The calculation performed in the preceding paragraph proceeds from the output back to the input. The signal would progress the other way, so let us think of the operation of the amplifier in time sequence. We apply the 1.050 V at the input at some instant of time. Electronic circuits respond quickly but not instantaneously; hence the output signal at this initial instant is zero. Thus the feedback voltage, v_f, will at first be zero and the entire 1.050-V input would appear at the main amplifier input, v_i. This signal is amplified by the main amplifier and the output voltage increases toward 200×1.050; but as the output increases, so does the feedback signal. Because the feedback signal subtracts from the 1.050-V input, the input voltage to the main amplifier diminishes as the output voltage rises. Thus we have competing effects: the higher the output voltage rises, the more voltage it feeds back and the more the input voltage to the main amplifier is reduced. In the end, the feedback signal subtracts 1.000 V from the input voltage of 1.050 V and the remaining 0.050 V is amplified by the main amplifier to yield an output voltage of 10 V.

8.5.2 The System Model

System Notation. A system representation of the feedback amplifier is shown in Fig. 8.49. This representation is characterized by the various blocks and circles connected with lines representing signal flow. Although our signals are voltages, customarily

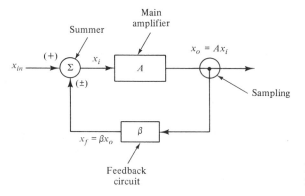

Figure 8.49 System representation of feedback.

denoted with v's, we have used a neutral symbol x to denote the signals on the system diagram in Fig. 8.49. Use of a neutral symbol is important because the signal, not the physical variable that represents the signal, is of primary importance in the system viewpoint. At some places in the system the signal might be represented by a voltage, at others a current, at yet others a pressure or the physical displacement of a mechanical component. The system representation embraces such hybrid systems.

The System Components.   We have shown four system components in Fig. 8.49. The main amplifier and the feedback circuit are obvious carryovers from the original formulation of the amplifier. Note that the direction of signal flow is indicated with arrowheads. At the output we have shown a "sampling" connection; clearly, this represents the parallel connection to the output of the main amplifier. In other feedback arrangements, the feedback network might be placed in series with the load and main amplifier output; with this arrangement the feedback network would sample the output current. Both possibilities are covered by the sampling symbol.

At the input we have drawn a circle containing a summation symbol to indicate the interaction of signals at the input of the system. In our previous example this circle represents KVL in Eq. (8.43), which can be put into the form

$$v_i = +v_{in} - v_f \tag{8.46}$$

where the $+$ and $-$ signs on the summer in Fig. 8.49 are associated with the input and the feedback signals, respectively. We may think of the summer as having a gain of $+1$ for v_{in} and a gain of -1 for v_f. The summer thus works as a "differencer" in this particular case, but the symbol is sufficiently versatile to cover all possibilities. In other contexts the summer at the input might be called a "comparator" because from another point of view it functions to compare the input with a sample of the output, furnishing the difference as the input to the main amplifier.

Analysis of the System.   The equations of the system are:

Main amplifier:

$$x_o = Ax_i$$

Feedback circuit:

$$x_f = \beta x_o$$

Summer:

$$x_i = x_{in} \pm x_f$$

In our equation for the summer, we have considered that the gain for the input signal is $+1$ but allowed for either $+1$ or -1 on the feedback signal. The main amplifier gain, A, may be positive or negative. As before, we can eliminate two of the variables and solve for the ratio of the output and input signals to obtain the gain with feedback.

$$A_f = \frac{x_o}{x_{in}} = \frac{A}{1 - (\pm 1)(\beta)(A)} \tag{8.47}$$

Equation (8.47) reduces to our earlier result [Eq. (8.44)] when the minus sign is used for the summer and $R_1/(R_1 + R_F)$, the gain of the voltage-divider circuit, is used for β.

The Loop Gain.   Another form for Eq. (8.47) is

$$A_f = \frac{A}{1 - L} \tag{8.48}$$

where $L = (\pm 1)(\beta)(A)$ is called the *loop gain*. The loop gain is an important concept in feedback theory and is defined as the product of all the system components around the feedback loop. The basic idea of the loop gain is suggested in Fig. 8.50. To calculate (or measure in the laboratory) the loop gain, break the feedback loop at some convenient point, insert a test signal, x_t, and calculate (or measure) the return signal, x_r. The loop gain is the ratio $L = x_r/x_t$, and is thus the gain around the loop. The loop may be broken, both in analysis and in the laboratory, only at a point where the function of the various system components are unimpaired. Care must be taken to terminate the break with the same impedance as the circuit saw before the break. For example, it would be inconvenient in the circuit in Fig. 8.48 to open the loop between R_F and R_1.

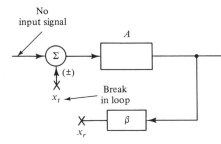

Figure 8.50   Derivation of loop gain.

The Importance of the Loop Gain.   The sign of the loop gain indicates the nature of the feedback. Negative loop gain indicates negative feedback. With negative feedback, the feedback signal subtracts from the input signal and the gain is reduced. Positive loop gain indicates positive feedback and is rarely used. Positive feedback increases the amplifier gain; indeed, for $L = +1$ the gain becomes infinite. This seeming anomaly is useful in the design of electronic oscillators, which can be considered amplifiers with output but no input.

The magnitude of the loop gain indicates the importance of feedback properties in influencing the system characteristics. When the loop gain is much larger than unity, the system properties are controlled largely by feedback considerations. We shall illustrate the importance of loop gain in the next section.

8.5.3 The Benefits of Negative Feedback

Improved Static Stability.   By *static stability* we mean the insensitivity of performance to changes in the system parameters. For instance, in our earlier example the gain of the main amplifier might decrease drastically due to the deterioration of a circuit component or replacement of a transistor. Let us say the gain of the main amplifier decreases from 200 to 100. Without feedback, this change would jeopardize the usefulness of the amplifier. With feedback, however, the gain becomes

$$A'_f = \frac{100}{1 - (-1)(0.1)(100)} = 9.09$$

Thus a 50% decrease in the gain of the main amplifier results in a 5% decrease in the overall gain of the system. We can see from Eq. (8.48) why this happens. When the loop gain is much greater than unity, the gain with feedback is approximately

$$A_f = \frac{A}{1 - L} \approx \frac{A}{-L} = \frac{A}{-(-1)(\beta)(A)} = \frac{1}{\beta} \qquad (8.49)$$

We see that as loop gain increases the gain approaches a value dependent on β. Thus the reduction of the main amplifier gain has little effect as long as the loop gain remains much larger than unity. The value of β could depend on resistors only, as in the specific example we are using, and thus be stable over long periods of time. In our example the reciprocal of β is 10, and we see that the gain of the system, both before and after the gain decrease, falls close to this value. Often in a practical system, the main amplifier merely provides sufficient gain to keep the loop gain much larger than unity, for in this case the β of the feedback circuit determines the overall gain with feedback.

We have used the word "static" to distinguish this type of stability from dynamic stability, the tendency of the system to vibrate under the influence of external stimuli. With a bridge, for example, the static stability might be good, meaning that the bridge footings are sound and the bridge members are sufficiently stiff to hold the bridge solidly in place. But under the influence of traffic or wind, the bridge might shake and even collapse, as did the Tacoma Narrows bridge in 1940, and hence exhibit poor dynamic stability.

Although negative feedback improves static stability, feedback can cause dynamic instability in a system. The problem is that time delays, or the frequency-domain equivalent (phase shift), can change negative feedback to positive feedback and hence cause dynamic instability or oscillations. This aspect of feedback theory lies beyond the scope of this text, but the study of dynamic stability is one focus of feedback theory at a more advanced level.

Improved Linearity.   Semiconductor devices such as transistors can be moderately linear in their active regions if signal variations are kept small, but large signals are often required. For example, the output amplifier in an audio system must produce a large signal to drive the speakers. Without feedback, such an amplifier would produce substantial distortion, but feedback techniques can be used to improve the linearity of the amplifier and thus reduce the distortion.

Equation (8.49) suggests how this is accomplished. The main amplifier introduces distortion into the system. If the loop gain is high, however, the gain is determined by the β of the feedback circuit, which depends on two resistors in this case. The resistors are linear components; hence the amplifier will produce minimal distortion if the loop gain is high. Thus a large amount of negative feedback is always used when good linearity is required for large-signal amplification.

Disturbance Reduction.   Feedback can reduce unavoidable disturbances that enter a system. As an example we shall consider the thermostatically controlled oven shown in Fig. 8.51. The input voltage to the system is compared with a voltage from a temperature sensor in the oven. The difference (v_d) is amplified and applied to a heater element (v_H). Because of the delay between the signal to the heater and the corresponding response from the sensor, the dynamic stability of this system is problematic. However, we shall ignore this issue and focus on the role of feedback in reducing disturbances causes by changes in the ambient temperature.

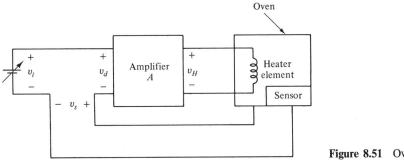

Figure 8.51 Oven controller.

A simplified system diagram is shown in Fig. 8.52. The heater element has been represented as a box having a gain A_H, which is defined as the change in box temperature divided by the change in the heater voltage. Although the relationship between heat (power) and voltage is nonlinear, we are for simplicity treating small changes as linear effects. The oven temperature is influenced both by the internal heater and by the ambient temperature, whose changes represent a disturbance in the system. This effect is represented with a summer. The sensor is also represented by a gain factor, A_S, which would represent changes in sensor output voltage divided by changes in oven temperature.

The loop gain of the system is

$$L = -AA_H A_S$$

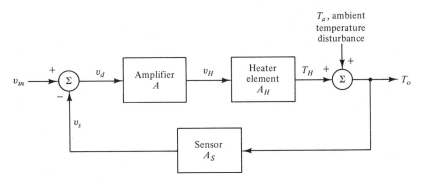

Figure 8.52 System diagram for oven controller.

Thus the relationship between oven temperature and input voltage would be

$$T_o = v_i \times \frac{AA_H}{1 - L}$$

The disturbance signal, T_a, represents a second input to the system. We redraw the system diagram in Fig. 8.53 to clarify this role. Considering the gain of the summer to be +1, we can determine a relationship between ambient and oven temperature:

$$T_o = T_a \frac{1}{1 - L}$$

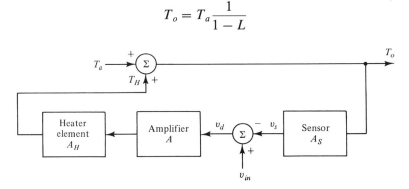

Figure 8.53 System diagram for controller with T_a as input.

Clearly, we want a large loop gain in order to reduce the effect of changes in the ambient temperature. For example, with a loop gain of 100, a 10° change in ambient temperature would cause a 0.1° change in oven temperature. Thus the effect of ambient temperature can be reduced to low levels, provided that the system can be stabilized dynamically.

Improved Response Time.   Negative feedback can improve the response time of a system. Consider, for example, an electrical relay, as pictured in Fig. 8.54. A relay is basically an electromagnet. When current passes through the coil, a magnetic force is produced at the gap and the lever closes against a spring. Such relays are used typically to activate switches and to operate locks and valves. We shall consider a relay having the following properties: current required to activate = 20 mA; maximum current = 30 mA; coil resistance (R_c) = 200 Ω; and coil inductance (L_c) = 2 H.

A suitable circuit for energizing the relay is shown in Fig. 8.55. When the relay is to be activated, the switch is closed. Current, initially at $i_0 = 0$, increases exponentially toward a final value of $i_\infty = V_s/R_c$. As current builds up, the relay will close when the current exceeds 20 mA. Let us assume that we wish the relay to close quickly as possible. In that case, we require the voltage source to be as large as possible, which in this case is limited by the maximum current to be 30 mA × 200 Ω = 6 V. With that value of voltage, the time required for the relay to close is easily shown to be $\tau \ln 3$, where $\tau = L_c/R_c$ is the time constant, 2H/200 Ω = 10 ms. Hence in this case the relay will close about 11.0 ms after the switch is closed.

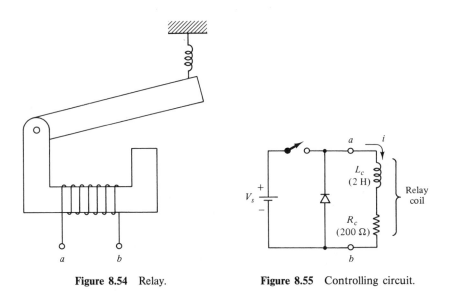

Figure 8.54   Relay.   **Figure 8.55**   Controlling circuit.

Throughout this period when the relay current is increasing, the diode remains OFF; but when the switch is opened, the inductor will react to keep the current going. This will cause the diode to conduct and the current will flow for a time through coil and diode, even with the switch open. The diode is thus placed across the relay coil to deenergize the inductance. Without the diode, a spark would appear at the switch contacts, as discussed on page 102.

Let us suppose that the 11-ms response time is too slow. We can use a feedback technique in the relay circuit to improve the relay speed. As shown in Fig. 8.56, we have added an amplifier and a resistor to the relay circuit. The resistor, R_s, produces a feedback signal which is proportional to the current in the relay coil. After the switch is closed, the equations of the system are: for the amplifier

$$v_o = A(V_s - v_f) = A(V_s - iR_s) \tag{8.50}$$

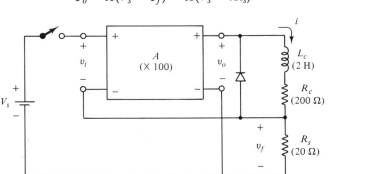

Figure 8.56   Feedback circuit to improve response time.

and for the coil and feedback resistor

$$v_o = L_c \frac{di}{dt} + (R_c + R_s)i \tag{8.51}$$

When we combine Eqs. (8.50) and (8.51), we obtain a differential equation describing the feedback system:

$$L_c \frac{di}{dt} + [R_c + (1 + A)R_s]i = AV_s$$

which can be put into the standard form

$$\frac{di}{dt} + \frac{i}{\tau_f} = \frac{AV_s}{L_c}$$

where τ_f would be $L_c/[R_c + (1 + A)R_s]$ and would clearly be shorter than the original time constant. With the numerical values we have shown in Fig. 8.56, the time constant drops from 10 ms to 0.91 ms and the response time would be reduced accordingly, from 11.0 ms to 1.0 ms. Other aspects to this example we will leave for a homework problem.

Improvement of Frequency Response.   As might be anticipated from the preceding example, and from the correspondence between speed of response in the time domain and bandwidth in the frequency domain, feedback can often improve the frequency response of a system. Let us reconsider the amplifier example in Fig. 8.48, except that we shall now introduce a bandwidth limitation to the main amplifier. We have shown the circuit in Fig. 8.57 (frequency-domain version), with the main amplifier now having a low frequency gain of A_o and a frequency cutoff of ω_c. The main amplifier frequency response has the form of a low-pass filter, the loss of the high frequencies resulting from a limitation within the amplifier. This limitation of the amplifier can be alleviated through negative feedback.

The equations of the system [Eqs. (8.41) to (8.43)] transform into the frequency

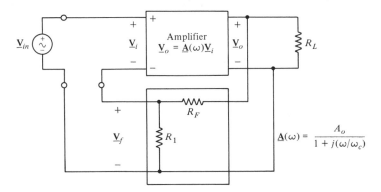

Figure 8.57   Improving bandwidth with feedback.

domain changed only in notation, and hence the amplifier characteristic with feedback becomes the frequency-domain version of Eq. (8.44):

$$\underline{\mathbf{A}}_f(\omega) = \frac{\underline{\mathbf{A}}(\omega)}{1 + [R_1/(R_1 + R_F)]\underline{\mathbf{A}}(\omega)} = \frac{A_0/[1 + j(\omega/\omega_c)]}{1 + [\beta]A_0/[1 + j(\omega/\omega_c)]}$$

where $\beta = \dfrac{R_1}{R_1 + R_F}$

This can be put into the form

$$\underline{\mathbf{A}}_f(\omega) = \frac{A_0/(1 + \beta A_0)}{1 + j(\omega/\omega_{cf})} \tag{8.52}$$

where $\omega_{cf} = (1 + \beta A_0)\omega_c$. Recognizing that the loop gain (L) is $-\beta A_0$, we see from Eq. (8.52) that the amplifier gain at low frequencies is reduced by the factor $1 - L$ and the cutoff frequency is increased by the same factor.

Bode plots for the amplifier with and without feedback are shown in Fig. 8.58. This application furnishes a good example of the claim which we made earlier that feedback trades gain for some other useful property. In this case gain is exchanged for bandwidth. The inherent bandwidth limitation of the amplifier has been overcome through the use of feedback.

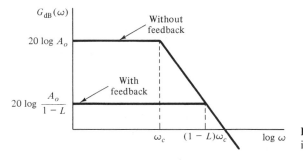

Figure 8.58 Feedback reduces gain but increases bandwidth.

When the bandwidth limitation is more complicated than we have used, the situation becomes more complicated. Feedback cannot overcome all bandwidth limitations. Nevertheless, the type of limitation we have used is a common one and hence the example is not trivial.

Impedance Control.   Feedback techniques can be used to change the effective impedance of an electronic circuit. This is useful for effecting maximum power transfer and for reducing loading effects. However, the system approach we have taken in this section precludes consideration of impedance levels. Impedance control is extremely important in electronic circuits, and feedback techniques are frequently used for this purpose. The details, however, rapidly become complicated and we shall not explore this subject in detail.

8.6 OPERATIONAL-AMPLIFIER CIRCUITS

8.6.1 Introduction

The Importance of OP Amps. An operational amplifier is a high-gain electronic amplifier which is controlled by negative feedback to accomplish many functions or "operations" in analog circuits. Such amplifiers were developed originally to accomplish operations such as integration and summation in analog computers for the solving of differential equations. Applications of op amps have increased until, at the present time, most analog electronic circuits are based on op amp techniques. If, for example, you required an amplifier with a gain of -10, rarely would you design a circuit of the type in Fig. 6.55; convenience, reliability, and cost considerations would dictate the use of an op amp. Thus op amps form the basic building blocks of analog circuits much as NAND and NOR gates provide the basic building blocks of digital circuits.

An OP-Amp Model and Typical Properties. The typical op amp is a sophisticated transistor amplifier utilizing a dozen or more transistors, several diodes, and many resistors. Such amplifiers are mass-produced on semiconductor chips and sell for less than \$1 each. These parts are reliable, rugged, and approach the ideal in their electronic properties.

Figure 8.59 shows the symbol and the basic properties of an op amp. The two input voltages, v_+ and v_-, are subtracted and amplified with a large voltage gain, A, typically 10^5 to 10^6. The input resistance, R_i, is large, 100 kΩ to 100 MΩ. The output resistance, R_o, is small, 10 to 100 Ω. The amplifier is often supplied with dc power from positive $(+V_{CC})$ and negative $(-V_{CC})$ power supplies. For this case, the output voltage lies between the power supply voltages, $-V_{CC} < v_o < +V_{CC}$. Sometimes one power connection is grounded (i.e., "$-V_{CC}$" $= 0$). In this case the output lies in the range, $0 < v_o < +V_{CC}$. The power connections are seldom drawn in circuit diagrams; it is assumed that one connects the op amp to the appropriate power source. Thus the op amp approximates an ideal voltage amplifier, having high input resistance, low output resistance, and high gain.

The high gain is converted to other useful features through the use of strong negative feedback. All the benefits of negative feedback are utilized by op-amp circuits. To

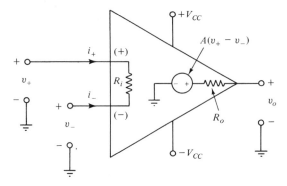

Figure 8.59 Op-amp model.

those listed earlier in this chapter, we would for op-amp circuits add three more: low expense, ease of design, and simple construction.

The Contents of This Section.   We begin by analyzing two common op-amp applications, the inverting and the noninverting amplifiers. We derive the gain of these amplifiers by a method that may be applied simply and effectively to any op-amp circuit. We then discuss active filters, which are op-amp amplifiers with capacitors added to shape their frequency response. We then deal briefly with analog computers and conclude by discussing some nonlinear applications of op amps.

8.6.2 Op-Amp Amplifiers

The Inverting Amplifier.   The inverting amplifier, shown in Fig. 8.60, uses an op amp plus two resistors. The positive $(+)$ input to the op amp is grounded (zero signal); the negative $(-)$ input is connected to the input signal (via R_1) and to the feedback signal from the output (via R_F). One potential source of confusion in the following discussion is that we must speak of two amplifiers simultaneously. The op amp is an amplifier which forms the amplifying element in a feedback amplifier which contains the op amp plus the associated resistors. To lessen confusion, we shall reserve the term "amplifier" to apply only to the overall, feedback amplifier. The op amp will never be called an amplifier; it will be called the op amp. For example, if we refer to the input current to the amplifier, we are referring to the current through R_1, not the current into the op amp.

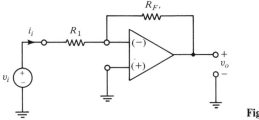

Figure 8.60   Inverting amplifier.

We could solve for the gain of the inverting amplifier in Fig. 8.60 either by solving the basic circuit laws (KVL and KCL) or by attempting to divide the circuit, in the style of Fig. 8.49, into main amplifier and feedback system blocks. We shall, however, present another approach based on the assumption that the op-amp gain is very high, effectively infinite. In the following, we shall give a general assumption, which may be applied to any op-amp circuit; then we will apply this assumption specifically to the present circuit. As a result, we will establish the gain and input resistance of the inverting amplifier.

1. We assume that the output is well behaved—does not try to go to infinity. Thus we assume that the negative feedback stabilizes the amplifier such that moderate input voltages produce moderate output voltages. If the power supplies are $+10$ and -10 V, for example, the output would have to lie between these limits.

2. Therefore, the input voltage *to the op amp* is very small, essentially zero, because it is the output voltage divided by the large voltage gain of the op amp:

$$v_+ - v_- \approx 0 \Rightarrow v_+ \approx v_-$$

For example, if $|v_o| < 10$ V and $A = 10^5$, then $|v_+ - v_-| < 10/10^5 = 100 \ \mu V$. Thus normally v_+ and v_- are equal within 100 μV or less, for any op-amp circuit. For the inverting amplifier in Fig. 8.60, v_+ is grounded; therefore, $v_- \approx 0$. Consequently, the current at the input to the amplifier would be

$$i_i = \frac{v_i - v_-}{R_1} \approx \frac{v_i}{R_1} \tag{8.53}$$

3. Because $v_+ \approx v_-$ and R_i is large, the current into the + and − op-amp inputs will be very small, essentially zero:

$$|i_+| = |i_-| = \frac{|v_- - v_+|}{R_i} \approx 0 \tag{8.54}$$

For example, for $R_i = 100$ kΩ, $|i_-| < 10^{-4}/10^5 = 10^{-9}$ A.

For the inverting amplifier, Eq. (8.54) implies that the current at the input, i_i, flows through R_F, as shown in Fig. 8.61. This allows us to compute the output voltage. The voltage across R_F would be $i_i R_F$ and, because one end of R_F is connected to $v_- \approx 0$,

$$v_o = -i_i R_F = -\frac{v_i}{R_1} \times R_F$$

Thus the voltage gain would be

$$A_v = \frac{v_o}{v_i} = -\frac{R_F}{R_1} \tag{8.55}$$

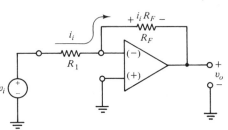

Figure 8.61 The input current flows through the feedback circuit.

The minus sign in the gain expression means that the output will be inverted relative to the input: a positive signal at the input will produce a negative signal at the output. Equation (8.55) shows the gain to depend on the ratio of R_F to R_1. This would imply that only the ratio and not the individual values of R_1 and R_F matter. This would be true if the input resistance to the amplifier were unimportant, but the input resistance to an amplifier is often critical. The input resistance to the inverting amplifier would follow from Eq. (8.53):

$$R_i = \frac{v_i}{i_i} \approx R_1 \tag{8.56}$$

For a voltage amplifier, the input resistance is an important factor, for if R_i were too low the signal source (of v_i) could be loaded down by R_i. Thus in a design, R_1 must be sufficiently high to avoid this loading problem. Once R_1 is fixed, R_F may be selected to achieve the required gain. Thus the values of the individual resistors become important because they affect the input resistance to the amplifier.

Let us design an inverting amplifier to have a gain of -8. The input signal is to come from a voltage source having an output resistance of 100 Ω. To reduce loading, the input resistor, R_1, must be much larger than 100 Ω. For a 5% loading reduction, we would set $R_1 = 2000\ \Omega$. To achieve a gain of -8 (actually 95% of -8, considering loading), we require that $R_F = 8 \times 2000 = 16\ \text{k}\Omega$.

Feedback effects dominate the characteristics of the amplifier. When an input voltage is applied, the value of v_- will increase. This will cause v_o to increase rapidly in the negative direction. This negative voltage will increase to the value where the effect of v_o on the $-$ input via R_F cancels the effect of v_i through R_1. Put another way, the output will adjust itself to withdraw through R_F any current that v_i injects through R_1, since the input current to the op amp is extremely small. In this way the output depends only on R_F and R_1.

The Noninverting Amplifier.   For the noninverting amplifier shown in Fig. 8.62 the input is connected to the $+$ input. The feedback from the output connects still to the $-$ op amp input, as required for negative feedback. To determine the gain, we apply the assumptions outlined above.

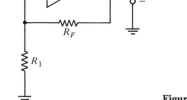

Figure 8.62   Noninverting amplifier.

1. Because $v_+ \approx v_-$, it follows that

$$v_- \approx v_i \tag{8.57}$$

2. Because $i_- \approx 0$, R_F and R_1 carry the same current. Hence v_o is related to v_- through a voltage-divider relationship

$$v_- = v_o \frac{R_1}{R_1 + R_F} \tag{8.58}$$

Combining Eqs. (8.57) and (8.58), we establish the gain to be

$$v_i = v_o \frac{R_1}{R_1 + R_F} \Rightarrow A_v = +\left(1 + \frac{R_F}{R_1}\right) \tag{8.59}$$

The + sign before the gain expression emphasizes that the output of the amplifier has the same polarity as the input: a positive input signal produces a positive output signal. Again we see that the ratio of R_F and R_1 determines the gain of the amplifier.

When a voltage is applied to the amplifier, the output voltage increases rapidly and will continue to rise until the voltage across R_1 reaches the input voltage. Thus little input current will flow into the amplifier, and the gain depends only on R_1 and R_F. The input resistance to the noninverting amplifier will be very high because the input current to the amplifier is also the input current to the op amp, i_+, which must be extremely small. Input resistance values exceeding 1000 MΩ are easily achieved with this circuit. This feature of high input resistance is an important virtue of the noninverting amplifier.

8.6.3 Active Filters

What Are Active Filters?   An active filter combines amplification with filtering. The RC filters we investigated earlier are called passive filters because they provide only filtering. An active filter uses an op amp to furnish gain but has capacitors added to the input and feedback circuits to shape the filter characteristics.

We derived earlier the gain characteristic of an inverting amplifier in the time domain. In Fig. 8.63 we show the frequency-domain version. We may easily translate the earlier derivation into the frequency domain:

$$v_i \Rightarrow \underline{\mathbf{V}}_i(\omega) \qquad v_o \Rightarrow \underline{\mathbf{V}}_o(\omega)$$

$$A_V = -\frac{R_F}{R_1} \Rightarrow \underline{\mathbf{F}}_V(\omega) = -\frac{\underline{\mathbf{Z}}_F(\omega)}{\underline{\mathbf{Z}}_1(\omega)}$$

The filter function, $\underline{\mathbf{F}}_V(\omega)$, is thus the ratio of the two impedances, and in general will give gain as well as filtering. We could have written the minus sign as $1\underline{/180°}$, for in the frequency domain the inversion is equivalent to a phase shift of 180°.

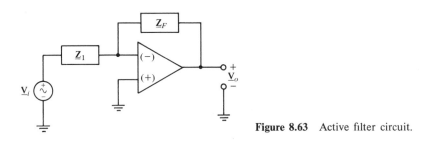

Figure 8.63   Active filter circuit.

Low-Pass Filter.   Placing a capacitor in parallel with R_F (Fig. 8.64) will at high frequencies tend to lower $\underline{\mathbf{Z}}_F$ and hence the gain of the amplifier; consequently, this capacitor converts an inverting amplifier into a low-pass filter with gain. We may write

$$\underline{\mathbf{Z}}_F(\omega) = R_F \,\|\, \frac{1}{j\omega C_F} = \frac{1}{(1/R_F) + j\omega C_F} = \frac{R_F}{1 + j\omega R_F C_F} \qquad (8.60)$$

Thus the gain would be

$$\mathbf{F}_V = -\frac{R_F}{R_1}\frac{1}{1 + j\omega R_F C_F} = A_V\frac{1}{1 + j(\omega/\omega_c)} \tag{8.61}$$

where $A_V = -R_F/R_1$, the gain without the capacitor, and $\omega_c = 1/R_F C_F$ would be the cutoff frequency. The gain of the amplifier is approximately constant until the frequency exceeds ω_c, after which the gain decreases with increasing ω. The Bode plot of this filter function is shown in Fig. 8.65 for the case where $R_F = 10$ kΩ, $R_1 = 1$ kΩ, and $C_F = 1$ μF. The shape is identical to that of the low-pass filter in Fig. 8.26, except for the increase in gain due to the op amp.

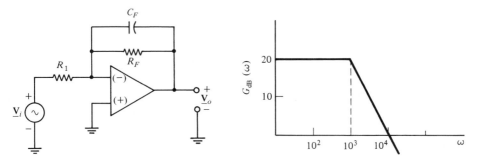

Figure 8.64 Low-pass filter circuit.   **Figure 8.65** Bode plot for active low-pass filter.

High-Pass Filter.   The high-pass filter shown in Fig. 8.66 uses a capacitor in series with R_1 to reduce the gain at low frequencies. The details of the analysis will be left to a problem. The gain of this filter is

$$\mathbf{F}_V(\omega) = -\frac{R_F}{R_1}\frac{j(\omega/\omega_c)}{1 + j(\omega/\omega_c)} = A_V\frac{j(\omega/\omega_c)}{1 + j(\omega/\omega_c)}$$

where $A_V = -R_F/R_1$ is the gain without the capacitor and $\omega_c = 1/R_1 C_1$ is the cutoff frequency, below which the amplifier gain is reduced. The Bode plot of this filter characteristic is shown in Fig. 8.67. This Bode plot is identical to that of the high-pass filter given in Fig. 8.28 except for the increase in gain due to the op amp.

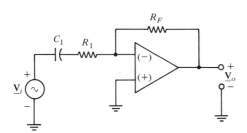

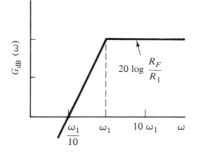

Figure 8.66 High-pass filter circuit.   **Figure 8.67** Bode plot for active high-pass filter.

Other Active Filters. By using more advanced techniques, one can simulate RLC narrowband filters and, by using additional op amps, many sophisticated filter character-istics can be achieved. Discussion of such applications lies beyond the scope of this text, but there exist many handbooks showing circuits and giving design information about active filters.

8.6.4 Analog Computers

Often a differential equation is solved by integration. The integration may be accom-plished by analytical methods or by numerical methods on a digital computer. Integra-tion may also be performed electronically with an op-amp circuit. Indeed, op amps were developed initially for electronic integration of differential equations.

An Integrator. The op-amp circuit in Fig. 8.68 uses negative feedback through a capacitor to perform integration. We have charged the capacitor in the feedback path to an initial value of V_1, and then removed this prebias voltage at $t = 0$. Let us examine the initial state of the circuit before investigating what will happen after the switch is opened. Since v_+ is approximately zero, so will be v_-, and hence the output voltage is fixed at $-V_1$. The input current to the amplifier, v_i/R, will flow through the V_1 voltage source and into the output of the op amp. Thus the output voltage will remain at $-V_1$ until the switch is opened.

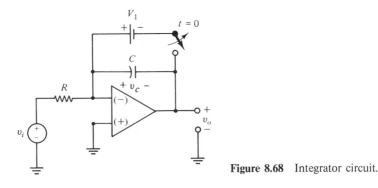

Figure 8.68   Integrator circuit.

After the switch is opened at $t = 0$, the input current will flow through the capaci-tor and hence the v_C will be

$$v_c(t) = v_c(0) + \int_0^t \frac{v_i(t')}{RC} \, dt'$$

Thus the output voltage of the circuit is

$$v_o(t) = -v_c(t) = -V_1 - \frac{1}{RC} \int_0^t v_i(t') \, dt' \qquad t \geqslant 0 \qquad (8.62)$$

Except for the minus sign, the output is the integral of v_i scaled by $1/RC$, which may be made equal to any value we wish by proper choice of R and C.

Scaling and Summing.   We need two other circuits to solve simple differential equations by analog computer methods. Scaling refers to multiplication by a constant, such as

$$v_2 = \pm K v_1$$

where K is a constant. This is the equation of an amplifier, and hence we would use the inverting amplifier in Fig. 8.60 for the $-$ sign or the noninverting amplifier in Fig. 8.62 for the $+$ sign.

A summer produces the weighted sum of two or more signals. Figure 8.69 shows a summer with two inputs. We may understand the operation of the circuit by applying the same reasoning we used earlier to understand the inverting amplifier. Since $v_- \approx 0$, the sum of the currents through R_1 and R_2 is

$$i_i = \frac{v_1}{R_1} + \frac{v_2}{R_2} \tag{8.63}$$

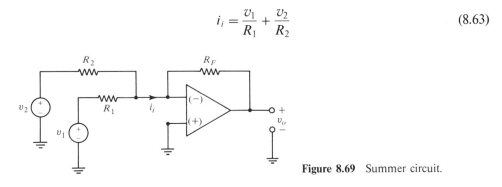

Figure 8.69   Summer circuit.

The output voltage will adjust itself to draw this current through R_F, and hence the output voltage will be

$$v_o = -i_i R_F = -\left(v_1 \times \frac{R_F}{R_1} + v_2 \times \frac{R_F}{R_2} \right)$$

The output will thus be the sum of v_1 and v_2, weighted by the gain factors, R_F/R_1 and R_F/R_2, respectively. If the inversion produced by the summer is unwanted, the summer can be followed by an inverter, a scaler with a gain of -1. Clearly, we could add other inputs in parallel with R_1 and R_2. In the example to follow, we shall sum three signals to solve a second order differential equation.

Solving a DE.   Let us design an analog computer circuit to solve the differential equation

$$2\frac{d^2v}{dt^2} + 4\frac{dv}{dt} + v = 6\cos 10t \qquad t > 0 \tag{8.64}$$

$$v(0) = -2 \qquad \text{and} \qquad \frac{dv}{dt} = +3 \quad \text{at } t = 0$$

Moving everything except the highest-order derivative to the right side yields

$$\frac{d^2t}{dt^2} = -2\frac{dv}{dt} - \frac{v}{2} + 3\cos 10t \tag{8.65}$$

The circuit which solves Eq. (8.64) is shown in Fig. 8.70. The circuit consists of two integrators to integrate the left side of Eq. (8.65), a summer to represent the right side, and two inverters to correct the signs. The noninverting inputs are grounded, and the inputs and feedback are connected to the inverting input of the op amps. Hence we have shown only the inverting inputs. With d^2v/dt^2 the input to the integrators, the output of the first integrator will be $-dv/dt$ [with the battery giving the initial condition of 3 V, as in Eq. (8.62)], and the output of the second integrator will be $+v$ (with an initial condition of -2 V). This output is fed into the summer, along with dv/dt after inversion, and the driving function $\cos 10t$, which must also be inverted to cancel the inversion in the summer. The input resistors connecting the three signals into the summer produce the weighting factors in Eq. (8.65), and hence the output of the summer represents the right side of Eq. (8.65). We therefore connect that output to our "input" of d^2v/dt^2 to satisfy Eq. (8.64). To observe the solution to Eq. (8.64), we merely open the switches at $t = 0$.

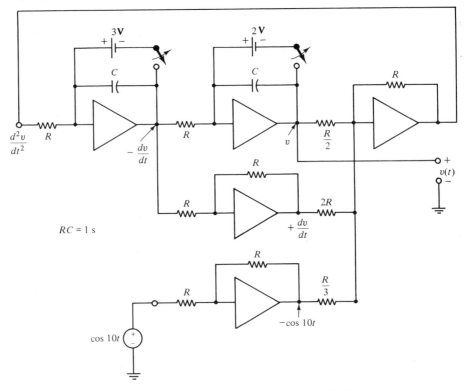

Figure 8.70   Analog computer to solve Eq. (8.64).

Clearly, these techniques can be applied to higher-order equations. Sophisticated use of analog computers requires a variety of refinements. Often, the equations being solved are scaled in time (time is sped up or slowed down on the computer) to accommodate realistic resistor and capacitor values. Also, voltage and current values can be scaled to bring the unknowns within the allowable range of the computer. In the next section we show how nonlinear operations can be introduced to solve nonlinear differential equations by analog methods.

8.6.5 Nonlinear Applications of Op Amps

Op amps can be combined with nonlinear circuit elements such as diodes and transistors to produce a variety of useful circuits. Below we discuss a few such applications. Many more circuits are detailed in standard handbooks and manufacturers' application literature for their products.

An Improved Half-Wave Rectifier.   The op amp in Fig. 8.71 drives a half-wave rectifier. When the input voltage is negative the output of the op amp will be negative and the diode will be OFF; hence the output will be zero. When the output is positive the diode will turn ON and the output will be identical to the input, because the circuit will perform as a non-inverting amplifier shown in Fig. 8.62 with $R_F = 0$. Use of the op amp effectively reduces the diode turn-on voltage. If the input voltage is greater than $0.7/A$, where A is the voltage gain of the op amp, the output voltage will exceed 0.7 V and turn on the diode. Hence the turn-on voltage is effectively reduced from 0.7 to $0.7/A$.

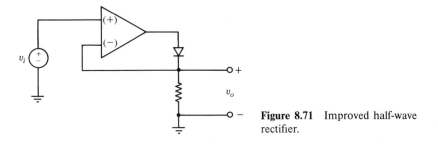

Figure 8.71 Improved half-wave rectifier.

This circuit would not be used in a power supply circuit; rather, it would be used in a detector or other circuit processing small signals, where the turn-on voltage of the diode would be a problem. An AM detector similar to that shown in Fig. 8.41 could be made by placing a suitable capacitor across the resistor.

A Logarithmic Amplifier.   By placing a diode in the feedback path, as shown in Fig. 8.72, we create an amplifier whose output is proportional to the logarithm of the input. Following our standard line of reasoning, we see the input current to be v_i/R. The output voltage will assume a value to draw this current through the diode. The diode characteristic is well represented by the ideal *pn* junction equation in Eq. (6.7); hence

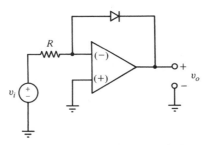

Figure 8.72　Logarithmic amplifier.

the output voltage will be

$$v_o = -\eta V_T \ln\left(\frac{v_i}{RI_o} + 1\right) \approx -\eta V_T \ln v_i + \eta V_T \ln RI_o \qquad (8.66)$$

The output thus contains a factor proportional to the logarithm of the input voltage. The circuit in Fig. 8.72, can be followed by another op-amp circuit which subtracts the constant term and adjusts the gain of the log amplifier to the required value. For example, the gain can be adjusted to produce an output in dB relative to 1 V. The circuit in its present form cannot deal with signals which go negative.

An Antilog Amplifier.　With a diode in the input circuit as shown in Fig. 8.73, the output voltage can be made proportional to the antilog function, which is the exponential function. We leave the derivation of the amplifier characteristics for a homework problem.

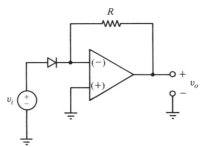

Figure 8.73　Antilog amplifier.

The existence of log and antilog amplifiers makes possible analog circuits for multiplication and powers of analog signals and thereby expands analog computer techniques to nonlinear problems. For example, if we wished a circuit to produce the product of two analog signals, we could take the log of each, add the logs, and take the antilog. If one or both of the inputs could go both positive and negative, they would have to be put through circuits that produce the absolute value, with the signs handled by a separate logic circuit.

Excellent analog multipliers and other nonlinear circuits are commercially available. Such circuits operate according to the principles outlined above but contain more complicated circuits to compensate for temperature effects and generally improve performance. Our purpose here has been to show some representative nonlinear applications of op amps.

PROBLEMS

Practice Problems

SECTIONS 8.2.1–8.2.2

P8.1. In Austin, Texas, the maximum daily average temperature occurs in mid-July and is 84.5°F. The minimum daily average temperature is 49.1°F and occurs in mid-January. Based on this information, and assuming that the daily average temperature $T(t)$ is well represented by the dc and first harmonic terms of a Fourier series such as Eq. (8.3), find C_0, C_1, θ_1, and ω_0. Let t be the time in months, with $t = 0$ at January 1, $t = 1$ at February 1, and so on. From your result, estimate the expected average temperature for Christmas Day.

Ans: $T(\text{Christmas}) = 46.2°$

P8.2. Consider a square wave similar to the one in Fig. 8.5a, except that it goes from $+2$ to -2 V and has a pulse width of 2.5 ms. Find the frequency (in hertz), the amplitude, and the phase (in degrees) of the eleventh harmonic.

Ans: $C_{11} = 0.231$, $\theta = \pm 180°$, $f_0 = 2200$ Hz

P8.3. What would be the Fourier series for the half-wave rectified sinusoid in Fig. 8.6a if the origin were drawn through the beginning of the pulse? *Hint:* Change variables from $t \longrightarrow t' - T/4$, where t' is the origin in the new time system.

P8.4. Show that the powers in the time and frequency domain are equal for the half-wave rectified sinusoid (Fig. 8.6a), whose Fourier series is indicated in Eq. (8.5). *Hints:* The time-domain power must be one-half that of an unrectified sinusoid; the nth harmonic amplitude is $2V_p/\pi(n^2 - 1)$, n even, with the signs alternating; sum up enough terms on your calculator to show approximate equality between the two domains. This series is difficult to sum mathematically.

SECTION 8.2.3

P8.5. Show that Eq. (8.9) for $\tau = T/2$ gives the same harmonic amplitudes and phases as those in Eq. 8.4.

P8.6. The nominal bandwidth of a standard telephone line is 4 kHz. What is the approximate duration of the shortest pulse which can be sent over such a telephone line? Given that the width between pulses should be equal to the pulse width, how many pulses per second can be sent over a phone line?

Ans: 0.25 ms, 2000 pulses/s or 2000 baud

SECTIONS 8.2.4–8.2.5

P8.7. For the half-wave rectified sinusoid in Fig. 8.6a, what fraction of the total power is carried by the dc component? *Hint:* For the ac power, just sum up the terms given. Note that the magnitudes diminish rapidly. See P8.4 for additional information.

Ans: 40.5%

P8.8. The "power" spectrum of a random signal is shown in Fig. P8.8.

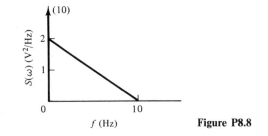

Figure P8.8

(a) What is the dc value of the signal?
(b) What power would this signal give to a 5-Ω resistor?
(c) What is the approximate correlation time for this signal?

Ans: (a) ± 3.16 V; (b) 4 W; (c) 0.1 s

P8.9. The effective bandwidth (BW) of standard AM radio is 5000 Hz. If such a radio were tuned off a station, receiving only static, what would be the approximate correlation time for the static?

SECTION 8.3.1

P8.10. A 60-Hz half-wave rectified sinusoid with a peak value of 10 V is filtered by a low-pass filter, as shown in Fig. P8.10. Calculate the dc and the peak value of the fundamental at the output.

Ans: 3.18 V, 0.133 V peak

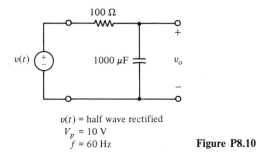

$v(t)$ = half wave rectified
$V_p = 10$ V
$f = 60$ Hz

Figure P8.10

P8.11. Show that the *RL* filter in Fig. P8.11 has a low-pass filter characteristic identical in form to that given in Eq. (8.22). Derive the value of the cutoff frequency, f_c, in terms of *R* and *L*.

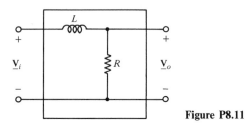

Figure P8.11

SECTION 8.3.2

P8.12. Make Bode plots of the following filter functions:

(a) $\underline{F}(\omega) = \dfrac{j\omega}{10 + j2\omega}$

(b) $\underline{F}(f) = \dfrac{1 + jf/10}{1 + jf/100}$

P8.13. Show that, at a frequency one octave (factor of 2) below the cutoff, the loss of the low-pass filter in Fig. 8.17 is approximately 1 dB.

SECTION 8.3.3

P8.14. Estimate the half-power frequencies of the narrowband filter function shown in Fig. 8.32.

SECTIONS 8.4.1–8.4.2

P8.15. A nonlinear circuit has input frequencies at 300 and 200 Hz. How many output frequencies are there below 1000 Hz, and what are these frequencies?

Ans: 10 frequencies, counting dc but not counting 1 kHz

P8.16. For the simplest design of the local oscillator (LO) in a radio, the ratio of the maximum to minimum LO frequencies should be as low as possible (i.e., the percent tuning range of the LO should be minimum). To show why the LO is placed above the RF band, calculate the ratio f_{max}/f_{min} for the LO both above and below the RF in the standard AM radio.

P8.17. The AM radio band of carrier frequencies is 540–1600 kHz and the standard IF is 455 kHz. Calculate the image frequency when the radio is tuned to the bottom of the band to receive a station broadcasting with a carrier of 540 kHz. Is this image in the AM band?

SECTIONS 8.5.1–8.5.3

P8.18. Work out the signal levels with $v_o = 10$ V for the amplifier in Fig. 8.48 with the reduced gain, $A = 100$. Make a table comparing with those calculated for $A = 200$ in Section 8.5.1.

P8.19. How large does the amplifier gain, A, have to be in the feedback amplifier in Fig. 8.48 before the gain with feedback is within 1% of $1/\beta$? What is the loop gain for this value of A? Based on this result, guess what loop gain would be required for a 0.1% difference between the exact gain and $1/\beta$.

P8.20. The feedback amplifier shown in Fig. P8.20 uses a current amplifier and has a single resistor, R_F, as a feedback circuit. In your analysis, assume that $R_F \gg R_L$, such that the output voltage is very nearly $v_o = i_o R_L$. Assume that the input impedance of the current amplifier is zero.

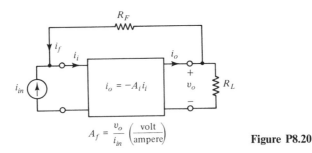

$$A_f = \frac{v_o}{i_{in}} \left(\frac{\text{volt}}{\text{ampere}}\right)$$

Figure P8.20

(a) Derive the gain with feedback, $A_f = v_o/i_{in}$, by using KCL, Ohm's law, and the gain equation for the current amplifier.
(b) For $A_i = 500$ and $R_L = 10 \, \Omega$, find R_F for an overall gain of $10^4 \, \Omega$.
(c) Put the results from part (a) in the form of Eq. (8.47) and draw the system diagram corresponding to this amplifier.
(d) What is the loop gain of the amplifier?
(e) If $A_i \longrightarrow \infty$, what is the limiting form for A_f?

P8.21. Assume that the amplifier in Fig. 8.48 has a severe nonlinearity, as shown by the input–output characteristic in Fig. P8.21. Without feedback this would cause serious distortion in the amplifier output. Derive and sketch the overall (with feedback) characteristics of v_o versus v_{in} with feedback to demonstrate the benefits of feedback in reducing the degree of nonlinearity. *Hint:* The final results will consist of straight-line segments. Assume selected values at the output and work back to the input, as was done in Section 8.5.1.

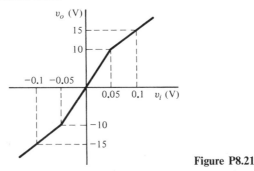

Figure P8.21

P8.22. For the relay circuit with feedback shown in Fig. 8.56:
(a) What would be the maximum input voltage to keep the diode current within the stated limit?
(b) Assume the relay opens at a current of 14 mA. In the original circuit, how long after the switch opens will it take for the relay to open, assuming that the relay has been closed a long time. Assume an ideal diode.
(c) Would the feedback circuit in Fig. 8.56 affect the time calculated in part (b)?

SECTIONS 8.6.1–8.6.2

P8.23. Design an op-amp circuit to provide gain of -3 and an input resistance of 5 kΩ. Draw the circuit and specify the component values.

P8.24. Determine the input resistance and voltage gain of the op-amp circuit in Fig. P8.24.

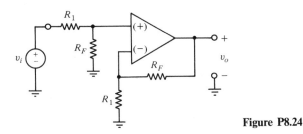

Figure P8.24

P8.25. Design an integrator with an input impedance of 10 kΩ which will provide an output

$$v_o(t) = 10 - 2 \int_0^t v_i(t') \, dt'$$

when the switch across the capacitor is opened at $t = 0$.

SECTION 8.6.3

P8.26. Derive the filter function, $\mathbf{F}(\omega) = \mathbf{V}_o/\mathbf{V}_i$, for the high-pass filter shown in Fig. 8.66. Confirm the formula for the critical frequency, ω_c. What would be the input impedance in the band of frequencies above the critical frequency, ω_c?

P8.27. Design an active high-pass filter with a cutoff frequency of 500 Hz, an input impedance of 1000 Ω in the high-frequency region, and a gain of $+10$ in the passband. *Hint:* Use a passive high-pass filter at the input of a noninverting amplifier.

P8.28. An op-amp amplifier is required to pass pulses of 1 ms width but to eliminate unwanted frequencies higher that those required for the pulse. The gain must be -2. Assume that you have available a 0.003-μF capacitor. Give the circuit and component values.

SECTION 8.6.4

P8.29. Design an analog computer circuit to solve the differential equation

$$10 \frac{dx(t)}{dt} + 3x(t) = SQ(t)$$

where $SQ(t)$ is a 5-V square wave with a frequency of 10 Hz (assume that you have a function generator to generate this waveform). The initial value of $x(0)$ is 2 V.

Application Problems

A8.1. The op-amp amplifier in Fig. A8.1 is arranged to amplify the difference between v_2 and v_1. That is, its output should be of the form $V_o = A_f(v_2 - v_1)$. Given that R_F and R_1 are fixed, find R_2 and R_3 such that only the difference between the inputs is amplified and the input resistance is identical for the two inputs.

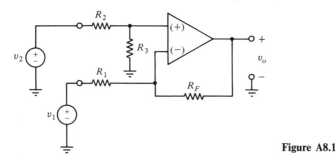

Figure A8.1

A8.2. In an audio system an amplifier is required which will give a gain of 10 (+ or −, it does not matter) and have upper and lower cutoff frequencies to exclude signals outside the normal audio region, 30 to 20,000 Hz. Combine high-pass and low-pass active filters and design a circuit that will accomplish this. Assume that the largest capacitor you have available is 100 μF.

A8.3. The small signal amplifier in Fig. 6.64 has an input impedance

$$Z_i = R_B \| r_\pi \qquad \text{where } R_B = R_1 \| R_2$$

which often is unacceptably low for a voltage amplifier because r_π is small. A circuit in which feedback improves the input impedance is shown in Fig. A8.3a. The small signal equivalent circuit for the amplifier is shown in Fig. A8.3b. This circuit provides an example of the use of feedback to control impedance. The feedback is proportional to the emitter current and is subtracted from the input voltage at the base of the base–emitter pn junction. Show that the input impedance to the transistor is raised from r_π to $r_\pi + (1 + \beta)R_E$.

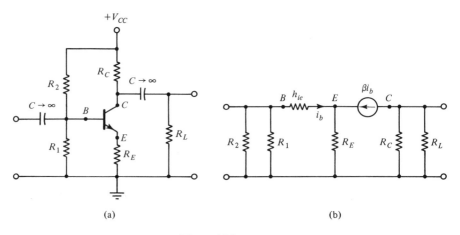

(a)   (b)

Figure A8.3

A8.4. Using the circuit in Fig. A8.4, design an amplifier that will produce an output of $20 \log (v_i/1 \text{ V})$, that is, the output is the input voltage in dB compared with 1 V. Assume that the diode characteristic is described by Eq. (6.7) for $\eta = 1.5$, $I_o = 10^{-10}$ A, and $V_T = 0.026$ V. Specify R_F for the required gain and V_B to remove the constant term in Eq. (8.66). (*Note:* V_B could be obtained with a voltage divider connected to the negative power supply.)

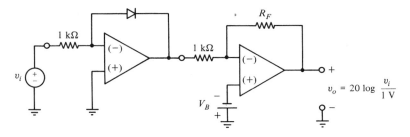

Figure A8.4

A8.5. The analysis of the passive low-pass filter in Section 8.3.1 ignores the loading at input and output. Consider now the circuit in Fig. A8.5, which contains a source resistance R_s and a load resistor R_L. The filter function $\underline{\mathbf{F}}(\omega) = \underline{\mathbf{V}}_o/\underline{\mathbf{V}}_i$ will be identical in form to that in Eq. (8.22), except that the gain will no longer be unity in the region below the critical frequency, ω_c, and the equation for ω_c is changed because it depends on the source and load resistances. Determine the revised expressions for ω_c and $\underline{\mathbf{F}}(\omega)$.

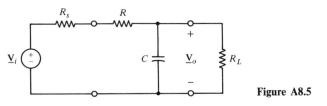

Figure A8.5

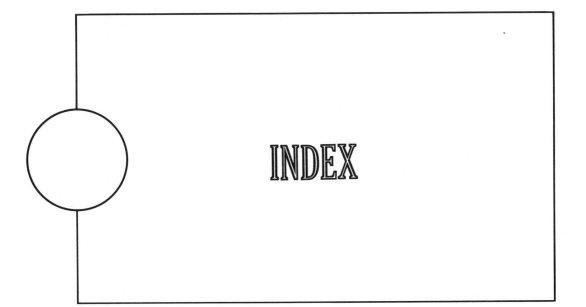

INDEX